Technology, Peace and Security | Technologie, Frieden und Sicherheit

Series Editor

Christian Reuter, Darmstadt, Germany

Technology plays a crucial role in various aspects of peace and security. This book series explores the intersection of computer science with peace and security studies. With a focus on cyber security and privacy, human-computer interaction as well as peace and conflict studies, it addresses topics such as peace informatics and technical peace research (cyber war, peace, arms control, and dual use), crisis informatics and information warfare (social media and collaborative technologies in conflict and crisis situations, misinformation and opinion manipulation), as well as usable safety, security, and privacy (resilient digital infrastructures, security and privacy enhancing technologies).

Technologie spielt eine entscheidende Rolle in verschiedenen Aspekten von Frieden und Sicherheit. Diese Buchreihe befasst sich mit Fragen an der Schnittstelle der Informatik mit der Friedens- und Sicherheitsforschung. Mit einem Fokus auf Cybersicherheit und Privatheit, Mensch-Computer-Interaktion sowie Friedens- und Konfliktforschung werden Themen wie Friedensinformatik und technische Friedensforschung (Cyberkrieg, Frieden, Rüstungskontrolle und Dual-Use), Kriseninformatik und Informationskrieg (soziale Medien und kollaborative Technologien in Konflikt- und Krisensituationen, Desinformation und Meinungsmanipulation) sowie nutzbare Sicherheit und Privatheit (resiliente digitale Infrastrukturen, Technologien zur Verbesserung der Sicherheit und des Datenschutzes) behandelt.

Jasmin Haunschild

Enhancing Citizens' Role in Public Safety

Interaction, Perception and Design of Mobile Warning Apps

Jasmin Haunschild
Darmstadt, Germany

Approved dissertation, Department of Computer Science, Technical University of Darmstadt
First review: Prof. Dr. Dr. Christian Reuter
Second review: Prof. Dr. Frank Fiedrich
Disputation on 17.11.2023

ISSN 3004-9318　　　　　　　ISSN 3004-9326　(electronic)
Technology, Peace and Security | Technologie, Frieden und Sicherheit
ISBN 978-3-658-46488-2　　　ISBN 978-3-658-46489-9　(eBook)
https://doi.org/10.1007/978-3-658-46489-9

Parts of this work were supported by funds of the German Federal Ministry of Education and Research and the Hessian Ministry of Higher Education, Research, Science and the Arts within their joint support of the National Research Center for Applied Cybersecurity ATHENE and by the LOEWE initiative (Hesse, Germany) within the emergenCITY centre.

This Springer Vieweg imprint is published by the registered company Springer Fachmedien Wiesbaden GmbH, part of Springer Nature.
The registered company address is: Abraham-Lincoln-Str. 46, 65189 Wiesbaden, Germany

Foreword

For many years, we believed that crises were phenomena that occurred frequently around the world but would become increasingly rare as the level of development increased. This was still believed in 2019, when Jasmin Haunschild began her dissertation. In recent years, however, we have been taught better and have not been spared major crises in Europe. These include the COVID-19 pandemic (since 2020) as a long-term crisis with contact restrictions and many warnings for the population, the Ahr Valley flood (2021) as an acute crisis with enormous, localized destruction, where warnings did not work that well, and the Ukraine war (since 2022) with energy shortages and supply bottlenecks. Doing the right thing in such crises as a person directly or indirectly affected is essential, as is obtaining the necessary information.

Jasmin Haunschild's dissertation contributes to human-computer interaction (HCI), especially crisis informatics: the application of computing and social science knowledge in the context of crisis situations. Her thesis is characterized by the application of methods of empirical social research and HCI in the context of population warning with mostly empirical findings, but also conceptual approaches in the context of warning apps. Notably, this dissertation deals with the role of crisis apps in the information ecosystem. This thesis thus examines a relevant topic.

This investigation is initially based on quantitative representative studies that analyse the current use of digital information channels, primarily crisis apps, warning apps and social media, citizens' expectations of them, future use and preferences with regard to functions. In particular, the thesis examines the use of information channels and warning apps during the COVID-19 pandemic. This is followed by design-oriented studies, in which persuasive mechanisms are examined in the context of warning apps, and a warning channel with bot-driven personalization is implemented and evaluated.

The studies conducted as part of the dissertation show that warning apps continue to close a significant information gap, which consists in the direct, reliable transmission of information by authorities, which particularly appeals to people who value direct contact with authorities, as well as in the reduction of information, thereby decreasing the risk of information overload. The analyses also show that users want additional functions that serve to establish direct contact, provide police and other emergency-related information, as well as functions that enable users to play a more active role as information providers. The dissertation also shows how different types of nudging can be used in the context of warning apps to increase crisis preparedness. The studies included in this PhD thesis have been published as eight peer-reviewed papers. In addition to working on her 24 publications and her dissertation, Jasmin was involved in project management (ATHENE-SecUrban) and research-oriented teaching of Information Technology for Peace and Security. Jasmin also played a key role in the continuous selection process for new student assistants, thus contributing to the future development of PEASEC.

Jasmin Haunschild has proven that she is capable of independent scientific work. Thus, in November 2023, her dissertation was accepted by the Department of Computer Science at the Technical University of Darmstadt for the degree of Dr. rer. nat. – as the fourth PhD thesis in our research group PEASEC. I would be delighted if research like this could contribute to improving population warning and thus contribute to reducing casualties. Jasmin, thank you for your contribution and for allowing me to accompany you on your way to your PhD. I wish you all the best and every success for the future.

Prof. Dr. Dr. Christian Reuter
Professor for Science and Technology
for Peace and Security (PEASEC)
and Dean of the Department
of Computer Science at Technical
University of Darmstadt
Darmstadt, Germany

Acknowledgements

The writing of this dissertation would not have been possible without the support of many people. First of all, I am deeply grateful for the invaluable council I have received throughout the course of this dissertation. Without the unwavering support and guidance of my supervisor, *Prof. Dr. Dr. Christian Reuter*, this work would not have come to fruition. His insightful advice, steadfast guidance, empathic and constructive feedback have been instrumental in shaping this research. I am truly grateful for this opportunity to grow academically and personally.

I extend my heartfelt appreciation to *Prof. Dr. Frank Fiedrich* for generously dedicating his time and expertise to serve as the second reviewer for this dissertation.

Within the dynamic environment of PEASEC at the Technical University of Darmstadt, I have been fortunate to be surrounded by a community of colleagues whose support, feedback, and humour are immeasurable. It is a privilege to witness your passionate commitment to addressing crucial societal issues. In particular, I want to acknowledge the contributions of my co-authors, especially *Dr. Marc-André Kaufhold, Selina Pauli, Prof. Dr. Verena Zimmermann, Dr. Nina Gerber, Philipp Kühn* and *Dr. Rolf Egert* whose collaborative efforts have played a pivotal role in the research presented in this dissertation. I would also like to thank *Dr. Thea Riebe, Laura Guntrum, Jonas Franken, Kilian Demuth, Dr. Thomas Reinhold, Leon Jung, Stefka Schmid, Dr. Marita Unden, Dr. Paul Gerber, Dr. Alina Stöver, Sofía Cerrillo, Franziska Bujara, Gerbert Roitburd, Henri–Jacques Geiß, Christian Richter, Felix Divo* and *Matthias Lang* for collaborating with me on topics outside of the scope of this dissertation.

The dedication and hard work of the research assistants have directly contributed to the depth and quality of the research findings. Their contributions have been invaluable, and I am truly appreciative of their many efforts. In particular, I would like to thank *Chantal Keller, Anja-Liisa Gonsior* and *Clarissa Neder* for their help editing the dissertation.

Working alongside students on their study projects has been a rewarding experience. I would like to extend my gratitude to *Elifnur Sukran Doˇgan, Glib Grozin, Lauren Pfluger, Grzegorz Jan Rolka, Ibrahim Ethem Saciak* and *Jakob Speitkamp* for their outstanding work in implementing the warning messenger bot. Your dedication and creativity have been indispensable.

I would also like to thank my colleagues at the Chair of International Relations at the Technical University of Braunschweig for supporting me in my first years in academia. I am deeply grateful to *Prof. Dr. Anja P. Jakobi* for many years of collaboration, for mentoring me and for encouraging me to accept the challenge of a dissertation. I have also particularly profited from insightful debates with *Dr. Bastian Loges, Sebastian Heidrich* and *Malte Mock* – thank you for this formative time.

I am thankful to the anonymous reviewers who provided valuable feedback that substantially improved the papers presented in this dissertation. The work in this dissertation was funded by the German Federal Ministry of Education and Research and the Hessian Ministry of Higher Education, Research, Science and the Arts within their joint support of the National Research Center for Applied Cybersecurity ATHENE and by the LOEWE initiative (Hesse, Germany) within the emergenCITY centre.

In addition to my academic circle, I owe a debt of gratitude to my family and friends. Their unwavering support, words of encouragement, and belief in me have been a source of strength throughout this journey. This endeavour would not have been possible without them.

Jasmin Haunschild

Abstract

Recent crises like the COVID-19 pandemic and the European Flood highlight the need for increased disaster resilience through a culture of preparedness. Engaging individuals in disaster readiness remains challenging, yet mobile warning apps can significantly enhance public safety by providing reliable information. This dissertation investigates the potential of mobile crisis apps to boost citizen engagement in disaster prevention and response. By examining user perspectives and developing design interventions, the study explores how warning apps can be more widely adopted and perceived as useful. The findings reveal that citizens find warning apps helpful and prefer them over social media during crises. Enhancing these apps with preparedness features can further improve their usefulness, though simpler solutions like messenger bots might increase adoption by reducing effort and resource costs.

Recent crises such as the COVID-19 pandemic and the European Flood have underscored the potential for low-probability safety and security hazards to occur and the need to increase disaster resilience. This requires a culture of preparedness – the engagement of individuals and communities in activities and practices that enhance their readiness for potential disasters. However, many citizens do not feel well-prepared and it is often difficult to engage citizens in public safety. Mobile warning apps can be an important channel for reliable information from disaster management agencies, helping citizens achieve situational awareness and take effective measures. While warning apps are offered in several countries, users' expectations of these tools remain partly unclear and different designs have rarely been explored.

Therefore, the aim of this dissertation is to investigate mobile crisis apps' potential for enhancing citizens' role in public safety. While studies in crisis informatics have investigated warning channels and e-government studies have investigated technology adoption and citizen engagement, the two have not been brought together. Examining mobile warning channels, particularly warning apps, from the user perspective, the dissertation seeks to increase user engagement in disaster prevention and response by increasing warning app adoption. This is done by investigating the use and perception of such apps, as well as by developing design interventions that increase citizens' contributions to public safety. To do this, firstly, representative and qualitative surveys were conducted, exploring the user perspective on warning apps, identifying trends and changes over time and comparing their usefulness to other sources. By investigating the COVID-19 pandemic and changes in risk cultures following severe crises since 2020, the studies analyse the impact of crisis experience on personal crisis responsibility and trust in emergency management through a comparative survey. Secondly, to enhance crisis preparedness, different versions of preparedness features are developed and evaluated in an experiment.

The results show that citizens believe that warning apps are useful, and those who used a warning app in a crisis report it to be one of the most helpful sources of information. In line with the theory of risk culture, the apps are preferred to social media in crises in Germany. Warning app adoption increased over time in Germany, while feature and design preferences remained largely stable. Analyses of the impacts of particular crises show that the lack of concise and reliable agency information about regulations put in place to manage the COVID-19 pandemic outbreak left an informational gap, that was filled only slowly by warning apps. With regard to involving citizens in crisis response through crisis preparedness, the results show that a persuasively designed preparedness feature can increase the taking of preparedness measures. In addition, extending warning apps by preparedness features can increase warning apps' perceived usefulness. However, using an app incurs costs in terms of effort and resources that prevent users from adoption, which can be avoided by using a messenger bot, thus promising a wider reach of warnings to citizens.

Author's Publications

In sum, 19 publications have been published in the context of the author's work. The following 8 publications are published as chapters in part II of this thesis:

1. Kaufhold, M.-A., Haunschild, J., & Reuter, C. (2020). Warning the Public: A Survey on Attitudes, Expectations and Use of Mobile Crisis Apps in Germany. *Proceedings of the 28th European Conference on Information Systems (ECIS)*, 1–16. https://aisel.aisnet.org/ecis2020_rp/84
2. Haunschild, J., Kaufhold, M.-A., & Reuter, C. (2022b). Perceptions and Use of Warning Apps – Did Recent Crises Lead to Changes in Germany? In M. Muhlhauser, C. Reuter, B. Pfleging, T. Kosch, A. Matviienko, K. Gerling, S. Mayer, W. Heuten, T. Doring, F. Muller, & M. Schmitz (Eds.), *Proceedings of Mensch und Computer 2022* (pp. 25–40). Association for Computing Machinery. https://doi.org/10.1145/3543758.3543770
3. Haunschild, J., & Reuter, C. (2021a). Bridging from Crisis to Everyday Life – An Analysis of User Reviews of the Warning App NINA and the COVID-19 Information Apps CoroBuddy and DarfIchDas. *Companion Publication of the 2021 Conference on Computer Supported Cooperative Work and Social Computing*, 72–78. https://doi.org/10.1145/3462204.3481745
4. Haunschild, J., Pauli, S., & Reuter, C. (2021). Citizens' Perceived Information Responsibilities and Information Challenges During the COVID-19 Pandemic. *Proceedings of the Conference on Information Technology for Social Good*, 151–156. https://doi.org/10.1145/3462203.3475886
5. Haunschild, J., Kaufhold, M.-A., & Reuter, C. (2020). Sticking with Landlines? Citizens' and Police Social Media Use and Expectation During Emergencies. *Proceedings of the International Conference on Wirtschaftsinformatik (WI) (Best Paper Social Impact Award)*, 1–16. https://doi.org/10.30844/wi_2020_o2-haunschild

6. Egert, R., Gerber, N., Haunschild, J., Kuehn, P., & Zimmermann, V. (2021). Towards Resilient Critical Infrastructures – Motivating Users to Contribute to Smart Grid Resilience. *i-com – Journal of Interactive Media, 20*(2), 161–175. https://doi.org/10.1515/icom-2021-0021

7. Haunschild, J., Pauli, S., & Reuter, C. (2023). Preparedness Nudging for Warning Apps? A Mixed-Method Study Investigating Popularity and Effects of Preparedness Alerts in Warning Apps. *International Journal of Human-Computer Studies, 172.* https://doi.org/10.1016/j.ijhcs.2023.102995

8. Haunschild, J., Henkel, M., & Reuter, C. (2025). Breaking Down Barriers to Warning Technology Adoption: Usability and Usefulness of a Messenger App Warning Bot. *i-com - Journal of Interactive Media, 24*

The following 11 papers are not included in the thesis, although their findings are supplementary to it:

9. Haunschild, J., & Reuter, C. (2021b). Perceptions of Police Technology Use and Attitudes Towards the Police – A Representative Survey of the German Population. In C. Wienrich, P. Wintersberger, & B. Weyers (Eds.), *Mensch und Computer 2021 – Workshopband* (pp. 1–12). Gesellschaft fur Informatik e.V. https://doi.org/10.18420/muc2021-mci-ws08-255

10. Zimmermann, V., Haunschild, J., Stover, A., & Gerber, N. (2023). Safe AND Secure Infrastructures? – Studying Human Aspects of Safety and Security Incidents with Experts from both Domains. In P. Frohlich & V. Cobus (Eds.), *Mensch und Computer 2023 – Workshopband,* (pp. 1–8). Gesellschaft fur Informatik e.V. https://doi.org/10.18420/MUC2023-MCI-WS01-225

11. Zimmermann, V., Haunschild, J., Unden, M., Gerber, P., & Gerber, N. (2022). Sicherheitsherausforderungen fur Smart City-Infrastrukturen. *Wirtschaftsinformatik & Management, 14*(2), 119–126. https://doi.org/10.1365/s35764-022-00396-5

12. Haunschild, J., Demuth, K., Geiß, H.-J., Richter, C., & Reuter, C. (2021). Nutzer, Sammler, Entscheidungstrager? Arten der Burgerbeteiligung in Smart Cities. *HMD Praxis der Wirtschaftsinformatik, 58,* 1129–1147. https://doi.org/10.1365/s40702-021-00770-8

13. Reuter, C., Haunschild, J., Hollick, M., Muhlhauser, M., Vogt, J., & Kreutzer, M. (2020). Towards Secure Urban Infrastructures: Cyber Security Challenges to Information and Communication Technology in Smart Cities. In C. Hansen, A. Nurnberger, & B. Preim (Eds.), *Mensch und Computer 2020 – Workshopband* (pp. 1–7). Gesellschaft fur Informatik e.V. https://doi.org/10.18420/muc2020-ws117-408

14. Riebe, T., Haunschild, J., Divo, F., Lang, M., Roitburd, G., Franken, J., & Reuter, C. (2020). Die Veranderung der Vorratsdatenspeicherung in Europa. *Datenschutz und Datensicherheit – DuD, 44*(5), 316–321. https://doi.org/10.1007/s11623-020-1275-3

15. Haunschild, J., Jung, L., & Reuter, C. (2023). Dual-Use in Volunteer Operations? Attitudes of Computer Science Students Regarding the Establishment of a Cyber Security Volunteer Force. In N. Gerber & V. Zimmermann (Eds.), *International Symposium on Technikpsychologie (TecPsy)* (pp. 66–81). sciendo. https://doi.org/10.2478/9788366675896-006

16. Haunschild, J., Kaufhold, M.-A., & Reuter, C. (2022a). Cultural Violence and Fragmentation on Social Media: Interventions and Countermeasures by Humans and Social Bots. In M. D. Cavelty & A. Wenger (Eds.), *Cyber Security Politics: Socio-Technological Transformations and Political Fragmentation* (pp. 48–63). Routledge. https://doi.org/10.4324/9781003110224-5

17. Haunschild, Jasmin, Guntrum, Laura Gianna, Cerrillo, Sofia, Bujara, Franziska, & Reuter, Christian. (2024). Towards a Digitally Mediated Transitional Justice Process? An Analysis of Colombian Transitional Justice Organisations' Posting Behaviour on Facebook. *Peace and Conflict Studies, 30*(2). https://nsuworks.nova.edu/pcs/vol30/iss2/4

18. Reuter, C., Riebe, T., Haunschild, J., Reinhold, T., & Schmid, S. (2022). Zur Schnittmenge von Informatik mit Friedens- und Sicherheitsforschung: Erfahrungen aus der interdisziplinaren Lehre in der Friedensinformatik. *Zeitschrift fur Friedens- und Konfliktforschung, 11*(2), 129–140. https://doi.org/10.1007/s42597-022-00078-4

19. Haunschild, J., Burger, F., & Reuter, C. (2024). Understanding Crisis Preparedness: Insights from Personal Values, Beliefs, Social Norms, and Personal Norms. *Proceedings of the 21st International Conference on Information Systems for Crisis Response and Management (ISCRAM) (Best Paper Award)*. https://doi.org/https://ojs.iscram.org/index.php/Proceedings/article/view/19

Contents

List of Figures

Part I
Synopsis

Introduction

1

This dissertation examines digital multi-hazard warning systems, in particular warning apps, from an interdisciplinary perspective combining crisis informatics and e-government. This introductory chapter presents the motivation for the research (Sect. 1.1) and the problems to be addressed (Sect. 1.2). The aims and research questions of the dissertation are then derived (Sect. 1.3). Section 1.4 presents the structure of the dissertation and a description of the research publications included in it.

1.1 Motivation

Recent global events such as the COVID-19 pandemic since 2020, the European floods in 2021 and the Russian war in Ukraine since 2022 have brought the potential of low-probability threats and their severe impacts to the forefront of public awareness. While these threats have long been recognised by experts, they are often ignored or overshadowed by other concerns in the eyes of the public and policy makers. The negative impacts of disasters are particularly pronounced in the Global South (Bündnis Entwicklung Hilft, 2022), which is particularly affected by rising hazards from both natural and human-made disasters, limited resources, the enduring legacy of colonialism, displacement, and rapid urbanisation. Yet, even in wealthy nations with vigilant risk management, perfect safety and security remain elusive. The *vulnerability paradox* suggests that infrequent disruptions lead to more severe consequences when they do occur, compounding the challenges for less experienced states (BMI, 2009; Fekete & Sandholz, 2021). Therefore, countries that are seldom impacted by emergencies may encounter significant repercussions and deficiencies in providing essential infrastructure during a crisis due to insufficient training and

© The Author(s), under exclusive license to Springer Fachmedien Wiesbaden GmbH, part of Springer Nature 2025
J. Haunschild, *Enhancing Citizens' Role in Public Safety*, Technology, Peace and Security | Technologie, Frieden und Sicherheit, https://doi.org/10.1007/978-3-658-46489-9_1

3

readiness. In such contexts, individuals may lack awareness of crisis protocols and fail to appreciate the importance of preparedness. Residents of lower socio-economic status are particularly vulnerable to the negative impacts of disasters (Shapira et al., 2018). Even in areas where crises are rare, it is therefore important to ensure that people know how to respond and are empowered to prevent negative outcomes from crises and to help themselves and others in crises (Goersch, 2013).

In Europe, the concept of civil protection has undergone significant changes since the Cold War era (Cronqvist et al., 2022). Protective infrastructure, such as sirens and bunkers, has been reduced due to various cost considerations. Recent crises in Europe have prompted a reassessment of crisis preparedness, disaster warning and civil protection measures, reigniting discussions on investing in these elements (Gather, 2022). *Resilience* has emerged as a core concept, referring to the ability of "system, community or society exposed to hazards to resist, absorb, accommodate to and recover from the effects of a hazard in a timely and efficient manner, including through the preservation and restoration of its essential basic structures and functions" (UN International Strategy for Disaster Reduction, 2009, p. 24). This concept emphasises the interconnectedness of sociotechnical systems, where individuals and communities intersect with technological resources, pointing towards the importance of each of these pillars. Ample research points towards the importance of community resilience (Mayer, 2019) and the resiliency enhancing effect of trusted communities (e.g. Tackenberg et al., 2022), emphasising the role of non-state actors for managing and overcoming crises.

The United Nations Sendai Framework for Disaster Risk Reduction prioritises the promotion of a "culture of prevention" and disaster risk education to foster resilient communities and an inclusive approach to disaster risk management (UN Office for Disaster Risk Reduction, 2015, p. 23). The call for a "culture of preparedness" echoes this sentiment (Appleby-Arnold et al., 2021). Swift and reliable communication is the bedrock of effective crisis response, as crises can escalate rapidly and require coordinated and timely action.

Citizen crisis preparedness entails possessing the knowledge, resources and plans to mitigate the impact of crises and ensure the safety and well-being of citizens. It involves an array of actions, such as devising emergency protocols, assembling emergency kits, participating in training programmes, and staying informed about potential hazards. The preparedness of citizens increases security and resilience and minimises the workload of emergency services and crisis management, because well-prepared citizens respond appropriately and make well-informed decisions in the event of an emergency. In addition, knowledge and preparedness can empower citizens to take an active role in their own safety and that of their families and neighbours. During crises, citizens turn to public authorities' channels for information.

Agency websites are the second most frequently accessed source of information during crises (Park et al., 2019), emphasising the significance of agency information in such circumstances. In the 2010 strategy for safeguarding the population in Germany, the Federal Office for Civil Protection and Disaster Assistance (BBK) identifies a deficit in local warning due to the reduction of sirens in Germany (BBK, 2010, p. 39–40). It is further noted that there is a deficit in the population's ability to protect itself, recognising that citizens rely too heavily on the state and ignore the fact that assistance may be delayed, unavailable to every citizen due to limited resources, or obstructed. Additionally, there is a lack of communication of information and material to the public: "Although comprehensive and didactically good information, self-help and training material are available also for the local levels and the local population, provided by the state government, this material is hardly known and not openly advertised" (BBK, 2010, p. 40–41, translation by the author). As crises and disasters are volatile and can strike quickly (Karutz et al., 2017, p. 55–50), rapid and reliable communication is a prerequisite for enabling people to respond and coordinate. This includes disseminating reliable information, improving situational awareness and promoting engagement in disaster preparedness.

In the digital age, crisis communication extends to various digital platforms such as websites, mobile broadcasts, alert apps and social media (Hauri et al., 2022; Reuter & Kaufhold, 2018). This digital transformation is in line with the concept of e-government, where digital technologies are used to deliver government services, increase citizen participation, and improve government operations (Bertot et al., 2012). In particular, crisis communication is evolving into a digital public service as sirens decrease (Distel & Lindgren, 2019). A number of countries have developed multi-hazard warning applications as part of their digital public warning systems (Hauri et al., 2022; Tan et al., 2017). Research suggests that the implementation of a warning app by a state is an indication to the public that the governing body is open to sharing control (Appleby-Arnold et al., 2019).

These apps offer features such as location-based notifications, risk maps, information updates, preparedness advice and the ability to store evacuation plans (Groneberg et al., 2017). However, their efficacy relies on a functional internet connection and the proactive uptake and downloading of the apps. Consequently, some countries, including Germany, have introduced cell broadcast (CB) as a more reliable crisis communication solution that, unlike apps, can reach mobile phones regardless of internet access and without the need to download an app (Hauri et al., 2022). However, alert apps also have unique advantages: they allow multimedia content such as maps, personalisation of alerts, multiple views of a notification, location-based alerts and two-way communication. Some disaster and emergency apps provide assistance with prevention and response, as well as features for citizens to conveniently coordinate (Tan et al., 2017). These apps are also employed to receive safety and security-related local information that does not reach severe warning levels. Alerts are especially valuable for individuals residing in border regions who may be connected to the other country's tower. As CB is solely text-based, design variations primarily concern wording and message framing. In contrast, the design of multi-hazard warning apps exhibit a broad range of discrepancies among countries, encompassing their appearance and functionality. As such, from a technical and human-computer interaction (HCI) perspective, warning apps are the more relevant field of study. In navigating the balance between information overload and scarcity, and combating the dissemination of false information, alert apps present a promising avenue for reliable, agency-sanctioned information within the field of crisis communication. Increasing citizen engagement through these channels could enhance citizen awareness and potentially enhance crisis response.

1.2   Problem Statement

A widespread deficiency in preparedness and engagement with disaster readiness is evident globally. According to a report in Canada, although 80% state having no obstacles in preparing, approximately two-thirds dedicated a maximum of one hour in the previous year to crisis preparedness (Canadian Red Cross, 2021). Similarly, a recent survey in the United States shows that the majority of people has not moved from a contemplation to an action and maintenance phase of preparedness (FEMA, 2022). In China, the picture appears to look similar, with a survey stating that "lack of motivation, negative attitude to preparedness and knowledge shortfall are major but remediable barriers for household preparedness" (Chen et al., 2019, p. 1) and another finds that only 28% are fully or partly prepared (Ning et al., 2021). It also shows that people are more likely to take action if they are confident in their ability to prepare and believe in the impact of preparedness. Registering for alerts and warnings was the most swiftly executed preparedness measure (FEMA, 2022). In contrast, around 60% of people in New Zealand state that they are likely to prepare, but perceive a lack of knowledge as the main barrier (Civil Denfence NZ, 2020).

In various European countries, there are significant concerns about citizens' lack of information and preparedness for disasters. According to Appleby-Arnold & Brockdorff (2018), approximately 72% of the population in European countries feel either insufficiently informed or not informed at all, while two thirds feel largely or totally unprepared.

There is a strong discrepancy between interest and readiness, with 92% of people expressing a strong desire for preparedness information. Intentions to prepare for disasters vary widely, with 91% of Portuguese intending to prepare or prepare a lot, compared to only 21% of Germans (Appleby-Arnold & Brockdorff, 2018). Interestingly, despite showing a high level of interest in disaster preparedness, Germans paradoxically show low intentions to prepare. This issue extends to the use of mobile phone apps for crisis management, where the use of multipurpose warning apps remains remarkably low in all European countries except Finland (Hauri et al., 2022), despite the widespread use of smartphones, social media and agency websites in crises (Ohme et al., 2020; Park et al., 2019; Reuter & Kaufhold, 2018). At the same time, however, citizens also say that they plan to use an app in the future (Reuter et al., 2019), indicating an interest in the medium.

Research suggests that a country's risk culture influences, to some extent, the overall motivation to take an active role in crisis preparedness and response (Cornia et al., 2016). Particularly in fatalistic risk cultures, people tend to believe that disasters are inevitable and that not much can be done to prevent or cope with them. In state-centred risk cultures, on the other hand, citizens perceive state authorities as responsible and capable of managing crises, which reduces the perceived responsibility to take the initiative. Research also suggests that this also affects the extent to which information and communication technologies, such as alert apps and social media, are used in crises (Reuter et al., 2019). However, the paradoxical position of German citizens with regard to preparedness points to a large gap in moving towards a culture of preparedness. This disparity may be attributed to the practices of disaster management agencies, as Germans display a high level of confidence in emergency management authorities but perceive a lack of trust from authorities towards them in terms of responding appropriately or being prepared (Appleby-Arnold & Brockdorff, 2018). The adoption of digital channels to engage citizens poses a challenge that extends beyond crisis management. Previous research has revealed a general adoption gap for e-government services in Germany (Distel & Ogonek, 2016). An important challenge relates to the relatively infrequent use of alert apps (Hauri et al., 2022), which may lead to a comparatively low perceived usefulness of these apps. Despite their infrequent use, these applications can be resource intensive: Background activity can drain the device's battery, thus reducing its longevity. Additionally, apps occupy storage space on the device, which may lead to insufficient storage. Another point to consider is that the app ought not to be automatically put into sleep mode due to long periods of inactivity as this may cause the users to miss important alerts thereby resulting in frustration. All of these factors can lead to a non-competitive cost-benefit ratio of alerting apps compared to other apps and to discontinuation of use (Tan et al., 2020c, 2020a). Therefore, analyses investigating the usefulness of warning apps, the conditions under which they are perceived to be useful, and the inhibitors can be used to improve the app, leading to greater adoption.

As smartphone users download very few new apps (Lella & Lipsman, 2017), each app should make a significant contribution to users—a particular challenge for warning apps in state-centred risk cultures where general preparedness motivation is low. However, little is currently known about how warning apps can be designed to be a useful warning component that increases the spread of reliable information and citizens' self-efficacy in crisis preparedness and response. Another open question is how to motivate users to contribute to security as co-producers through the use of mobile crisis apps. Citizen co-production refers to the collaboration between citizens and government or other organisations to provide public services or address public problems. It is about recognising that citizens are not just passive recipients of public services, but actively involved in the design and delivery of those services. In the case of security, this means involving citizens as active partners in the development and implementation of measures to improve security and reduce risks. By working together, governments and organisations can respond more effectively to threats while building trust and cooperation between citizens and government. In the context of crisis preparedness and response, alerting applications could be used to engage citizens in the co-production of security. In this dissertation, the commonly used HCI term *user* is replaced by *citizen*, as this frame includes both users and non-users of technologies, thus giving more space to *potential users*. Whilst some research delves into the demands of citizens for utilising alert applications and social media during crises (Dallo & Martí, 2021; Reuter et al., 2019), these technologies are rapidly evolving and their usage is influenced by various factors, including cultural ones. Therefore, nationally representative research is needed that allows conclusions and recommendations to be drawn for specific local contexts. In addition, research should examine current use and its relationship to specific crises in order to provide a complete and up-to-date picture.

Summary—Problem Statement

Engaging citizens in crisis preparedness and response is a challenge, especially in fatalistic and state-oriented risk cultures. To motivate citizens to participate in crisis preparedness, digital alert channels must be designed to provide a sufficient amount of useful and user-friendly features, as this will increase the likelihood of adoption.

1.3   Aims and Research Questions

To address the limited citizen engagement with public safety and crisis preparedness, the dissertation analyses user perceptions of warning apps, aiming to understand the role that this channel plays in the crisis communication and information ecosystem. Evaluating design interventions, it aims to deduce implications for effective design in increasing citizen engagement with warnings and crisis preparedness.

To address the issues at hand, the present dissertation seeks to answer the following main research question:

RQ How can warning apps be used to engage citizens in public safety?

This question is divided into the following sub-questions:

RQ_1 What are users' interactions, perceptions and expectations of mobile crisis apps?

Information seeking and information exchange is increasingly taking place in the digital realm. Due to the often fast onset of crises and the complexity and uncertainty that is inherent to them, the mobile warning component is particularly relevant. RQ1 thus explores how citizens use mobile crisis apps and how they complement other channels, paying particular attention to their usefulness. Thus, the question reveals users' needs in their particular usage contexts, environments and culture. Answering RQ1 gives insights into user preferences with regard to mobile crisis apps.

RQ_2 How can mobile crisis apps be designed and improved to enhance users' role in public safety?

The second research question takes the insights from RQ1 to explore how smartphone-based mobile crisis tools can be designed to a) allow users to stay informed in crises and b) to be motivated to take proactive precautionary and preventive measures. Therefore, design interventions are implemented that seek to increase engagement with warning apps and crisis preparedness.

1.4   Content, Structure and Underlying Publications and Contributions

This dissertation is organised into two parts, a synopsis presenting and discussing the overall research question, and the empirical findings that the synopsis is based on, which have been published as scientific papers.

Part I, the Synopsis, introduces the topic and presents the contribution and structure of the dissertation (Chap. 1), before discussing related work and deriving the existing research gaps (Chap. 2). Then, the overall research design of this dissertation is portrayed (Chap. 3), followed by a summary of the findings (Chap. 4), which are discussed in Chap. 5. Finally, the dissertation is summarised in the conclusion (Chap. 6).

In part II, peer-reviewed and published papers are presented that contribute to answering the research question. This dissertation consists of the works that have been previously published as journal articles and conference papers and were often the results of joint work and co-authorship. I was the first and corresponding author of six of these papers, second author of one paper, whereas one was ordered alphabetically. In the following, I describe my contribution to each of the papaers. All co-authors have kindly given permission to include the works in this dissertation.

CHAPTER 7: *Kaufhold, M.-A., Haunschild, J., & Reuter, C. (2020). Warning the Public: A Survey on Attitudes, Expectations and Use of Mobile Crisis Apps in Germany.* Proceedings of the 28th European Conference on Information Systems (ECIS), *1–16.* https://aisel.aisnet.org/ecis2020_rp/84

The chapter shows how warning apps are used during crises, how this use compares to data from two years earlier, what features citizens value, and what implications this has for design. The findings suggest that design preferences are largely unaffected by sociodemographic factors and highlight a preference for features promoting active citizen engagement (e.g. contributing to locating missing persons). However, the analysis indicates that warning apps are perceived as very important, but only used by a small fraction of the German population. Moreover, citizens' most crucial demand is for a single app that encompasses all significant emergencies and functionalities.

AUTHORSHIP STATEMENT The paper constitutes a joint work of Jasmin Haunschild, Marc-André Kaufhold and Christian Reuter. Jasmin Haunschild led the management and writing process of the paper. The final research design, literature review and interpretation of the results were done jointly. Jasmin Haunschild was responsible for data presentation and data analysis, conducting quantitative and descriptive analyses and interpreting the results. She was also responsible for presenting the paper at the conference. Marc-André Kaufhold proposed the initial research design of the study and was responsible for the survey design and data collection. Christian Reuter was a general advisor of this work and contributed with continuous feedback during all phases of the paper writing process.

CHAPTER 8: *Haunschild, J., Kaufhold, M.-A., & Reuter, C. (2022b). Perceptions and Use of Warning Apps—Did Recent Crises Lead to Changes in Germany? In M. Muhlhauser, C. Reuter, B. Pfleging, T. Kosch, A. Matviienko, K. Gerling, S. Mayer, W. Heuten, T. Doring, F. Muller, & M. Schmitz (Eds.),* Proceedings of Mensch und Computer 2022 (pp. 25–40). *Association for Computing Machinery.* https://doi.org/10.1145/3543758.3543770

The chapter presents findings related to how warning app use has changed between 2017, 2019 and 2021 and the role COVID-19 apps play in this context. Although COVID-19 contact tracing apps have surpassed warning apps in popularity, the usage of warning apps has also seen an increase. The study shows that preferences have remained largely stable and the increase in warning app use has been linear. Additionally, the chapter investigates recent developments, such as citizens' openness to using messenger channels for alerts, their desire for pandemic-related updates within warning apps, and the success of a regional warning app.

AUTHORSHIP STATEMENT The paper constitutes a joint work of Jasmin Haunschild, Marc-André Kaufhold and Christian Reuter. The research design and discussion of the results was done jointly. As corresponding and leading author, Jasmin Haunschild was responsible for the review of the literature, data presentation and data analysis, conducting quantitative and descriptive analyses and interpreting the results. The questionnaire design was done jointly by Jasmin Haunschild and Marc-André Kaufhold. Marc-André Kaufhold was responsible for the data collection through a representative survey and contributed to the literature review and discussion. Christian Reuter was a general advisor of this work and contributed with continuous feedback during all phases of the paper writing process.

CHAPTER 9: *Haunschild, J., Kaufhold, M.-A., & Reuter, C. (2020). Sticking with Landlines? Citizens' and Police Social Media Use and Expectation During Emergencies.* Proceedings of the International Conference on Wirtschaftsinformatik (WI) (Best Paper Social Impact Award), *1–16.* https://doi.org/10.30844/wi_2020_o2-haunschild

The chapter reports the results of a representative survey among the German population in 2021 which asked about citizens' use of social media in crises, their perception of advantages and disadvantages compared to other media, particularly warning apps. It further reports how current usage compares with data from two years earlier. The study shows that different usage types can be identified, with one groups preferring direct agency contact through, e.g. warning apps, and another group that relies more on social media. The findings suggest that citizens increasingly both search for, but also share information in crises and that YouTube plays a larger role than other research suggests. Age, largely, does not impact the use of social media, suggesting that in crises, citizens enlarge their media repertoire to include channels that are not typically used. While direct communication with agencies via emergency hotlines is regarded as very important, scepticism towards social media is decreasing compared with previous years.

Authorship Statement The paper constitutes a joint work of Jasmin Haunschild, Marc-André Kaufhold and Christian Reuter. The research design, literature review and interpretation of the results were done jointly. As corresponding and leading author, Jasmin Haunschild was responsible for data presentation and data analysis, conducting quantitative and descriptive analyses and interpreting the results. She was also responsible for presenting the paper at the conference. Marc-André Kaufhold focused on the survey design (with support by Christian Reuter) and did the data collection through a representative survey. Christian Reuter was a general advisor of this work and contributed with continuous feedback during all phases of the paper writing process.

Chapter 10: *Haunschild, J., Pauli, S., & Reuter, C. (2021). Citizens' Perceived Information Responsibilities and Information Challenges During the COVID-19 Pandemic.* Proceedings of the Conference on Information Technology for Social Good, *151–156.* https://doi.org/10.1145/3462203.3475886

The chapter shows how citizens navigated the information ecosystem during the first wave of the COVID-19 pandemic in Germany. It reports a one-week qualitative repeated survey among Germans, in which citizens reported on the media and ICT used during each day, how important information was received, whether it was shared, and which sources were the most helpful ones. In addition, challenges and perceived information gaps were reported. The chapter shows that a section of the population employed an ordered news consumption strategy, often relying on local newspapers, radio and websites of local administration. However, many also suffered from information overload and perceived staying up to date as strenuous. The federal and local differences were also a great challenge. Warning apps were hardly used and not perceived as very helpful. The chapter identifies gaps in the information ecosystem and derives design implications for agency information channels that could reduce information overload and ensure reliability of the information.

AUTHORSHIP STATEMENT The paper constitutes a joint work of Jasmin Haunschild, Selina Pauli and Christian Reuter. As corresponding and leading author, Jasmin Haunschild led the overall research design, data collection process, analytical framework, literature review, management and writing process of the paper. Coding and interpretation of the collected data and the visual design of the results was done jointly by Jasmin Haunschild and Selina Pauli. Christian Reuter was a general advisor of this work and contributed with continuous feedback during all phases of the paper writing process.

CHAPTER 11: *Haunschild, J., & Reuter, C. (2021a). Bridging from Crisis to Everyday Life—An Analysis of User Reviews of the Warning App NINA and the COVID-19 Information Apps CoroBuddy and DarfIchDas.* Companion Publication of the 2021 Conference on Computer Supported Cooperative Work and Social Computing, *72–78.* https://doi.org/10.1145/3462204.3481745

The chapter shows the emergence of regulation information apps as a particular type of app that resulted from crisis information needs that arose during the COVID-19 pandemic. A market analysis revealed that only one state-run warning app and two privately run apps depicted the different rules and regulation that apply to fight the COVID-19 pandemic in Germany. Analyses of user reviews of these apps reveal persistent information needs and affordances from the user perspective. They also revealed that maintenance and keeping information correct and updated were challenges and lead to incorrect information, disappointment and diminishing trust. The chapter discusses the inclusion of regulatory information in warning apps and the need for machine-readable information for efficient crisis communication.

AUTHORSHIP STATEMENT The paper constitutes a joint work of Jasmin Haunschild and Christian Reuter. As corresponding and leading author, Jasmin Haunschild led all stages of the research and writing process of the paper. In particular, Jasmin Haunschild was responsible for the research design, data collection, data analysis through descriptive and qualitative analyses, interpretation of results and writing of the paper. Christian Reuter was a general advisor of this work and contributed with continuous feedback during all phases of the paper writing process.

CHAPTER 12: *Egert, R., Gerber, N., Haunschild, J., Kuehn, P., & Zimmermann, V. (2021). Towards Resilient Critical Infrastructures—Motivating Users to*

Contribute to Smart Grid Resilience. i-com—Journal of Interactive Media, 20(2),
161–175. https://doi.org/10.1515/icom-2021-0021

The chapter presents various approaches for motivating people to contribute to
infrastructure resilience. It shows that many strategies have been established to
motivate users to contribute to defined aims and goals. However, these are rarely
or never applied to fostering resilience of critical infrastructures. Using the energy
sector as an example, the chapter applies these insights to infrastructure resilience,
showing how citizens might be motivated to contribute to the resilience of urban
infrastructures and to the provision of public safety and security more broadly.

AUTHORSHIP STATEMENT The paper constitutes a joint work of Rolf Egert,
Nina Gerber, Jasmin Haunschild, Philipp Kühn and Verena Zimmermann. As corresponding and leading author, Jasmin Haunschild led the overall management and
writing process of the paper. The authors contributed the introduction and literature
review together. The research design was done jointly. Analysis of the motivational
strategies and their application to the energy sector (Chap. 2) was mainly done by
Verena Zimmermann, Nina Gerber and Rolf Egert. The evaluation and discussion
were mainly done by Jasmin Haunschild, Nina Gerber and Verena Zimmermann.
The participation strategies (Chap. 4) were mainly developed by Jasmin Haunschild,
Philipp Kühn and Rolf Egert. The conclusion (Chap. 5) was done mainly by Jasmin
Haunschild.

CHAPTER 13: *Haunschild, J., Pauli, S., & Reuter, C. (2023). Preparedness Nudg-
ing for Warning Apps? A Mixed-Method Study Investigating Popularity and
Effects of Preparedness Alerts in Warning Apps.* International Journal of Human-
Computer Studies, 172. https://doi.org/10.1016/j.ijhcs.2023.102995

The chapter presents the design and experimental evaluation of persuasive design
and nudging towards crisis preparedness in warning apps. The study set out to engage
users in crisis preparedness information beyond immediate warnings via interactive
nudges conveying crisis preparedness information and reminders through push-
notifications. The acceptance of nudging and nudge preferences are first explored
through a representative study. Then, the nudges are co-designed with users, before
being tested in an experiment where users engaged with an app that sent them preparedness notifications. One group received the notification modelled on a widely
used current warning app with an improved design and two groups received different nudges (a social nudge and a confrontational nudge) in the improved design.
The study reports the effect of the different nudges on implemented preparedness

measures and the acceptance of this type of interaction and alert. Statistical analyses investigate whether some groups were more or differently engaged with the notifications. Short post-interaction surveys and a post-study survey explored participants' perceptions, openness to the notifications, their thought processes and attitudes towards continuing usage of an app with such preventative and nudging notifications.

AUTHORSHIP STATEMENT The paper constitutes a joint work of Jasin Haunschild, Selina Pauli and Christian Reuter. As corresponding and leading author, Jasmin Haunschild led all stages of the research process and was responsible for the conceptualisation, methodology and writing of the original and review draft and the editing, as well as data curation and visualisation. Selina Pauli assisted in the development of the conceptualisation, methodology and data curation, as well as the writing of the original draft and visualisation, all pertaining to the experiment. Christian Reuter was a general advisor of this work and contributed with continuous feedback during all phases of the paper writing process. In addition, he was responsible for project administration and funding acquisition.

CHAPTER 14: *Haunschild, J., Henkel, M., & Reuter, C. (2024). Breaking Down Barriers to Warning Technology Adoption: Usability and Usefulness of a Messenger App Warning Bot. i-com – Journal of Interactive Media, 24*

The chapter shows the exploration of delivering warnings via messaging apps. Presenting the design science research approach, the study aimed to create a warning channel with low adoption barriers, while maintaining good usability throughout the set-up process. The study iteratively implemented a warning bot based on a literature review of requirements for warning apps and on two rounds of user feedback. These were then combined with requirements for a conversational interface design. For the evaluation, users were asked to perform a set of tasks to familiarise themselves with the bot and to complete the personalisation process. Then, open feedback was gathered and the bots' usefulness and utility were evaluated. The study shows that with the extensive use of buttons as rail-guards, the bot received a good usability rating. As anticipated, the function of automatic warning notification was seen as the most relevant. While, on the one hand, the restricted design options in a messaging app were perceived as a limitation, the fact that it does not require downloading an app were seen as assets. The results suggest that a warning bot could be a valuable addition to the warning mix.

AUTHORSHIP STATEMENT The paper constitutes a joint work of Jasmin Haunschild, Markus Henkel and Christian Reuter. As corresponding and leading author, Jasmin Haunschild led the research, implementation and writing process of the paper. In particular, Jasmin Haunschild devised the research design and user evaluation studies, instructed a student group in the implementation, supervised the data collection, analysed the data, interpreted the results and wrote the paper. Markus Henkel revised the paper especially with regard to the integration of further literature. Christian Reuter was a general advisor of this work and contributed with continuous feedback during all phases of the paper writing process.

Related Work

2

The following chapter introduces relevant research related to the use of information and communication technologies to inform, involve and engage citizens in crises. To understand the use of apps in the context of crises, Sect. 2.1 introduces relevant terms related to crisis informatics and the research concerning warning apps. Sect. 2.2 presents different crisis communication models. This is followed by Sect. 2.3, which introduces work from human computer interaction (HCI) and information systems (IS) investigating the acceptance of technologies and the role of technology design. In addition, the section introduces the e-government literature that contributes insights on the adoption and non-adoption of digital government services. Finally, Sect. 2.4 introduces the interdisciplinary work focusing on psychological and cultural factors related to risk perception and engagement in crisis management, particularly the theory of risk cultures.

2.1 Crisis Informatics and Disaster Apps

The terms crisis and disaster are often used interchangeably. While the former can also relate to private and corporate crises, the latter is more closely linked to public safety threats. They both share an element of urgency: *Crises* are a "significant threat to operations that can have negative consequences if not handled properly" (Coombs, 2007, p. 1), while disasters incur "widespread human, material, economic or environmental losses which exceed the ability of the affected community or society to cope using its own resources" (UN International Strategy for Disaster Reduction, 2009, p. 4). *Hazards* are potential sources of harm that can escalate into disasters if

J. Haunschild, *Enhancing Citizens' Role in Public Safety*, Technology, Peace and Security | Technologie, Frieden und Sicherheit, https://doi.org/10.1007/978-3-658-46489-9_2

they are not managed effectively, or if they put vulnerable systems or populations at risk. A term that is broader and less connected to natural disasters is "public safety threats". The term describes risks to the safety and well-being of the public, including natural and human-made disasters, as well as less dramatic occurrences that nonetheless affect the public's safety negatively, such as cybersecurity incidents, infrastructure failures, public health emergencies or large-scale transportation accidents. In this dissertation, the terms crisis and disaster will be used to describe extreme events that jeopardise public safety. The term public safety hazard will be used to describe less dramatic safety and security relevant occurrences.

Public safety is commonly seen as one of the main governance services provided by the state (Gerhold et al., 2021). In Germany, the federal states (Länder) are responsible for warnings in the event of a disaster (civil protection), the municipalities for warning in everyday hazards (fire protection, technical assistance and public safety). In this context, the Modular Warning Systems (MoWaS) serves at all federal levels as a uniform technical platform. Yet, despite a good functioning of the crisis management agencies in Germany, gaps exists with regard to the public warning system, and safety and security are never perfect. Therefore, the German strategy includes the aim of strengthening the population's self-help capabilities as well as improving the warning system (BBK, 2010; BMI, 2022).

Crisis informatics "is a multidisciplinary field combining computing and social science knowledge of disasters; its central tenet is that people use personal information and communication technology to respond to disaster in creative ways to cope with uncertainty" (p. 224 Palen & Anderson, 2016). Relevant internet enabled technologies include websites, social media and apps. Inspired by the research community of computer-supported collaborative work (CSCW), crisis informatics stresses the importance of citizens' contributions to situational awareness and the role of social media as a two-way communication channel (Kaufhold, 2021; Reuter & Kaufhold, 2018). A review of the research of crisis informatics shows crisis communication can be differentiated into inter-agency communication for internal coordination (agency to agency), top-down information to citizens (agency to citizens), bottom-up information from citizens to agencies, e.g. providing information for situational awareness, and citizen to citizen communication for coordination and collaboration (Reuter & Kaufhold, 2018). Fig. 2.1 depicts these communication modes and differentiates between crisis communication and self-organisation.

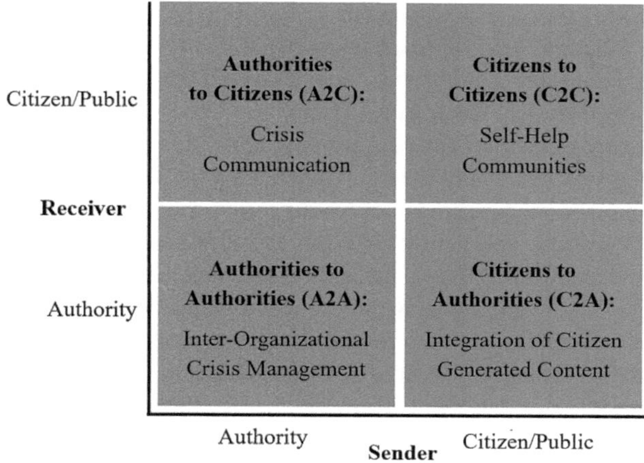

Fig. 2.1 Crisis communication modes by receivers and senders. (Source: Reuter & Kaufhold (2018))

The review shows the various research efforts directed at analysing the roles that social media play in disaster contexts and design interventions seeking to use social media for situational awareness and to build community relations (Reuter & Kaufhold, 2018). Aiming to understand the online activities of people in cases of crises, research has found that online activity in response to disasters centres on *helping, being anxious, returning, supporting* and *exploiting* (Hughes et al., 2008). One branch of research seeks to harness the information that citizens provide through their social media channels, often to increase emergency managers' situational awareness (Kaufhold, 2021; Kaufhold, Bayer, & Reuter, 2020). For example, a study shows how volunteered geolocation information can be used to optimise disaster response (Tzavella et al., 2018). Recently, the research on social media in crises has also included the use of social bots for information gathering on Facebook (Stieglitz et al., 2022). Another branch looks at the way technology shapes and enables a more active role of citizens, as volunteers and helpers (Fathi et al., 2020; Kaufhold & Reuter, 2016). Some research has also analysed the role of apps in relation to disaster management, showing that some apps used for other purposes are also used to fulfil functions during disasters, e.g. the use of weather apps which show extreme weather (Tan et al., 2017). In contrast, apps that are built for disaster purposes have functions that are specifically tailored to emergency situations, but they are infrequently used and thus less wide-spread (Tan et al., 2017). Tan et al. (2017) suggest that to increase the usage of disaster apps, they need to be promoted and citizens need to be motivated to use these apps, or regular apps can be augmented with disaster functions—an approach taken e.g. in a study where a neighbourhood app provides day-to-day features, but becomes a resilient communication network in case of crises (Haesler, Mogk, et al., 2021).

A subcategory of disaster apps are warning apps. As the mobile component of public warning systems, one of their key functions is to alert the population to disasters. An international comparison of warning apps shows that they offer various functionalities to assist users in times of crises and disasters (Groneberg et al., 2017). Identifying and analysing 59 apps, the authors find that the providers of these apps includes a range of organisations, with government entities being the most frequent contributors (38 instances), closely followed by humanitarian organisations such as the Red Cross (8 instances). Additionally, private sector entities, such as weather reporting apps, leverage data from diverse sources to contribute to the pool of information. The apps predominantly address multiple scenarios. There is a discernible correlation between specific scenarios and region-specific crises and disasters.The public safety apps provide three main sets of functionalities: information, communication and preparedness. The most widespread feature is the ability to receive push notifications during crisis events, followed by the integration of mapping systems and the provision of general news and information. With regard to communication, various functions incorporate social media integration. Users are also given the option to initiate emergency calls or send emergency messages via email, text, or SMS. Finally, common functionalities for crisis preparedness include emergency planning checklists, behavioural guidelines, and hazard descriptions (Groneberg et al., 2017).

Both qualitative and quantitative research have been used to explore usage patterns, usability requirements for warning apps and user demands (Dallo & Martí, 2021; Reuter & Spielhofer, 2017; Tan et al., 2019). This research shows that when it comes to warning apps, users strongly prefer a multi-purpose warning app that combines many relevant disaster types in one app, rather than being specific to one hazard (Dallo & Martí, 2021), and some research has analysed these apps in particular (Hauri et al., 2022; Reuter, Kaufhold, Leopold, & Knipp, 2017). A study elaborates the usability guidelines for disaster apps, which centre around the challenges of making critical information salient, considering users' cognitive load, and building trust while anticipating the level of interaction with users (Tan et al., 2020b). Another study has investigated the usability of warning apps by analysing user reviews (Tan et al., 2019) and shows that warning apps have specific design requirements that differ from those of other warning apps, e.g. a simple design, sparing phone resource usage or noticeable audio output with a "wake-up function". This leads the authors to suggest an additional category of usability design requirements for disaster apps, namely *app dependability* (Tan et al., 2019)—an aspect that was also mentioned in German app reviews (Kotthaus et al., 2016). Some research has investigated aspects of warning app (non-)adoption (Fischer et al., 2019; Reuter, Kaufhold, Leopold, & Knipp, 2017) and reasons for (dis-)continuing using a warning app (Fischer et al., 2019; Kotthaus et al., 2016; Tan et al., 2020c). With regard to continuance of warning app use, app utility, app dependability, and user interface output positively influence the continued intention to use disaster apps (Tan et al., 2020c). In particular, users must perceive that the app delivers its intended function and does not deviate from it (Tan et al., 2020c). Conversely, user interface input and user interface graphics have negative influences, suggesting that users do not want to have to provide inputs to use the app. In addition, the study suggests that the serious usage context of the app results in users preferring a very simple design, rather than complex graphics which might distract from the main purpose and use up cognitive capacity in a crisis (Tan et al., 2020c). Concerns about data security have been shown to have negative effects on warning app use intention (Fischer et al., 2019).

This suggests that not only usability, but also usefulness as an important factor (see Sect. 2.3). While usability refers to the ease of using a system or its usage being effortless, a useful system "is one for which a user believes in the existence of a positive use-performance relationship" (Davis, 1989). This leaves the open question how multi-purpose warning apps can be designed to be more relevant for users. Two strategies are possible, which have hardly been explored. On the one hand, the inclusion of new features could be explored that lead to the app being used more in daily life. This is a strategy that is also suggested by the Finnish warning app's relative success compared to other apps and its inclusion of a larger set of emergency functions, such as a display of defibrillators and of general emergency numbers (Hauri et al., 2022). This could include the extended use of warning apps to engage citizens in crisis preparedness, an activity that is also relevant in the pre-disaster state. Crisis knowledge and preparedness advice and support in implementing preparedness measures could be conveyed with a trusted warning app. Taking cognitive biases and intention-behaviour gaps into account, crisis preparedness features should consider the inclusion of persuasive technologies or nudging elements to help users implement preparedness advice (Caraban et al., 2020; Oinas-Kukkonen & Harjumaa, 2009). However, so far, persuasive elements have hardly been investigated with regard to preparedness or warning apps (Kotthaus et al., 2016; Mol et al., 2021), with one study examining household preparedness against floods as an insightful exception (Mol et al., 2021). Instead, related research is heavily focused on behaviour change with regard to health (Dennison et al., 2013; Taj et al., 2019). Information about crisis preparedness via warning apps could also be understood as an element of e-learning. However, the few studies that investigate e-learning and crisis preparedness target an audience that holds crisis management capacities due to being responsible for vulnerable populations, such as crisis responders, hospital staff or caregivers (Ingrassia et al., 2014) or education facilities and the integration of crisis knowledge in school curricula (How et al., 2020). Some studies investigate the use of serious games and virtual simulations for crisis preparedness and risk perception (Chittaro et al., 2017; Hatayama & Nakai, 2020; Kim et al., 2021; Markwart et al., 2019; Meesters & vande Walle, 2013; Toyoda et al., 2014). However, considering that cognitive biases limit citizens' willingness to engage with risks and crisis preparedness, it is unclear how users would get in touch with such content. Therefore, it seems paramount to investigate their potential inclusion in warning apps, where citizens would be more likely to find such content and engage with it.

2.2   Citizen Engagement and Crisis Communication

Crisis communication is relevant at all stages of the crisis cycle, from pre-crisis to evaluation. Different crisis communication models relate to the public in different ways. The Crisis and Emergency Risk Communication (CERC) Model by the US Center of Disease Control identifies five stages, described in Fig. 2.2, was the

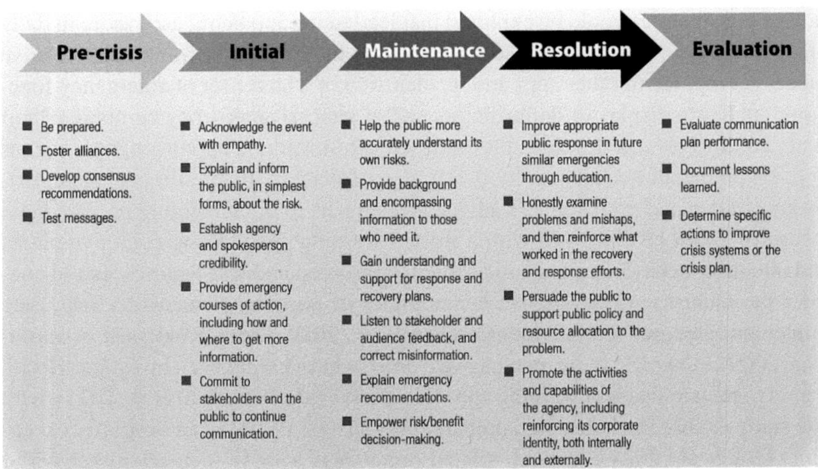

Fig. 2.2 Crisis communication model. (Source: Centers for Disease Control and Prevention (2020))

first one to combine the more long-term oriented risk communication and acute crisis communication (Bean et al., 2015). The goals of communication encompass enabling citizens to respond effectively to crises, through preemptive measures and enhanced self-efficacy during such times; enhancing public awareness of risks and knowledge about effective crisis response measures; and building trust between citizens and crisis management.

Studies related to crisis communication and public warnings have long explored the design of warning texts and the role of risk communication, e.g. (Mileti & Sorensen, 1990; Wachinger et al., 2013). Some of this work has focused on understanding recipients' reactions to warnings and on reducing milling (Bean et al., 2016; Wood et al., 2018), showing that people need to be able to identify the source of the warning and trust it, and typically seek to verify the information through further information gathering or through social cues. A study dedicated to public safety makes a number of recommendations aimed at crisis communication (Gerhold et al., 2021). It emphasises the importance of timely and effective communication between authorities responsible for disaster response and the general population. The text recommends that these communications should include information about the agency's awareness of the event, the actions being taken to manage the situation, and where and when the public can expect further updates. Specific information needs of the population, such as those gathered through media monitoring or citizens' helplines, should also be addressed. This approach aims to establish the emergency services as a trusted and relevant source of information for the public. It describes social media as a means of two-way communication and a technology that can help move away from the dominant top-down style of communication that sees citizens as passive consumers of information. This top-down style should be replaced by "crisis communication 3.0", which means actively involving citizens in the communication of crisis-related information, e.g. via a warning app and other one-to-many communication channels (Tan et al., 2017). As digital channels, the study mainly recommends using *Facebook* and *Twitter* (now called *X*), somewhat neglecting other social media and technology trends, such as the increasing use of messaging apps (Newman et al., 2022). Agencies are encouraged to establish and promote these channels in advance to maximise their reach and trustworthiness during emergencies. In disaster-prone regions, it is also suggested that separate social media channels be established for the sole purpose of disseminating situational information during events. For web-based communication, the text suggests presenting situation information using infographics and visual aids, which can make current response measures more comprehensible. Utilising geospatial data and creating an online information portal that provides details about damage events, danger zones, deployed response teams, and mitigation actions is also recommended.

A study analysing the communication after the Brisbane flood in 2011 suggests that crisis communication can be differentiated along two dimensions: strategic vs. operational, and reputation-oriented vs. resilience-oriented (Olsson, 2014). Figure 2.3 shows the resulting communication types and their definitions.

Fig. 2.3 Dimensions of crisis communication. (Source: Olsson (2014))

An analysis of the German crisis communication by the federal office for civil protection and disaster assistance holds that the top-down many-to-one communication used in mainstream media and via warning apps can be classified as operational and reputation-oriented (Wahl & Gerhold, 2021, p. 67). With regard to citizens' information needs and social media use, Dailey and Starbild find that information that is shared by agencies can be differentiated into "need-to-know" information that is actionable and essential for affected populations for decision-making. By contrast, "good-to-know" information is related to informing the general population, serving to raise awareness (Dailey et al., 2016). Summarising studies related to information needs and communication behaviour on social media in crises, Wahl and Gerhold (2021, p. 103) identify the following relevant clusters of information types: information about the well-being and whereabouts of loved ones, damage/consequences/impacts, potential impacts/consequences, description of the event or affected area, information about the affected geographical area, information about the event in general, past similar events, recommendations for action, emergency response measures by authorities, emergency response measures by the community or other private persons, dealing with other affected persons with the event, weather information or warnings, photos or videos of eyewitnesses, references to additional information about the event.

Research on social media use shows that citizens act as co-producers of crisis information through their online activities on (micro-)blogs such as Twitter or as contributors of photos and videos (Dailey et al., 2016; Gui et al., 2017; Meijer, 2014). Some work sees citizens as co-producers of public safety particularly after a crisis has occurred, investigating the contributions that their online activities make for sense-making, information sharing and problem solving (Gao et al., 2011; Haesler, Schmid, et al., 2021; Liu et al., 2016). In addition, citizens' engagement is also investigated with regard to their contributions as volunteers (Douglasdotter, 2023). However, a gap exists when it comes to investigating the roles citizens can play in co-producing safety *before* a crisis occurs. While no current study gives a comprehensive list of the state of crisis preparedness, past studies show that very few households meet the preparedness recommendations (Menski et al., 2015). According to the study, in 2015, around half of German households judged that they could encounter serious problems if the supply of telecommunication infrastructure or the water provision failed for more than one day (Menski et al., 2015). A similarly limited level of household preparedness and disaster prevention knowledge can be found in many countries across the globe (see Chapter 1). In studies on crisis preparedness, gender has been found to have ambiguous effects, at times leading to greater or less crisis response knowledge and responsibility for household preparedness (Cvetković et al., 2018). A study of fire and medical emergency preparedness in Germany comes to the conclusion that collective factors (such as number of people and children in the household, marital status) has a greater effect than individual characteristics such as gender and age (Knuth et al., 2017). Whether men and women have different attitudes regarding the domain of crisis ICT is currently an open question. Few studies investigate aspects that are related to motivating people to implement crisis preparedness measures.

A study focusing on citizen engagement in e-government and on marketing strategies suggests that groups differ according to their current engagement (Simintiras et al., 2014). To convert non-users to users, marketing should focus on the benefits (i.e., value) of adopting e-government services ("transactional" orientation). Secondly, limited use should be enhanced through the integration of additional services, allowing for seamless usage of various services ("optimisation" orientation). Finally, to foster continuous engagement, e-governance should allow users to engage in debate and public policy formulation ("representation'" orientation) (Simintiras et al., 2014). Crisis communication through digital channels can be regarded as part of e-government, which involves the use of digital technologies to deliver government services, engage citizens, and improve the efficiency and transparency of government operations. E-government services in this domain include access to information, alert systems, online training and resources, digital communication channels, crisis mapping and data sharing or public engagement in policy making. However, a literature review on e-government and crises shows a strong focus on websites for information provision and on social media for two-way communication (Roztocki et al., 2023).

Internet-enabled technologies have the potential to improve crisis communication. Wogalter and Mayhorn (2005) suggests that new and emerging technologies increase the effectiveness of warnings through warning interactivity, dynamic modification and personalisation. Key advantages are that technologies can serve as sensors that either notice a hazard or inform about it via personalised or adaptive visual and sound displays. In addition, they can help structure the response complex processes or use multimedia to display and tailor information so that it is better processed and increases compliance (Wogalter & Mayhorn, 2005). In particular, apps offer the possibility to freely design warning tools that realise these potentials. Schrock (2015) finds that distinctive communicative affordances of mobile media are their portability, availability, locatability, and multimediality. Similarly, political science has investigated the promises of new technologies to foster participation, access to information and services and to enhance deliberative processes (Coleman & Blumler, 2009). The e-government discipline is also investigating the nexus between app use and citizenship. A recent example of this nexus has been shown with the Ukrainian app *Diia*, which was built to offer e-government services and then became important during the war, ensuring the provision of services even for displaced people and those living in occupied territories.

About half of the apps that were built for disaster purposes surveyed in an English-language review study (2016-2020) allowed to gather information from citizens for the purpose of processing by an agency or organisation, e.g. for situational awareness via crowdsourcing (Tan et al., 2017). However, only two apps allowed for direct one-to-one two-way communication, with individual citizens contacting agencies via apps. Twenty of the 57 apps surveyed included community features that allowed citizens to share information with other citizens. Seventeen apps were alert or notification apps, or apps that allowed individuals to send emergency information to their contacts (Tan et al., 2017). The survey shows that the apps envision and enable citizens to have different underlying roles, often combined. People take on public roles as victim, as information receiver, as in-situ sensor and as offsite volunteer (Tan et al., 2017, p. 302).

In crises, people particularly turn to agencies as sources of information (Park et al., 2019). Designed to offer reliable agency warnings and information about disasters (Tan et al., 2017), adoption rates of warning apps in Europe remain relatively low. For example, the Swiss warning app is used by only around 12% of the population (Hauri et al., 2022), while exceptionally high usage by approximately 30% of the population is seen in Finland where the app offers a more comprehensive set of crisis functions (Hauri et al., 2022). Protection motivation, technology adoption and e-government adoption models, and user experience metrics can be used to explain the adoption of warning apps.

2.3 Technology Acceptance

Several models and theories attempt to explain technology adoption in general, and may be helpful in understanding the adoption of warning apps. Originally designed to understand technology adoption in the workplace, at its core, the models build on the theory of planned behaviour (TPB) (Ajzen, 1991) aiming to understand the intention to use a technology and the resulting usage behaviour (Lai, 2017). The theory of planned behaviour focuses on beliefs and their influence on attitude, subjective norm and perceived behaviour control, which then result in a behaviour or its intention. The technology acceptance model (TAM) (Davis, 1989) and its extension (TAM2) (Venkatesh & Davis, 1996) focus on two main aspects that influence acceptance: perceived ease of use and perceived usefulness. Ease of use addresses system characteristics which make it effortless or effortful to use and learn to use a system (Venkatesh & Davis, 1996). Perceived ease of use also influences the perceived usefulness, which indicates the extent top which a technology is helpful and contributes value for the user. Further extensions of the model have mainly incorporated various factors as independent variables. For example, the Unified Theory of Acceptance and Use of Technology (UTAUT) (Venkatesh et al., 2003) incorporates four key determinants: performance expectancy, effort expectancy, social influence, and facilitating conditions and four moderators: age, gender, experience, and voluntariness. In the original validation of UTAUT, it was able to explain 70% of the variance of behavioural intention (Venkatesh et al., 2003).

Figure 2.4 shows the various, widely used models to explain behaviour, particularly the technology adoption intention and behaviour.

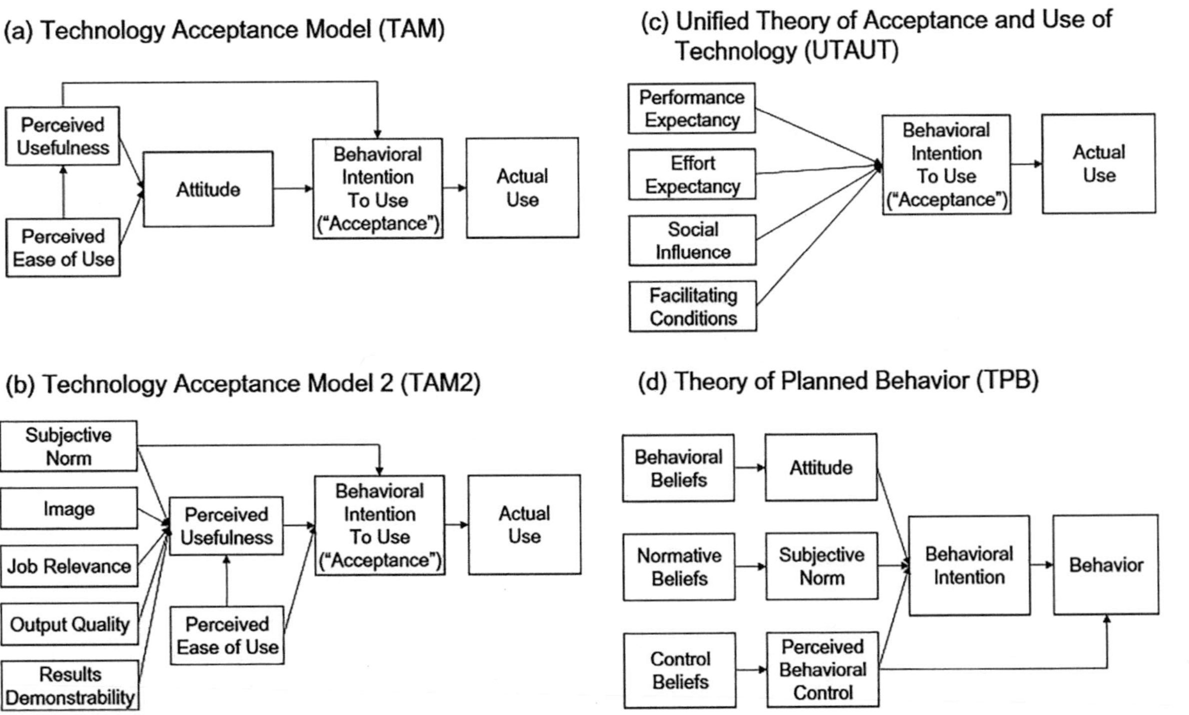

Fig. 2.4 Models explaining technology adoption. (Source: Holden and Karsh (2010), showing a) TAM (Davis, 1989), b) TAM2 (Venkatesh & Davis, 1996), c) UTAUT (Venkatesh et al., 2003), d) TPB (Ajzen, 1991))

Showing the various models and their application in different disciplines, Momani (2020) demonstrates that one strand of research is used for the development of technologies, and the other in the scientific field by social psychology, social sciences and computer science. Thus, some models focus more on understanding the psychology and context of behaviour, while others focus more on the characteristics of the system (Momani, 2020). Indeed, a study of the use of the TAM and UTAUT in health informatics shows various adoptions of the model to better fit to the health care domain. A review of the models' use in health care domain indicates mixed results, with some studies finding no effect of ease of use (Holden & Karsh, 2010) and other stressing the importance of social factors. Applying the UTAUT construct to the context of e-government adoption, Chen and Aklikokou (2020) discover that citizens' inclination to use e-government services is significantly influenced by perceived usefulness and perceived ease of use. Notably, social influence impacts perceived ease of use, while perceived usefulness remains unaffected. Factors such as facilitating conditions and trustworthiness also play a pivotal role in this scenario (Chen & Aklikokou, 2020). An analogous study on an e-government tool for citizens' incident reporting arrives at a similar conclusion (Susanto et al., 2017).

Despite the demonstrated need to adapt technology acceptance models to the domain and to identify independent variables that influence perceived usefulness and ease of use, studies of the acceptance of warning apps or other digital means for engaging citizens in crisis preparedness are absent. Protection motivation theory has been shown to be particularly successful in explaining the intention to adopt and—to a lesser extend—to keep using a warning app. It explains 69% and 45% of the intention to start using a warning app or to continue using a warning app (Fischer-Preßler et al., 2021). For both users and non-users, social influence and trust in the quality of the information provided are central, while maladaptive rewards, i.e. the prospect of saving battery and capacity by not using an app, deter both groups (Fischer-Preßler et al., 2021). For current non-users' of warning apps, perceived vulnerability and response efficacy (the perceived effect of using a warning app) positively influence the adoption intention (Fischer-Preßler et al., 2021). For users of warning apps, additional factors negatively influence the intention to continue using a warning app: The perceived costs associated with using a warning app (in terms of effort and time), self efficacy (feeling capable of using the functions of warning apps) and prior experience of an emergency (Fischer-Preßler et al., 2021). Overall, the theory and its variables only explains about half of the continuance intention (Fischer-Preßler et al., 2021).

Nonetheless, the challenge of low technology adoption is not limited to warning apps; agencies struggle with the perplexing underutilisation of e-government services (Distel & Ogonek, 2016), which is particularly pronounced in Germany

(Distel, 2020). Research has revealed that, beyond technological barriers, five distinct categories of hindrances exist: socioeconomic, communication, cultural, individual, and service-related barriers (Distel & Ogonek, 2016). These investigations are rooted in the Inhibitor Theory, asserting that potential losses often outweigh potential gains (Kahneman & Tversky, 1984), underscoring how perceived losses can impede technology acceptance. A new approach for improving user satisfaction with e-government centers around "proactivity". It involves identifying and addressing potential issues, providing services, and disseminating information before they are requested or problems arise (Ayachi et al., 2016; Scholta & Lindgren, 2019). As warning apps' information provision is based on location information or personalisation, this can be regarded as such a service. Indeed, a study on the use of messaging apps for news dissemination identifies as set of motivations and affordances that these channels offer via-a-vis other channels. People were often motivated to use messaging apps for news consumption by efficiency, which includes the containment of information overload, the convenient and effortless access via incidental new consumption without having to perform any additional steps to access the news (Lou et al., 2021). The study finds that the affordances that allow for the advantages to play out include customisability of subscriptions; increased accessibility through push notifications (Lou et al., 2021). Particularly the personalisation and the convenience of push notifications are related to the idea of proactivity and suggests that this can be a helpful aspect of digital information provision.

Moving to the measurement of technology success, usability takes center stage as a key aspect of user experience. Defined as the extent to which a product enables specified users to achieve predefined goals effectively, efficiently, and satisfactorily within a specific context DIN EN ISO 9241-11, 2018, p. 4, usability is commonly assessed through the widely used System Usability Scale (SUS) (Bangor et al., 2008; Brooke, 1995; Lewis, 2018). However, SUS alone explains just 39% of user recommendations, highlighting the relevance of other factors in shaping user perception (Sauro & Lewis, 2016). As usability predominantly addresses task-related attributes, aspects like visual appeal, utility, and trust come into play for future adoption and recommendations. In this context, the user engagement scale emerges, identifying dimensions like perceived usability, aesthetic appeal, reward, and focused attention (O'Brien et al., 2018), complementing SUS by emphasising aesthetics and likelihood of recommendation. Beyond the technology itself, users may encounter various barriers such as resistance to change, unfamiliarity with the technology, and organisational or cultural factors (Distel & Ogonek, 2016). Moreover, the platform on which a technology operates also impacts barriers. For instance, downloading specific apps incurs costs with regard to learning and not taking an action offers advantages ("maladaptive rewards" (Hassandoust & Techatassanasoontorn, 2020)), e.g. saving data and battery. Another factor relevant to non-adoption of technology is a bias towards maintaining the status quo. This has also been shown in relation to the adoption of e-government services, where it has a strong negative impact on adoption (Distel, 2020). Therefore, familiarity, current use of the system and the absence of maladaptive rewards could increase the adoption of alert channels.

In many domains, socio-demographic characteristics influence technology acceptance. Therefore, age, gender, education and income are considered by many technology acceptance and digital divide studies and they increase the explanatory power of technology adoption models (Niehaves & Plattfaut, 2014). Due to their single-item measurements, they are effective and feasible variables (Niehaves & Plattfaut, 2014). Research has found that gender, viewed as a social construct, also matters for technology acceptance. A study looking at technology adoption in the workplace indicates that stereotypically masculine individuals were significantly influenced only by attitude, whereas for stereotypically feminine participants only subjective norm and perceived behavioural control were significant predictors of behavioural intention (Venkatesh et al., 2004).

2.4   Individual Preparedness and Risk Culture

For technologies in the safety and security domain, safety-related personal attitudes such as risk perception, trust, and subjective norm are aspects that could increase usage intentions (Fischer et al., 2019). Risk perception and preparedness are complex phenomena (Wachinger et al., 2013; Weinstein & Klein, 1995). A review indicates that personal experience of a natural hazard and trust in authorities and experts have the most substantial impact on risk perception (Wachinger et al., 2013). However, the connections are not straightforward and a risk perception paradox has been identified: Perceiving a high risk does not necessarily lead to the adoption of preparedness measures. Instead, individuals with high risk perception often still choose not to prepare (Wachinger et al., 2013). Similarly, a study across several European countries indicates that, in most countries, even though worry and perceived risk correlate, only worry and concern about disasters—and not perceived risk—are related to interest in receiving information about preparedness measures (Appleby-Arnold & Brockdorff, 2018, p. 19). Intending to implement preparedness measures is strongly related to an interest in preparedness information and worry (Appleby-Arnold & Brockdorff, 2018). Germany appears to be an interesting exception, because in Germany, worry and concern about disasters is not related to preparedness intentions, which is the case in the other surveyed countries (Italy, Portugal and the Netherlands) (Appleby-Arnold & Brockdorff, 2018).

Interestingly, in some countries, feeling prepared and preparedness intentions correlate (Appleby-Arnold & Brockdorff, 2018). In some countries, feeling more informed leads to more (Italy) or less (the Netherlands) worry, and in the Netherlands also to less preparedness intentions (Appleby-Arnold & Brockdorff, 2018, p.23, 26). Again, things are different in the case of Germany. Figure 2.5 shows the generally perceived need for and willingness to receive more preparedness information across all countries. However, while citizens in most countries also state an intention to take preparedness measures (over 60% intent to prepare quite a lot or

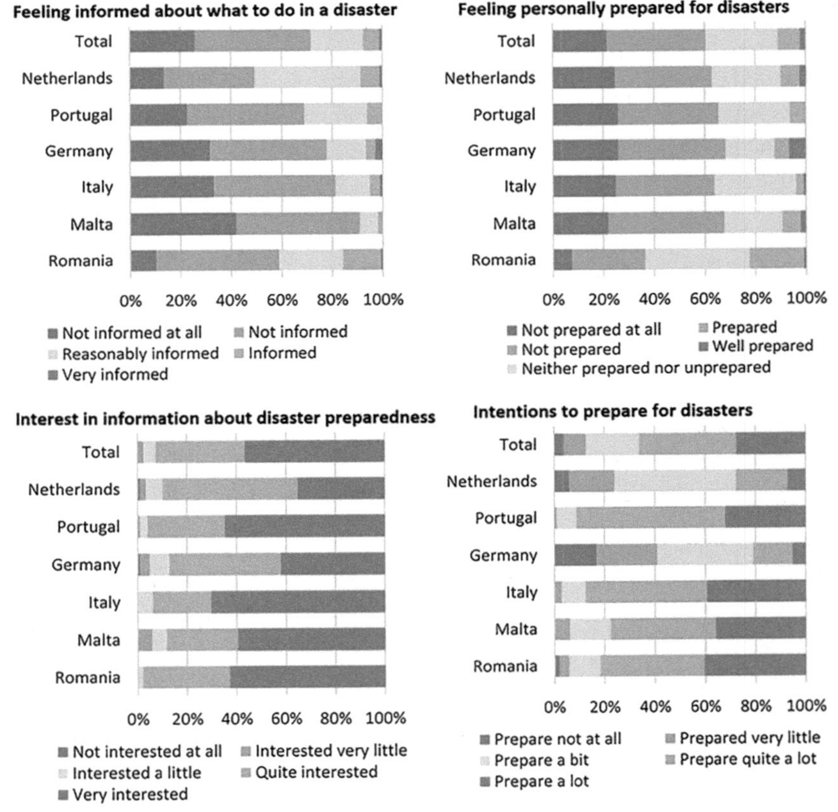

Fig. 2.5 European comparison of attitudes and perceptions regarding preparedness information and intentions. Compilation from Appleby-Arnold and Brockdorff (2018)

a lot for disasters), this is hardly the case in Germany, where almost 20% have no intention to prepare at all and another 20% intend to prepare only very little. Only around 20% intend to prepare at least quite a lot—compared to over 80% in all other countries except the Netherlands (Appleby-Arnold & Brockdorff, 2018, p. 15).

This is despite the fact that Germany has the greatest percentage of people feeling fully unprepared (Appleby-Arnold & Brockdorff, 2018, p. 14). Interestingly, interest in preparedness information is similar and high in all countries (Appleby-Arnold & Brockdorff, 2018, p. 15), while less than 25% across all countries feel at least reasonably informed about disasters (with the exception of Romania and the Netherlands, where this is the case for more than 40%).

While 10% would like to receive preparedness information only when there is an increased risk, 76% of the German sample would like to receive preparedness information at least once a year, of these 27% once every six months and only 13% once every three months (Appleby-Arnold & Brockdorff, 2018). This is similar to the Netherlands, but significantly less than in most other states (Appleby-Arnold & Brockdorff, 2018). These findings underline the impact of contextual factors such as culture and the particular challenge of motivating German citizens to prepare for crises.

The variations between the countries, particularly the lack of preparedness intentions in Germany, can also be explained by national *risk culture*. Risk culture in the context of disasters warnings (Cornia et al., 2016) consists of three dimensions: *Disaster framing* describes the interpretation of the "role of human agency, i.e. whether people frame disasters emphasising human mastery over hazards or, on the contrary, human subjugation to hazards" (Cornia et al., 2016, p. 4). Another dimension is trust in authorities (*trust context*), which refers to the extent to which people are willing to rely on state emergency management institutions, often related to past positive or negative experiences. Finally, the dimension *disaster blaming* describes the attribution of responsibilities to people or entities to explain the disaster. Taken together, risk culture explains the degree to which people perceive that it is necessary or adequate to prepare for a disaster. Cornia et al. (2016) identify three ideal types of risk cultures, which are depicted in Fig. 2.6.

Especially in fatalistic risk cultures, people tend to believe that disasters are unavoidable and that little can be done to prevent or counter them. In state-oriented risk cultures, on the other hand, citizens perceive that state agencies are responsible and capable of managing a crisis, which reduces the perceived responsibility to take the initiative. Research also suggests that this affects the extent to which information and communication technologies, such as warning apps and social media, are used in crises (Reuter et al., 2019). However, the paradoxical position of German citizens with regard to preparedness indicates a large gap in moving towards a culture of

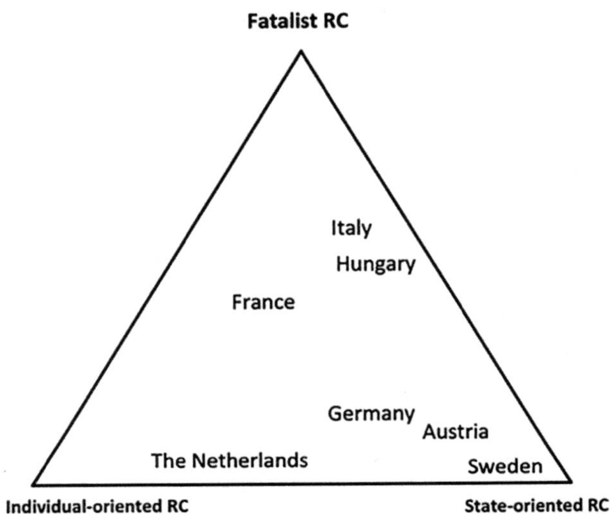

Fig. 2.6 Classification of countries by risk culture ideal type. (Source: Cornia et al. (2016))

preparedness. This may also be due to the strategies used by disaster management agencies: While themselves highly trusting in emergency management authorities, Germans largely perceived that authorities do not trust them to respond appropriately or to prepare (Appleby-Arnold & Brockdorff, 2018, p. 33). The key characteristics of each ideal type risk culture can be summarised as follows (Cornia et al., 2016):

- **Individual-oriented risk culture:** Disaster harm is caused by lack of preventive measures, which should be put in place *by the citizens*, who are responsible for putting in place preventive measures for themselves and their family. Citizens' trust in authorities can be either high or low, and disaster consequences are to be blamed on careless citizens.
- **State-oriented risk culture:** Disaster harm is caused by lack of preventive measures, which should be put in place *by public authorities*, who are responsible for protecting their citizens through preventive measures. Citizens' trust in authorities is high. but the state is to be blamed in case of harm caused by disasters.
- **Fatalistic risk culture:** Disaster harm cannot be avoided, therefore *nobody* is responsible for disaster prevention or can be blamed for disaster consequences. Disaster management focuses on rescue and post-event phase rather than prevention. Trust in state authorities is low.

Concerning the state-oriented risk cultures, as found in Germany, which neglect individual preparedness (Cornia et al., 2016; Reuter et al., 2019), warning apps could focus more on increasing crisis preparedness. Research shows that the ways in which crises are confronted is influenced by a country's risk culture (Cornia et al., 2016). In some cultures, the notion of individual responsibility is greater, leading people to take actions and to use social media more. In state-oriented risk cultures (such as Germany), trust in authorities is high and it is expected that the authorities will provide in case of a crisis, while trust in social media is lower (Appleby-Arnold et al., 2019; Reuter et al., 2019). However, studies focusing on warning app users' expectations in a state-oriented risk culture are so far rare and preparedness features have not been investigated. A report on how to measure individual preparedness shows that there is a lack of empirical evidence on the effectiveness of preparedness measures, such as maintaining a first aid kit (Kohler et al., 2020). The authors emphasise the opportunity to investigate the COVID-19 pandemic as a case of a slow-onset global crisis. In addition, they suggest that more research should investigate smartphones as preparedness tools, as smartphones commonly include relevant crisis features such as flashlights and the ability to receive radio frequencies, and they can be enhanced by crisis-relevant apps, digital storage of documents or offline preparedness advice (Kohler et al., 2020).

The importance of emotional rather than analytical assessments of disaster risk and the lack of preparedness behaviour across countries (Appleby-Arnold & Brockdorff, 2018) might to some degree be explained by psychological research, which shows that cognitive biases can lead people to make sub-optimal decisions (Kahneman et al., 1982; Siegrist & Árvai, 2020). As a result, even when knowledge about risk and preparedness measures exists, feasible measures are not always implemented. Citizens often perceive the likelihood of a disaster as too low to take measures, and assume that they will receive information soon enough through other channels, such as the media (Reuter, Kaufhold, Leopold, & Knipp, 2017). Many interventions exists to encourage behaviour change (Michie et al., 2013), which can be supported by persuasive technology design (Orji & Moffatt, 2018). Among these interventions, nudging seeks to alter "people's behaviour in a predictable way without forbidding any options or significantly changing their economic incentives" (Thaler & Sunstein, 2008, p. 6). *Nudges* are interventions that are made at the moment of decision-making to influence people to make wise decisions, for example when creating passwords (Sunstein, 2015; Zimmermann & Renaud, 2021). Therefore, in addition to increasing awareness and knowledge about disaster risks and intervention costs and benefits, motivational strategies should be explored to support citizens in acting according to their intentions and to contribute to greater public safety (Caraban et al., 2020).

2.5   Synthesis of Factors Relevant for Warning App Adoption

Protection motivation is a theoretical construct that was designed to explain the adoption of warning apps (Fischer-Preßler et al., 2021). It explains the intention to adopt a warning app and to a lesser extent the intention to continue using one. Its most significant constructs are social influence, trust in the information quality and maladaptive rewards (advantages derived from not using a warning app, such as saving capacity or battery on a smartphone) (Fischer-Preßler et al., 2021). In Germany, the perceived level of disaster risk influences the preparedness intention, whereas in other European countries disaster worry is more significant (Appleby-Arnold & Brockdorff, 2018). Furthermore, preparedness intention correlates with current preparedness, indicating that those who have already implemented preventive measures are more motivated to implement more measures (Appleby-Arnold & Brockdorff, 2018). This might be explained by the importance of the personal norm for crisis preparedness, which has been suggested to be relevant in flood preparedness (Mol et al., 2021). A personal norm indicates what individuals perceive to be appropriate behaviour or what they feel morally obliged to do (Schwartz & Howard, 1984). Additionally, trust and social norms influence the intention to use warning apps, with partner usage and social influences positively impacting adoption (Fischer et al., 2019; Tan et al., 2020c). State-oriented risk cultures, like Germany, emphasise trust in authorities but may neglect individual preparedness (Appleby-Arnold & Brockdorff, 2018; Cornia et al., 2016; Dallo & Martí, 2021; Reuter et al., 2019). Germans are more skeptical of social media use during crises and more likely to use a disaster app instead (Appleby-Arnold et al., 2019; Reuter et al., 2019).

In addition to these personal and cultural factors, the technology itself is also relevant for whether or not it is adopted. As indicated by the TAM, usefulness and usability are the main factors of adoption intention (Davis, 1989). With regard to warning apps, Tan et al. (2020c) find that users continue using apps that effectively perform their intended function, are reliable, and make critical information salient and easy to understand (Tan et al., 2020c). Usefulness increases with the centralisation of warning scenarios in one app and personalised alerts (Dallo & Martí, 2021). On the one hand, cultural and personal factors also likely influence the perceived usefulness. As citizens in different countries place more or less trust in authorities or feel to varying degrees personally responsible for preventing and managing disasters, they are also likely to judge warning apps and warnings differently (Cornia et al., 2016; Reuter et al., 2019). However, it is unclear whether state-oriented risk cultures lead to citizens adopting warning apps, because they feel that these deliver valuable information transmitted by trusted state agencies. On the other hand, citizens in state-oriented risk cultures may also refrain from using warning apps as they do not feel personally responsible for being informed and acting in case of an emergency. Another factor stressing the relevance of usefulness for continued use is the finding of a negative correlation between continuance intention and self-efficacy. Current users of warning apps who perceive that the apps are easy to use appear to want to continue using a warning app *less*, possibly because they perceive the available features as too limited (Fischer-Preßler et al., 2021). Finally, technology design and usability influence adoption and continuance of use. Usability guidelines focus on information salience, low cognitive load, building trust, simple design, minimal resource usage, clear audio output, graphical design and minimal user interface input (Tan et al., 2019, 2020b, 2020c). As a consequence, usability also influences discontinuation, with complex user interfaces and high resource use leading to disuse (Fischer-Preßler et al., 2021; Tan et al., 2020c). Research shows that warning apps have additional requirements compared to other mobile apps (Tan et al., 2020b).

> **Summary—Key Frameworks Related to Adoption and Use of Warning Apps**
>
> - **Protection motivation and risk perception:** Main constructs are social influence, trust in the information quality and maladaptive rewards (advantages derived from not using a warning app, such as saving capacity or battery on a smartphone). In Germany, the perceived level of disaster risk influences the preparedness intention (Appleby-Arnold & Brockdorff, 2018).
> - **Cultural and social norms:** State-oriented risk cultures, like Germany, emphasise trust in authorities but may neglect individual preparedness (Appleby-Arnold & Brockdorff, 2018; Cornia et al., 2016; Dallo & Martí, 2021; Reuter et al., 2019;). Germans are more skeptical of social media use during crises and more likely to use a disaster app instead (Appleby-Arnold et al., 2019; Reuter et al., 2019). Social norms influence the intention to use warning apps, with partner usage and social influences positively impacting adoption (Fischer et al., 2019; Tan et al., 2020c).
> - **Usefulness:** For warning apps, making critical information salient and easy to understand depends on effective and reliable alerts (Tan et al., 2020c), and on centralisation of warning scenarios in one app and personally relevant alerts (Dallo & Martí, 2021). Risk culture may influence perceived usefulness, as citizens place more or less trust in authorities or feel to

varying degrees personally responsible for preventing and managing disasters (Cornia et al., 2016; Reuter et al., 2019).

- **Usability:** Usability guidelines focus on information salience, low cognitive load, building trust, simple design, minimal resource usage, clear audio output, graphical design and minimal user interface input (Tan et al., 2019, 2020b, 2020c). Usability also influences discontinuation, with complex user interfaces and high resource use leading to disuse (Fischer-Preßler et al., 2021; Tan et al., 2020c).

2.6   Research Gap

Looking at the research on citizen engagement in crisis response and crisis preparedness, several gaps emerge: Firstly, the primary emphasis of the research centers on the capabilities of social media and their socio-technical innovation. This research has focused on citizen reporting and citizens' contributions to situational awareness (Kaufhold, 2021). This has led to an emphasis on the affordances of social media for crisis management from an agency perspective (Bullock et al., 2021; Eismann et al., 2021) and on the numerous ways that social media are used in crises for communication and situational awareness, emotional support and coordination (Gui et al., 2017; Haesler, Schmid, et al., 2021, Hughes et al., 2008; Kaufhold, 2021; Reuter, 2022; Reuter & Kaufhold, 2018; Vieweg et al., 2010), and to improve trust and community relations (Akkaya et al., 2019; Bullock, 2018; Denef et al., 2012). In addition, research has looked at the ways that social media enables distributed problem-solving and digital volunteerism (Palen et al., 2009; Reuter, 2022). These aspects play a crucial role in understanding how social media enable new forms of engagement, citizen contributions and improved situational awareness. However, not all people are willing and able to engage with social media during emergencies (Appleby-Arnold et al., 2019; Reuter, 2022). In addition, social media platforms are susceptible to the dissemination of false and misleading information (e.g. Kaufhold et al., 2020; Marchal & Au, 2020; Starbird et al., 2014) and information overload can occur (Park, 2019; Schmitt et al., 2018). Therefore, more studies in crisis informatics are needed that investigate the role of channels other than social media for reaching out and engaging citizens, especially given the decline in the use of public social media platforms for news consumption (Lou et al., 2021; Newman et al., 2023).

However, this leaves research gap regarding the limited number of studies examining warning apps and their potential to foster citizen engagement in the realms of crisis management and preparedness. This represents an important area for exploration, suggesting an opportunity to delve into the role that warning apps can play as a highly adaptable channel of direct citizen-agency communication. Bridging this gap contributes valuable insights into the role of different technologies in managing crises and in enhancing public participation, ultimately contributing to more effective crisis management strategies.

Few studies have investigated warning app usability (Tan et al., 2019, 2020a) and they have found that warning apps have design requirements that are specific to the crisis use context and the resulting infrequent use. Technology adoption has been studied extensively in other domains, especially mobile health (Holden & Karsh, 2010; Mackert et al., 2016; VanAhn et al., 2019), but research has only started to explore the adoption of warning channels (Fischer-Preßler et al., 2021; Tan et al., 2020a, 2020c). While this research has yielded important insights, these are mainly based on user reviews in app markets. Two large quantitative studies have investigated warning app usage from the user perspective: One study by the Canadian Red Cross in 2012 investigated social media use in crises (Canadian Red Cross, 2012). Another more recent study analysed representative samples from four European countries with regard to social media use and warning app use in crises (Reuter et al., 2019). Yet, this study contained only three questions directed at the extent to which people had experience with downloading and using warning apps and other disaster-related apps. Another qualitative study investigated similar aspects to compare attitudes and used crisis information sources in different countries (Appleby-Arnold et al., 2019, 2020). Only one study compares a large set of crisis communication channels for warnings, focusing on the features of warning apps in Europe (Hauri et al., 2022). Other recent studies have started to explore warning app adoption and (dis-)continuation of use (Fischer-Preßler et al., 2021; Tan et al., 2020c), which gives insights into user needs and user satisfaction. Therefore, a second research gap concerns the existence of little user-centred research on warning apps.

Thirdly, many studies have investigated behaviour change in domains such as health (Matthews et al., 2016; Nwafor et al., 2021), sustainability (Beermann et al., 2022; Zimmermann, Hein, et al., 2021) and education, with safety and security topics receiving limited attention (a notable exception being password creation (Zimmermann & Renaud, 2021)). However, only one study has focused on motivational aspects and nudging and disaster preparedness, looking at flood reduction investments (Mol et al., 2021). One study has analysed user reviews to identify user comments related to the persuasiveness of a warning app (Kotthaus et al., 2016). A third gap thus consists in a lack of studies dedicated to nudging in crisis preparedness and the use of persuasive elements in warning apps.

Summary—Research Gaps

- **Few Studies of Warning Apps**: Few studies have investigated the contribution of warning apps in the crisis information ecosystem. However, due to lower usage of social media in some socio-demographic groups, design constraints imposed by social media platforms that follow financial interest, more studies are needed that investigate warning apps as channels of verified information that, independent of social media platforms, offer singular warning and crisis preparedness capabilities and design possibilities.
- **Lack of User Centred Research on Warning Apps**: Studies in crisis informatics are often dedicated to improving the work processes of crisis managers. However, only scant attention has been paid to users' expectations and preferences regarding warning apps or to the factors affecting usability and continuance of use. Currently, no studies exist that have used the user-centred design process to explore and evaluate design solutions for warning apps.
- **Limited Research on Motivating Citizens for Safety**: While there is extensive research on citizen-generated data in crisis response, there is comparatively less focus on motivating citizens to actively contribute to public safety and to prepare for crises. In addition, research has investigated behaviour change in many domains, but they have largely omitted crisis preparedness and response. Finally, no studies have sought to use warning apps for digital nudging or persuasive system design that may enhance citizens' engagement with crisis management.

Research Design 3

The following chapter describes the research fields that this work is situated in (Sect. 3.1). It then discusses the contributions of empirical studies and design studies (Sect. 3.2), as well as how they complement and build on each other in a mixed methods approach (Sect. 3.3). Finally, this chapter outlines the research context in which the thesis was developed, with a focus on the resilience of smart cities in relation to security incidents (Sect. 3.4).

3.1 Research Field

Focusing on users' technology acceptance and engagement in crises, the dissertation is situated in the disciplines of crisis informatics, human-computer interaction, information systems and e-government. As a field with a particular focus on human behaviour in relation to technical systems, HCI is "concerned with the design, evaluation and implementation of interactive computing systems for human use and with the study of major phenomena surrounding them focuses on facilitating effective and intuitive communication between humans and computers by investigating the design, evaluation, and implementation of interactive computing systems in social, organisational and political environments" (Hewett et al., 1992, p. 5). HCI has developed over time, shifting its focus on computing technology (first wave), to cognitive psychology (second wave) to social science (third wave) (Harrison et al., 2007). While technology and human cognition remains at the core of HCI, the discipline now also stresses the importance of social and cultural context: *"The assumption that the needs and concerns of human users are an intrinsically important part of computer system design is central to HCI work. As this assumption leads researchers and practitioners to move beyond specific interface design questions toward the*

J. Haunschild, *Enhancing Citizens' Role in Public Safety*, Technology, Peace and Security I Technologie, Frieden und Sicherheit, https://doi.org/10.1007/978-3-658-46489-9_3

*consideration of larger contextual issues, societal and political questions neces-
sarily intrude"* (Hochheiser & Lazar, 2007, p. 339). This shift can be seen in the
way that HCI research has been engaging with the concept of utility (see Fig. 3.1).
Toomim et al. (2011) show that in the 1960s, goals were predefined and measures
via efficiency and accuracy, with a strong focus on operation. In the 1980s, the goals
were still predefined but users were free to choose their tool, leading to a focus on the
application, with measures of success becoming user-friendliness and learnability.
According to the authors, another shift in the 2000s lead to an understanding of users
choosing both the goal and the tool, and the design metric became utility. Indeed,
many researchers use the term *usefulness* to indicate this understanding of users'
free choice and their evaluation of a tool's overall value, as opposed to its value
related only to accomplishing a task: "Usefulness can be considered as a construction because the technological usefulness of software artefacts remains unknown,
or known only indirectly until they are in use" (Nocera et al., 2007, p. 155). Instead
of the practical acceptability of technology, Nocera et al. (2007) point out that the
sociology of technology (Bijker et al., 1987) understands usefulness as "actively
and socially constructed by users rather than merely perceived as a property of technology" (Nocera et al., 2007, p. 155). In the context of crisis communication tool,
the term usefulness is thus more instructive, as it opens up the analyses to cultural,
social and political contexts that shape norms and values. Research in information
systems (IS) is dedicated to improving the effectiveness and efficiency of organisations (Hevner et al., 2004). To achieve this, technology must be accepted by users,
which has lead to a great emphasis of user perceptions (Davis, 1989).

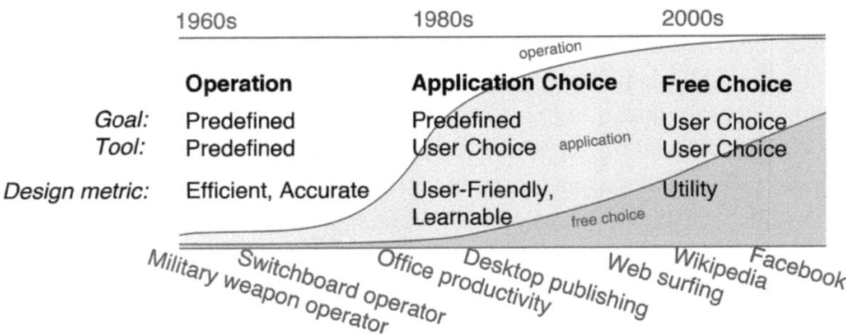

Fig. 3.1 Development of utility over time. (Source: Toomim et al. (2011))

The social and political context is highly relevant in the application domain of
civil protection and agency-citizen communication, as it related to the governance of

security and the provision of state services. The sub-discipline of crisis informatics does justice to this nexus, as it "views emergency response as an expanded social system where information is disseminated within and between official and public channels and entities" (Palen et al., 2009, p. 469).

Certain aspects, such as the dependability of a warning tool, can be measured objectively. when it comes to the interaction with users, judgement and interpretation become relevant. While aspects of perception can also be measured—e.g. as is done in the technology acceptance model, which measures *perceived* usefulness—interactions are influenced by aspects such as values, norms or experiences, which result in users *constructing* the reality of their experience. Epistemologically, this view of reality adheres to constructivist ideas (Wendt, 1992). This school of thought assumes that security and safety are not objective, but influenced by perception. Constructivism has therefore engendered analyses of framing, perceptions, culture and situated interactions. In addition, instead of seeing humans as the ideal *homo oeconomicus*, who is a fully rational agent who weights gains against costs (rational choice approach), the constructivist school stresses the importance of norms and values, and accounts for emotions and cognitive biases. Constructivism has also been applied to computer science research (Duarte & Baranauskas, 2016; Nordmann & Ripper, 2019), which has established theories, frameworks and methods that resonate with constructivist ideas. For example, value sensitive design seeks to ensure a reflection of which values and whose values are fostered by a technological intervention (Friedman & Kahn Jr, 2007). Other work has similarly considered the impact of culture on technology use and technology acceptance (Cornia et al., 2016; Reuter et al., 2019; Venkatesh & Davis, 2000), the role of habits (Limayem et al., 2007), or social norms, e.g. for privacy (Horne et al., 2015). Applied to computer science, the constructivist view stresses that technology and interactions mutually influence each other, that norms, values and framing matter for technology interactions and that humans make sense of technology in a situated context (Leonardi & Barley, 2010).

Crisis communication can be understood as a "wicked problem" due to the complex and multifaceted nature of managing and responding to crises effectively. The term was coined by social scientist Rittel (1973) and refers to problems that are difficult to define, have no definitive solution, are intertwined with numerous interconnected factors and are characterised by involved stakeholders having competing preferences. Crisis management and crisis communication fit this description, because they are characterised by complexity and uncertainty. Disasters cannot be predicted and their effects and management are highly context dependent (Lee, 2016). Therefore, there is always a lack of an identical precedent. As each experience is new, there is no clear playbook to follow, making crisis communication a challenging, evolving and culturally contingent process. While crises often provide opportunities for learning and improving in crisis communication strategies, the complex and unique nature of each crisis can make it difficult to effectively transfer lessons from one situation to another. Another challenge lies in the diffuse aim of achieving "preparedness". The questions of how prepared is well-prepared and what constitutes too much preparedness can never be answered objectively. Instead, it depends on the subjective value that is placed on safety and security and on the outcome of future events. A person who may be perceived as overly prepared may be proven right in the event of an emergency. A person investing in preparedness measures without ever needing them may conclude that they have been "too" prepared. Therefore, communicating the "right" measure of risk and the "right" amount of preparedness is a wicked problem.

As such, crisis communication does not have a clear one-size-fits-all solution. Instead, it requires a flexible, adaptable, and context-specific approach. The wicked problem paradigm was originally developed to explain planning and governance issues, but has also been applied to describe resilience (Lee, 2016) and information science (Kunz & Rittel, 1972). As such, it has been used in the discipline human computer interactions (HCI) (Agrawal et al., 2020; Kuznetsov & Tomitsch, 2018) and design thinking (Buchanan, 1992). In particular, the framework draws attention to the constraints and tensions that shape the search for solutions (Tatar, 2007). Regarding the design aspects of crisis communication, the complexity and constraints emerge from the crisis context as outlined above, but also from tension between the logic of government agencies' e-government approaches and citizens' technology use, from privacy concerns, from creating joyful interactions in a serious context that requires maintaining trust and from tensions between cultural norms and personal norms.

From the computer science disciplines, the dissertation applies the following approaches to interactions with technology in crises:

- **User-centred research:** User-centred research follows the user-centred design process, which is an iterative process based on understanding the user context, specifying user requirements, producing design solutions and evaluating them (DIN EN ISO 9241-11, 2018). The approach emphasises involving users throughout the design process and understanding their needs, goals, and preferences (Da Silva et al., 2011; Keinonen, 2008), e.g. through participatory design, requirements analyses or surveys (Vredenburg et al., 2002). Applied here, the approach ensures that crisis warning aligns with users' needs, requirements and interests, resulting in more usable interfaces and more useful tools. An important element of this is investigating users' experiences, including via prototypes (Kangas & Kinnunen, 2005).

- **Human factors and psychology:** By considering cognitive theories and models, HCI helps to understand how people perceive, process, and comprehend information and make decisions in crises. This is particularly relevant with regard to cognitive biases surrounding risk perception (Eiser et al., 2012) and effects such as milling or procrastination related to preparedness measures (Wood et al., 2018). Methods and design interventions in HCI have applied choice architecture models to technology design, resulting in the development of persuasive technology design and digital nudging (Bergram et al., 2022; Caraban et al., 2019).
- **Mobile computing:** This approach draws attention to the advantages and challenges associated with using mobile technology. In crisis informatics, mobile technologies are central, as they allow immediate notification and rely less on infrastructure which may be damaged in a disaster due to their battery use (Kaufhold et al., 2021). However, mobile technology use also comes with challenges, such as smaller interfaces, a reduced ease of typing and limited battery, space and data traffic resources (Huang, 2009).

As a domain that is mainly understood to be a responsibility of the state, providing information for crisis management can be regarded as an aspect of e-government. Due to its embeddedness in a state agency and the elements that make it akin to a "wicked problem", an understanding of the political structures and cultural-historical contexts is important to ensure that systems are compatible with the environment that they are designed for. As such, the dissertation applies several approaches from political science:

- **E-government adoption:** Looking at tools that are provided by emergency management organisations, we use public administrations' focus on e-government adoption to help understand the use and non-use of technologies provided for public information provision (Distel & Ogonek, 2016; Rana et al., 2015). In particular, in this dissertation, the service ideal is prioritised over the other e-government ideals (such as professional or efficiency) (Distel & Lindgren, 2019), focusing on „[improving] the availability, accessibility and usability of government services by providing them online" (Distel & Lindgren, 2019, p. 4).
- **Citizen engagement as polycentred governance:** Particularly, the dissertation uses the governance concept of co-production to manage disasters and risk (Djalante et al., 2013; Paniagua & Rayamajhee, 2022). It was conceptualised by Elinor Ostrom as „the process through which inputs used to produce a good

or service are contributed by individuals who are not "in" the same organisation. […] Coproduction implies that citizens can play an active role in producing public goods and services of consequence for them" (Ostrom, 1996). As such, she describes a mode of polycentred governance that goes beyond regulation either by the state or the market and includes invited (top-down) or citizen-initiated (bottom-up) activities (Khine et al., 2021).

3.2   Research Approaches

HCI researchers and practitioners employ various methods and approaches to study and improve the interaction between humans and computers. Research can be conducted in an artificial setting (i.e. a laboratory experiment) or in a natural setting (i.e. field studies, action research or case studies), or it can be independent of the environment, such as normative or survey research. Research purposes include understanding, describing, engineering, re-engineering and evaluating (Kjeldskov & Graham, 2003). Wobbrock and Kientz (2016), show that contributions in HCI can be clustered into the following seven categories: (1) empirical, (2) artefact, (3) methodological, (4) theoretical, (5) benchmark/dataset, (6) survey, and (7) opinion. Through the development of new artefacts, studies can empirically investigate the effect of designs and features on perceived usefulness and user engagement. Similarly, the discipline of information systems is based on the two paradigms of behavioural science and design science. Behavioural science "seeks to develop and justify theories (i.e., principles and laws) that explain or predict organisational and human phenomena surrounding the analysis, design, implementation, management, and use of information systems" (Hevner et al., 2004, p. 76). Design science, rooted in engineering, "is fundamentally a problem-solving paradigm [that] seeks to create innovations that define the ideas, practices, technical capabilities, and products through which the analysis, design, implementation, management, and use of information systems can be effectively and efficiently accomplished" (Hevner et al., 2004, p. 76). Based on pragmatist thought (Aboulafia, 1991), Hevner argues that technology and behaviour are not separate, but rather "truth (justified theory) and utility (artefacts that are effective) are two sides of the same coin and […] scientific research should be evaluated in light of its practical implications" (Hevner et al., 2004, p. 77).

Examining the connections between human behaviour and artefact design in HCI, Adam et al. (2021) argue that HCI makes three types of design-inspired research. In the *exterior mode*, the focus is on designing work systems and studying human behaviour by conducting mainly observational studies. The *interior mode* is focused on IT system design and technical studies examining technical improvements. Finally, the *gestalt mode* focuses on a "synergistic balance between IT systems and human behaviour" (Adam et al., 2021, p. 4) by combining technical and observational studies.

Following the *gestalt mode* of HCI (Adam et al., 2021), the dissertation combines empirical and design research studies, aiming for a synthesis of user behaviour and artefact design that, in combination, lead to a greater engagement with crisis information. The empirical studies almost exclusively employ methods that elicit data from the prospective users and users of warning apps. Therefore, they are fully in line with user-centred design. As the design studies which produce an artefact are iteratively designed based on user feedback and evaluated by users, they also adhere to this paradigm. Figure 3.2 shows the HCI contributions.

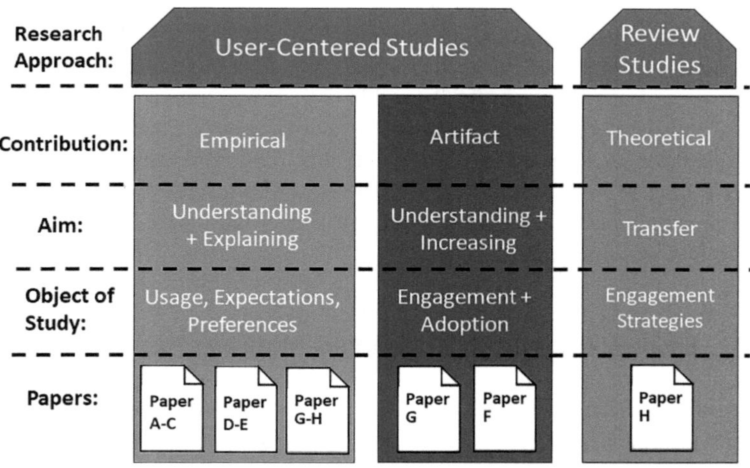

Fig. 3.2 HCI contributions of papers, based on contribution types by Wobbrock and Kientz (2016)

3.3   Research Methods

The research conducted for this dissertation is sensitive to ethical issues. With regard to researching warnings and crisis experiences, particular attention was paid to ensuring that there would never be any doubt about the truthfulness of real warnings. Additionally, we only chose research designs that would not induce fear or stress in participants. We strove to include diverse participants to design warning channels that would serve the needs of a large part of the citizenry (Hochheiser et al., 2010). For each research step, issues of privacy, data security and participant safety and well-being were considered. For each study that was conducted with research participants, the terms for the data acquisition were made transparent and all participants agreed by signing a consent form or agreeing to the terms of the experiment. Additionally, in order to further value the time spent on the research, participants could choose to be informed about the outcome of the studies. In paper G, participants were deceived concerning the true purpose of the study, in order to avoid biasing the participants. In addition, due to the potential impact of warning technologies in case of emergencies, the research approach ensured that participants could always fully discern the alerts that pertained to the study. Due to the taken precautions, the research was accepted by the ethics board of the Technical University of Darmstadt (EK 41/2021). After the study, participants were debriefed and informed about the true purpose of the research.

The data were stored on the devices of the involved researchers. Any online data and services were hosted on servers within the EU and compliant with EU data security and privacy law, or on the university's own servers. Qualitative data were transcribed and anatomised, and anonymity was ensured during data collection, e.g. by asking private data only as clusters, in order to ensure that participants would remain anonymous. The original voice recordings were deleted once the transcription was finished.

3.3.1   Empirical Methods and Mixed-Methods Approach

Empirical studies in HCI explore user interactions with systems and users' needs. In line with this, Papers A, B and C use mainly quantitative methods to inquire about user needs, experiences and preferences. Papers D and E use qualitative and quantitative methods to understand how users behave in a crisis and how they evaluate different information channels. However, a central aspect of the success of public warnings revolves around reaching a large section of the population. Therefore, the empirical studies target not only people who have experience with warning apps, but also those, who do not. Furthermore, the studies cover both people who have had experiences with crisis situations (about 60% of the samples in the representative studies) and those who have not. Empirical studies were also used to evaluate the user experience interacting with the prototype artefacts. The evaluation studies consist of a field experiment in which users interacted with a warning app prototype for one week. Iterative evaluations were also conducted via task-based interviews, in which users performed a series of tasks and commented on their experiences, supplemented by semi-structured interview questions. After another design itera-tion, the artefact was again evaluated through a survey, with users interacting with the tool on their own devices. Table 3.1 summarises the empirical studies conducted in the course of the dissertation.

Table 3.1 Empirical studies presented in this dissertation

Method, Date	Topic	Sample	Paper
Focus Group, Feb. 2019	Practitioners' view on warning app and social media use in crises	German central federal police agency employees: $N=12$ civil protection, $N=15$ police	Paper A, C
Rep. Survey, May 2019	Citizens' disaster app and social media use, expectations and preferences in crises	$N=1{,}219$ participants representative of German population	Paper A, C
Rep. Survey, Oct. 2021	Citizens' (new) disaster app usage patterns, expectations and preferences in crises	$N=1{,}090$ participants representative of German population	Paper B, G
Daily Online Survey, April 2020	Diary entries about crisis information channels, preferences and information gaps	Residents from Hesse, Germany, $N=133$ instances by $N=24\text{--}47$ participants	Paper D
App Store Reviews, March-May 2021	Helpfulness and gaps of COVID-19 regulation apps	$N=832$ reviews of three apps	Paper E
Design Workshop, June 2021	Development of persuasive nudges for disaster preparedness	$N=5$ participants	Paper G
Field Experiment, Dec. 2021	Test of the effectiveness of nudging for disaster preparedness	$N=76$ participants in three groups	Paper G
Semi-Structured Interviews, Feb. 2023	Usability of a messaging app warning channel	$N=14$ users	Paper H
Evaluation Survey, March 2023	Usability and usefulness of a messaging app warning channel	$N=26$ users	Paper H

The dissertation follows a mixed methods design, which can be chosen for many different purposes: (1) to corroborate findings (triangulation), (2) to enhance or clarify findings (complementarity), (3) to inform or develop another method (development), (4) to explore a topic with the aim of initiating another research question (initiation) and (5) to extend the breadth and range of inquiry (expansion) (Schoonenboom & Johnson, 2017). The answer to RQ1, concerned with usage and user expectations, is derived largely from representative surveys (Papers A, B and C). This quantitative survey design allows conclusions about majority opinions and identifies significant differences between socio-demographic groups. Additionally, by asking about crisis experiences, the surveys offer a large sample to analyse only the group of people who report having used a specific technology for crisis information in a crisis. Guidelines for valid item design were used, paying attention to phrasing positively, clearly, short, concisely, and understandably, limited to one statement per item and avoiding leading questions (Moosbrugger & Kelava, 2012). In each lengthy survey, quality questions were utilised to ensure that participants were paying attention and those not passing these questions were excluded. The samples represent the German population and match its distribution with regard to gender, age, formal education, income and, in some cases, residence in a federal state within Germany. Concerning quantitative data collection, in 2017, 2019 and 2021 ISO-certified panel providers were used to recruit the samples.

The analysis utilised various statistical tests, including Chi-Square tests, Pearson's Phi, Cramer's V, Kendall's tau-b, ANOVA, and Pearson's correlation coefficient, depending on the scaling of the dependent and independent variables. For categorical variables, such as gender, the t-test for independent samples was applied, paying attention to the assumption of homogeneity of variance through a Levene test. In some instances, non-linear correlations, were tested, e.g. to test for generational differences. To test these, data are recoded into categories to test group effects, such as binary categories for those under 25, over 45 and those over 60-years old. Effect sizes of Pearson's r of |0.10| as a small, of |0.30| as a moderate, and of |0.50| as a strong correlation (Cohen, 1988). For the statistical analysis, IBM SPSS Statistics 26 was used. When possible, the data from different years were compared. Details about the data collection for each chapter, such as the survey questions and items, can be found in the electronic supplementary material.

Quantitative surveys also have some disadvantages: In particular, a limitation of the surveys was that participants made only brief remarks in the open questions. Therefore, the surveys lacked some nuance. Additionally, because the surveys inquired about past behaviour in crises, the surveys relied on participants' memories, which are never entirely reliable. Following the multi-method and mixed-method approach, additional methods were combined to gain further insights and to add nuance. The surveys included open-ended questions to explore unforeseen aspects as part of an inductive approach (Mayring, 2014). For qualitative coding, coding schemes were developed, both deductively, using existing theory, and inductively, deriving new codes that emerged from the texts (Mayring, 2014). For the analyses, MAXQDA and RQDA were used. Intercoder reliability was tested using Cohen's Kappa k (Altman, 1991).

Papers A and B collected new representative data for the years 2019 and 2021 and compared these to data from previous years, analysing how warning technologies were currently used and how usage had evolved over time. It inquired about feature requirements and user preferences for warning apps. In addition, to view warning app use in crises in comparison with other two-way communication channels, social media use was investigated and compared with warning app use (Paper C) for complementarity. The first step in this research was a practitioner workshop in order to derive research questions that were relevant to practice (initiation). To expand the research and for triangulation, two studies (Papers D and E) focused on how users dealt with the COVID-19 pandemic. One study (Paper D) used a qualitative survey consisting mainly of open questions to explore how ICT contributed to information flow and the usefulness of the different channels and technologies. For this research, participants answered a daily questionnaire for one week, reporting their ICT usage for crisis information, as well as any information gaps they perceived. The qualitative study was conducted during the first wave of the COVID-19 pandemic, thus offering reports on users behaviour while the crisis was unfolding. The study also initiated Paper E, because it identified particular unmet information needs resulting from the quickly changing and locally different COVID-19 behaviour regulations in a federal country such as Germany. The paper therefore analysed user reviews of crisis information apps used during the pandemic. The reviews, which were gathered from the Apple, Android and Xaomi app stores, showed what users perceived as information needs that were met or not met by the apps. They also offered insights into users' expectations of the state-run multi-hazard warning app NINA, compared with users' expectations of apps that were designed and run pro-bono to help users understand local behaviour restrictions related to the pandemic.

3.3.2   Design Oriented Methods

The second type of contributions that this dissertation makes is artefact or design oriented: "Whereas empirical contributions arise from descriptive discovery-driven activities (science), artefact contributions arise from generative design-driven activities (invention). Artefacts, often prototypes, include new systems, architectures, tools, toolkits, techniques, sketches, mockups, or environments that reveal new possibilities, enable new explorations, facilitate new insights, or compel us to consider new possible futures. New knowledge is embedded and manifested in artefacts and the supporting materials that describe them" (Wobbrock & Kientz, 2016, p. 39). The artefact-driven research is helpful in analysing how mobile crisis apps can be designed and improved to enhance users' role in public safety (RQ2). This type of research centres around identifying relevant challenges form a user perspective, followed by designing, implementing and evaluating the solution. From this process, insight can be gleaned into user needs and behaviours, the development process, as well as implications for design. Following this process, this dissertation builds on the insights about users and the crisis context to develop and evaluate solutions. For the ideation step, paper F explored motivational strategies that could used to engage citizens in public safety by looking at critical infrastructure safety. The findings from this paper were used to devise and evaluate the motivational strategies used in paper G.

As such, paper G presents the design process of a persuasive warning app feature for preparedness notification. Based on the digital nudge design method (Mirsch et al., 2018), the processes consisted of the steps (1) defining the context, (2) ideation and design, (3) implementation, and (4) evaluation of the digital nudge. For step 1, a representative survey explored the general acceptance of nudging in the context of crisis preparedness and warning apps and preferences regarding different nudge types. Then, a design workshop explored ideas for implementing the two preferred types of nudges in warning apps. Following this, a feature for engaging users in crisis preparedness and motivating them to implement preparedness measures was developed with a focus on persuasive design. The feature and the nudge design variations were then tested in an experiment, with users interacting with the app for one week.

A second design study, presented in Paper H, presents the design science research (DSR) approach applied to the development of a bot-driven warning messenger channel. DSR holds that „[...] knowledge and understanding of a design problem and its solution are acquired in the building and application of an artefact" (Hevner & Chatterjee, 2010). DSR can be comprehended as comprising three distinct cycles (Hevner, 2007). The *relevance cycle*, encompassing user studies and an understanding of requirements, ensures the investigation addresses a pertinent problem, thereby establishing a connection between the environment and the design science research process. Within the DSR process, the *design cycle* iteratively links design with evaluation outcomes. The design cycle involves six steps: (1) identification and motivation of the problem, (2) establishment of solution objectives, (3) design and development, (4) demonstration, (5) evaluation, and

(6) communication (Peffers et al., 2007; vom Brocke et al., 2020). The DSR process connects with the knowledge base through the *rigor cycle*, which guarantees grounding in and augmenting the knowledge base (Hevner, 2007), thus contributing to scientific advancements. Gregor & Hevner (2013) describe different types of research contributions that can be made through DSR: Contributions can be made by developing new solutions for known problems (*improvement*), by inventing new solutions of new problems (*invention*) or by extending known solutions to new problems (*exaptation*). The authors state that for exaptation, "[t]he new field must present some particular challenges that were not present in the field in which the techniques have already been applied" (Gregor & Hevner, 2013, p. 347). Applying existing solution to crisis communication and warning app design, the study contributes an *exaptation*. Following the DSR, in Paper H, the problem of having mobile public warning channels with very low barriers to adoption was identified. Following this, objectives were defined which centred around offering high ease of use and usability of mobile warnings in safety critical contexts within a highly accessible messaging app. Based on this, requirements were collected based on a literature review, before the design was explored in a workshop to transfer the requirements to the implementation. A warning bot prototype was iteratively developed, which was then evaluated.

3.4   Research Context

The research context of this dissertation is safety and security in urban contexts. The safety of urban infrastructures is explored in the LOEWE centre "Emergency Responsive Digital Cities" (emergenCITY), whereas the security of urban infrastructures is explored in the ATHENE research area "Secure Urban Infrastructures" (SecUrban).

The National Research Center for Applied Cybersecurity (ATHENE, https://www.athene-center.de), formerly known as Centre for Research in Security and Privacy (CRISP, 2015–2019) is funded by the Federal Ministry of Education and Research (BMBF) and the Hessian Ministry of Science and Art (HMWK). Since 2019, ATHENE examines usable security, biometrics, cloud security, cryptography, cybersecurity analytics and defences, digital forensics, economics of IT security, industry 4.0, internet and infrastructure security, media security, mobile and cyber-physical system security, privacy and trust, secure software systems, security management, social and ethical aspects of IT security, threat modelling and security evaluation, as well as automotive security. ATHENE is an alliance of the Fraunhofer

Institutes SIT and IGD as well as the universities Technical University of Darmstadt, the Goethe University Frankfurt and Darmstadt University of Applied Sciences.

In addition, this research contributes to the LOEWE centre emergenCITY, funded by the Hessen federal state's initiative for the development of scientific and economic excellence (LOEWE, https://www.emergencity.de). Established in 2008, LOEWE is the research promotion program of the federal state of Hesse, Germany. LOEWE funds excellent research projects and especially the cooperation between academia, non-university research institutes and businesses. emergenCITY is funded by LOEWE from 2020 to 2023 and is dedicated to the research of ICT resilience and the resilience of digital cities, with the aim of enabling digital cities of the future to overcome safety crises. The centre focuses on the program areas communication, information, city and society and cyber-physical systems. It is a cooperation between the Technical University of Darmstadt, the Philipps University Marburg and the University Kassel, with the Federal Office for Civil Protection and Disaster Assistance and the city Darmstadt as associated partners.

Findings 4

Together, the papers presented in this dissertation give insights into the question of how warning apps can be designed to engage citizens in public safety (RQ). In the following, the findings are summarised by answering the research questions RQ1 and RQ2.

4.1 Users' Perceptions and Expectations of Warning Apps

In the following, RQ1 is answered, which addressed users' interactions and perceptions of mobile crisis apps. This is done by, first, summarising the findings regarding warning app usage and perceptions, second, by looking at the expectations and preferences, and third by analysing the effect of sociodemographic variables.

4.1.1 Warning App Usage and Perceptions

The three quantitative representative surveys (Papers A, B C, Chaps. 7–9) collectively show a moderate current use of warning apps in crises in Germany. In 2019, 26% of respondents had never used a warning app, of which 16,5% were currently using one. The gap between the two numbers shows that numbers of downloads of an app, as used in Hauri et al. (2022), are only a vague indicator because a significant group of people had stopped using such apps. The two multi-hazard warning apps on the German app market were by far the most used crisis apps, while the others had usage numbers of below 4% (see Paper A, Chap. 7). In 2021, 17% indicated that they did not know any multi-hazard warning apps, showing that further information about these apps could increase adoption to some degree. However, a large

J. Haunschild, *Enhancing Citizens' Role in Public Safety*, Technology, Peace and Security | Technologie, Frieden und Sicherheit, https://doi.org/10.1007/978-3-658-46489-9_4

share of non-adopters appear to be hampered by a status quo bias—20% indicated that they would like to use a warning app in the future, while currently not using one. One third (30%) had no intentions of using a disaster or warning app. In 2021, the number of people who had stopped using warning apps was 7% (see Paper B, Chap. 8). With these numbers, Germany appears to be ahead of other European states, some of which partly had had technical problems when they introduced their warning apps, but it remains below the usage found in Finland, where the number of downloads equals a third of the population (Hauri et al., 2022). However, between 2017 and 2021, the use of warning apps has continually increased, reaching 25% in 2021 (see Paper B, Chap. 8). The increase in usage of the two main multi-purpose warning apps (from 4% to 12% to 17% from 2017, 2019 to 2021 for *NINA* and from 6% to 9% to 10% for *KATWARN*) indicates that neither the COVID-19 pandemic nor the severe effects of the European flood lead to a disproportionately slow or quick uptake in these apps. Fig. 4.1 shows the usage developments for the two main warning apps *Nina* and *KATWARN* and the regional warning app *HessenWARN*. It shows that adoption and awareness of the apps has increased and the significance of the regional app in its area.

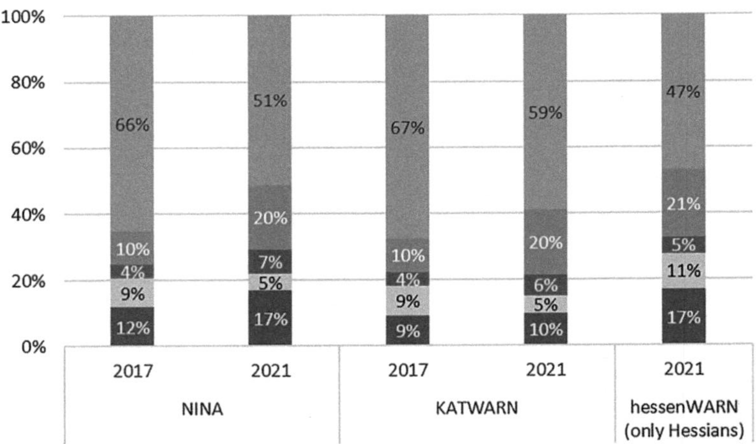

Fig. 4.1 Development of usage of main German warning apps over time (for the regional app *HessenWARN*, only residents of Hessia are depicted (n=83) (■ current use, ▨ planned future use, ■ past use, ▨ aware but no usage intention, ▨ unaware of this app), based on data presented in Chap. 8

Instead of warning apps, in the case of the COVID-19 pandemic, the tracing apps *Corona-Warn-App* and *Luca* were the apps of choice. Due to the long-term character of the health crisis, warnings were less relevant than tracing potentially infectious contacts. On the one hand, the separation of these crisis functions contradicts the user requirement of combining all crisis functions in one app and can be seen as an opportunity missed for multi-purpose warning apps. However, the contact tracing apps require sensitive data and thus very specific privacy-preserving technologies. Separating the warning app and the contact tracing apps in this case seems feasible, to increase user acceptance of the contact tracing app. In 2021, more people thought that they would likely use a warning app in the future for a number of purposes than in 2017. The main purpose lies in receiving warnings (78% would likely or very likely use a warning app for this purpose in the future), followed by receiving safety advice (64%). The use of a warning app to contact emergency services showed the biggest increase in interest (from 32% to 54%). But more than 60% also perceive it as likely to use such an app to find information (63%) and to share information with emergency services (61%). This indicates citizens' increased willingness to use warning apps for two-way communication. More research should therefore investigate two-way communication, especially since a recent study shows that around 50% of emergency managers are rather willing to include citizen-generated information in their situation assessment (Holzhüter et al., 2023).

Interestingly, user demands for warning types and relevant functions remained largely stable, including with regard to pandemic-related information and preventive advice (see Paper B, Chap. 8). This indicates that the experience of recent crises has not significantly affected the way that crises and personal responsibility for crisis response and preparedness are perceived. Following this reasoning, it appears that Germany has remained a state-oriented risk culture in which state organisations, rather than individuals, are perceived as responsible for crisis management and preparedness (Cornia et al., 2016; Reuter et al., 2019). In addition, warning app usage was characterised by a continuous rather than exponential growth. Despite recent crises, there seems to be neither an abandoning of warning apps, nor a stark up-take of them. However, as the connection between the worry, risk appraisal, intention and implementation of preparedness measures—adoption of warning app can be regarded as such a measure (FEMA, 2022)—are not straightforward (Appleby-Arnold & Brockdorff, 2018; Paton, 2019; Wachinger et al., 2013), more research into the dimensions of risk cultures and their possible changes in Germany is feasible.

In 2019, direct contact with emergency services, but also with personal contacts, were perceived as the most helpful sources of information, followed by traditional media, such as television and radio. This still held true in 2021. The findings are similar to those of a study that investigated the use of information sources in a health crisis (the Zika virus) and found that people were mainly using TV and personal conversations, in addition to consulting websites of health agencies (Chon & Park, 2021). Their study showed that the high-risk group, as well as those particularly motivated to comply with instructions, relied on additional social media sources. The authors suggest that instead of relying on social norms, this group of people tried to find more information on which to base their decisions (Chon & Park, 2021).

Among the online sources, warning apps are perceived as most helpful (67%) by the people who had actually used the respective sources, whereas print media and social media were perceived as the least helpful (by ca. 30%) (see Paper A and B, Chaps. 7 and 8). With regard to factors that influence the perceived helpfulness, age is a significant negative factor with regard to the perception of social media's helpfulness (Spearman's correlation coefficient $r_s = -0.207$, p < .001) in emergencies, while not significant for the evaluation of warning apps ($r_s = 0.037$, p = .323). Comparing crisis apps to social media use in crises in 2019 (see Paper C, Chap. 9), we find that the use of social media in crisis situations is wide-spread among all ages, with *WhatsApp* being most widely used, followed by Facebook. This use of a messaging app underlines the importance of investigating messaging apps, which are increasingly used for communication with larger groups of unknown people, in addition to the original use for chat communication with personal contacts (Newman et al., 2022). The data indicate that those who had used and found social media and warnings apps to be helpful also found other internet sources to be particularly helpful. However, people who appreciated warning apps also praised the helpfulness of direct contact to emergency services more. The correlations indicate that warning app users in emergencies are a group that particularly seeks out privileged information from emergency services, be it through an app, direct contact or websites, rather than using legacy media.

The use of social media in emergencies has increased, including the sharing of information through social media, with people sharing, in particular, information about the weather, traffic and increasingly reassuring others about their safety (see Paper C, Chap. 9). A majority perceives social media as faster and more accessible. In addition, reservations are decreasing: The number of people who find social media more accurate, more reliable and richer increased slightly and people saw fewer disadvantages of using social media in 2019 than in 2017. However, people increasingly perceive that emergency services are too busy to monitor social media. This indicates an understanding of the way emergency services operate and an understanding for the complexities required for filtering relevant social media posts and enhancing them (Kaufhold, 2021).

Looking at the case of crisis communication in the COVID-19 pandemic (see Papers D and E, Chaps. 10 and 11), we find that despite the long duration of the crisis, the issue of information insecurity persisted. Health crises differ drastically from a crisis such as a shooting or a natural disaster, as they unfold over a long time period and do not evolve as rapidly as most disasters. Despite the relaxed time frame and the abundance of information, there were unfulfilled information needs. While a large group of people was satisfied with their sources, particularly radio and TV, others felt that particularly due to different regulations in different federal states and even on the city-level, it was difficult to keep up to date with the regulations. As a result, several crisis apps emerged to close this gap. However, the apps were suffering from having to manually code the restrictions and translating the legal text into information that could be easily grasped. While some information, such as number of infections, vaccinated people and availability of intensive care hospital beds, were readily available, the legal regulations were not provided in a machine readable format (Moon, 2020).

4.1.2   The Impact of Socio-Demographic Factors

The representative studies in this dissertation show that socio-demographic factors play, at most, a minor role with regard to warning app interactions and preferences. Throughout the studies, no effect of education was found. With regard to preferences, age and gender are similarly largely irrelevant, which implies that needs in crisis and with regard to crisis communication from agencies do not differ. This indicates that warning apps, possibly due to their simplicity, are accessible to all people regardless of age and education.

Differentiating between the group of younger and older people, we find that age is largely not a strong factor with regard to technology use for crisis communication and warning app preferences. Table 4.1 shows the significant effects of age and lack thereof on aspects related to technology use for crisis communication. This is in line with a study investigating the protection motivation, which found that age, gender, and education were no relevant factors in influencing warning app adoption or continuance of use (Fischer-Preßler et al., 2021). User evaluation of warning apps, preferences, perceived helpfulness, nor social media use has a linear correlation with age.

Table 4.1 The effect of age on crisis communication, WA = warning app, SM = social media

Topic	Effect
WA use	No linear effect; slightly less use of WAs by under 25s[1]
Sharing	Increased trust in the secure transmission of GPS data to emergency services[2]
Features	Older people show stronger preference for easy use, inclusive usability, a seniors' mode, self-explanatory design, location specific information and control over type of information[3]
Download of WA	Fewer older people having ever downloaded a warning app[4]
Share information on SM	Very small negative correlation with age[5]
Content sharing on SM	Under 25s are more likely to share feelings and video; less likely to not have shared any type of content[6];
	Over 45 years old are less likely to share most types of content or any content at all[7]

[1] $\chi^2(1) = 8{,}175, p < 0.004, \phi < -0.104$
[2] Spearman's $\rho = -0.125, p < 0.001$
[3] $\rho = 0.191, \rho = 0.185, \rho = 0.172, \rho = 0.171, \rho = 0.12, \rho = 0.115$
[4] $\tau b = 0.06, p = 0.038$
[5] $\tau b = -0.15, < 0.001$
[6] $X(1) = 31.097, p < 0.001; \phi = 0.194; X(1) = 4.612, p = 0.032; \phi = 0.075); X(2) = 10.012, p = 0.002; \phi = -0.11$
[7] $X(2) = 13.222, p < 0.001; \phi = 0.127$

Interestingly, it is the group of younger people (under 25s) who have a slightly decreased likelihood of using a warning app. This could indicate that in the case of warning apps, aspects related to risk perception or disaster concerns may be more relevant than technology acceptance. In a similar vein, a study on the use of health technologies found that instead of age, health literacy affected technology adoption (Mackert et al., 2016). However, the effects are also very small and should not be exaggerated.

The insignificant difference regarding the older age group may also result from the methodology of the study. As it consisted of an online survey, it only reached those who access the internet and possibly have more online experience or a greater interest in using online technologies. In addition, research indicates that the elderly (65+ ears of age) are not a homogenous group, but rather age remains a significant explanatory factor even in this age group (Neves et al., 2018). The results may thus have been different, had the group of older adults been further differentiated. While, generally, age negatively affects the intention to use a technology, the elderly are more motivated by performance expectancy (Niehaves & Plattfaut, 2014). The results are similar when it comes to social media use in crises. They show that regardless of age, people refer to social media to gather information about crises. It seems that older citizens are prepared to use technologies that they may be using less in their daily lives, such as *YouTube*, in case of emergencies (see Chap. 9). These findings are supported by a recent study which finds only a small negative correlation between sharing of feelings on social media in crises, indicating that older people use social media somewhat more in crises for this purpose (Münscher et al., 2023). Overall, the study suggests that warning apps and social media use are not significantly affected by a digital divide effect. If anything, attention should be paid to spark interest in younger people in the use of warning apps. A study of the continuance of use of warning apps shows that, interestingly, those people with *higher* self-efficacy were more likely to discontinue using warning apps (Fischer-Preßler et al., 2021). The authors suggest that warning apps may be offering too few features for proficient users to continue using them. Whether this is an aspect that influences the younger user group should be investigated in future research.

Similarly, male and female warning app users did not differ significantly in their usage and demands. Slight variation exists, for example, with regard to preferences for warning types, with women more interested in a variety of topics, particularly ones related to policing and school cancellations, indicating a greater perceived insecurity and their childcare responsibilities. They are also more willing to grant permissions to the app to increase their safety. Interestingly, women are slightly more prepared to share videos, photos and live transmissions via warning apps. However, when it comes to social media, men are more willing to share similar content. This could indicate women's higher trust in warning apps and higher mistrust in social media. However, women may also believe that the sharing via warning apps is more effective. In addition, in 2019, women were somewhat more likely to have shared their feelings in relation to crises on social media. This finding differs from a recent German study that found that women were slightly less likely to do the same in crises (Münscher et al., 2023). The comparison of survey data suggests that the sharing of feelings on social media diminished significantly between 2017 and 2019, from 46% to 27%. Therefore, the changes could possibly result from further changes in attitudes since then, especially as social media platforms have also undergone significant changes., Table 4.2 summarises the findings of the representative surveys regarding the effect of gender or lack thereof. While some effects are significant, they were all very small and should therefore not be overstated.

Table 4.2 The effect of gender on crisis communication preferences and attitude, WA = warning app, SM = social media

Topic	Effect
WA use	Slightly less WA use by women[1]
WA preferences	Women are less prepared to have several crisis apps, oppose apps replacing other sources more strongly, and favour more strongly the pre-installation of WAs[2]
Sharing	Women are slightly more prepared to share videos and photos, live transmission, being asked before transmitting GPS data[3]
Features	Gender plays a small role for usability and accessibility features. Women are more interested in search of missing persons, direct messages to ES, telephone, and GPS-enhanced emergency calls, prepared to grant access to their remote location[4]
Warning types	Women are slightly more interested in emergency warnings, police messages, cybercrime, school cancellations[5]
Perceived helpfulness	Slightly more positive evaluation by women, e.g. helpfulness of contact with emergency services[6]
Content sharing on SM	Women are more likely to share feelings; men are more likely to share their location, videos, witness accounts[7]

[1] $\chi^2(1) = 12.97, p < 0.001, \phi = 0.14$

[2] $t(1119) = -3.204, p < 0.001; t(1113) = -1.88, p < 0.06; t(1132) = 2.776, p < 0.006$

[3] Cramer's V = 0.108; $p < 0.01$; Cramer's V = 0.095; $p < 0.037$; Cramer's V = 0.115; $p < 0.005$

[4] Cramer's V between 0.1 and 0.15; Cramer's V = 0.113, $p < 0.006$; Cramer's V = 0.144, $p < 0.001$; Cramer's V = 0.127, p 0.001; Cramer's V = 0.137, $p < 0.001$; Cramer's V = 0.104, $p < 0.017$

[5] Cramer's V = 0.133, $p < 0.002$; Cramer's V = 0.121, $p < 0.002$; Cramer's V = 0.105, $p < 0.014$; Cramer's V = 0.102, $p < 0.019$

[6] Cramer's V $< 0.107, p < 0.003$

[7] X(2) = 24.907, $p < 0.001, \phi = 0.174$; X(2) = 10.36, $p = 0.006, \phi = 0.112$; X(2) = 16.23, $p < 0.001, \phi < 0.14$; X(2) = 6.928, $p < 0.001, \phi = 0.088$

Summary—RQ1: Users' interactions and perceptions of mobile crisis apps

- **Reach of Warning Apps:** Usage in Germany increased to 25% in 2021.
- **Crisis Apps:** The warning apps *Nina* and *KATWARN* are most widely used, while other crisis apps are not widely adopted.
- **Acceptance:** Around 60% find warning apps feasible; 70% expect information from emergency services via this channel.
- **User Group:** The evaluation of warning app helpfulness particularly correlates with helpfulness of direct contact with emergency services and of other internet sources.
- **Sociodemographics** Age, gender, and education have no or minimal impact on warning app usage and preferences.
- **Design Preferences:** People wish for warning apps that combine a large number of security and safety incidents. Preferences have been largely stable over time.
- **Impact of Crisis Experiences:** Despite recent crisis experiences, the use of warning apps has continued in a linear fashion. The only significant changes exist with regard to an increased readiness to contact emergency managers directly through the apps.
- **Helpfulness:** Warning apps rated as the most helpful online source by its users.
- **Importance and Lack of Communication About Regulations During COVID-19 Pandemic:** Citizens felt strong responsibility for staying informed during COVID-19. Agency communication, mainly through TV, radio, and local news channels, was considered crucial, but frustration was voiced regarding the lack of direct agency information. Lack of communication of up-to-day, correct and concise information about local, regional and national regulation.

4.2   Enhancing Citizens' Role in Public Safety Through Warning Apps

RQ2 of this dissertation asked how warning channels can be designed and improved to allow users to contribute to public safety. This question can be answered by analysing the studies with regard to citizens' willingness to engage with crisis preparedness and to make contributions to public safety. Citizens' contributions can be increased through the following strategies: Firstly, as a precondition to empowering citizens to take on personal responsibility in crisis management, the adoption of warning apps and other warning channels can be increased. Secondly, citizen contributions can be increased by implementing functions that allow citizens to share their knowledge and to empower them to take on more active roles. Thirdly, persuasive system design elements and nudging can be used to increase engagement with crisis preparedness.

Mobile Warning Channel Adoption As described in Sec. 4.1, the use of warning apps is generally perceived as sensible. In the studies presented in this dissertation, a large set of functions have been investigated that can increase the usefulness of warning apps to citizens. However, as has been shown in Chap. 2, despite a general interest in disaster preparedness, it can be difficult to motivate people to take preparedness actions. Therefore, a second strategy can be to increase ease of use and decrease adoption costs, making it particularly easy to start engaging with crisis preparedness.

Following this strategy, a warning bot within a messaging app was explored. The bot allows users to query the warning status of a location, to set up a personalised subscription for specified warnings and severity levels at a selected geographical location. In addition, the bot provides disaster preparedness recommendations. The analysis of the requirements showed that delivering warnings through messaging apps, the apps impose some limitations that reduce the usability of the warning function and they way information can be presented. Auditory output is important to alert people to severe crises, and warning apps can override other phone setting, allowing to send alerts despite the phone being muted (Tan et al., 2020a). While this is currently not possible in messaging apps, the auditory output of channels can be individualised, allowing users to chose a distinctive sound. However, this would require additional effort and familiarity with the settings in messaging apps. In addition, messaging apps user interfaces consist of conversation streams, making it more difficult to display information permanently. For example, crisis preparedness information could be sent as text messages, quickly leading to cluttering of the interface, it can be provided as a link or downloadable file. In the implemented bot, sending users a file to download was chosen. While this offers no interactive or personal content, it allows users to store the information on their phone, making it accessible even without an internet connection.

The conversational interface also leads to a lack of visual feedback about the status of the bot. Therefore, the conversational interface needs to be designed carefully to provide guidance and feedback. In the implemented bot, this was done by using buttons to display the next available actions and by posting short text feedback about any changes that users had made to the settings of the bot. Figure 4.2 shows the interface design of the implemented bot.

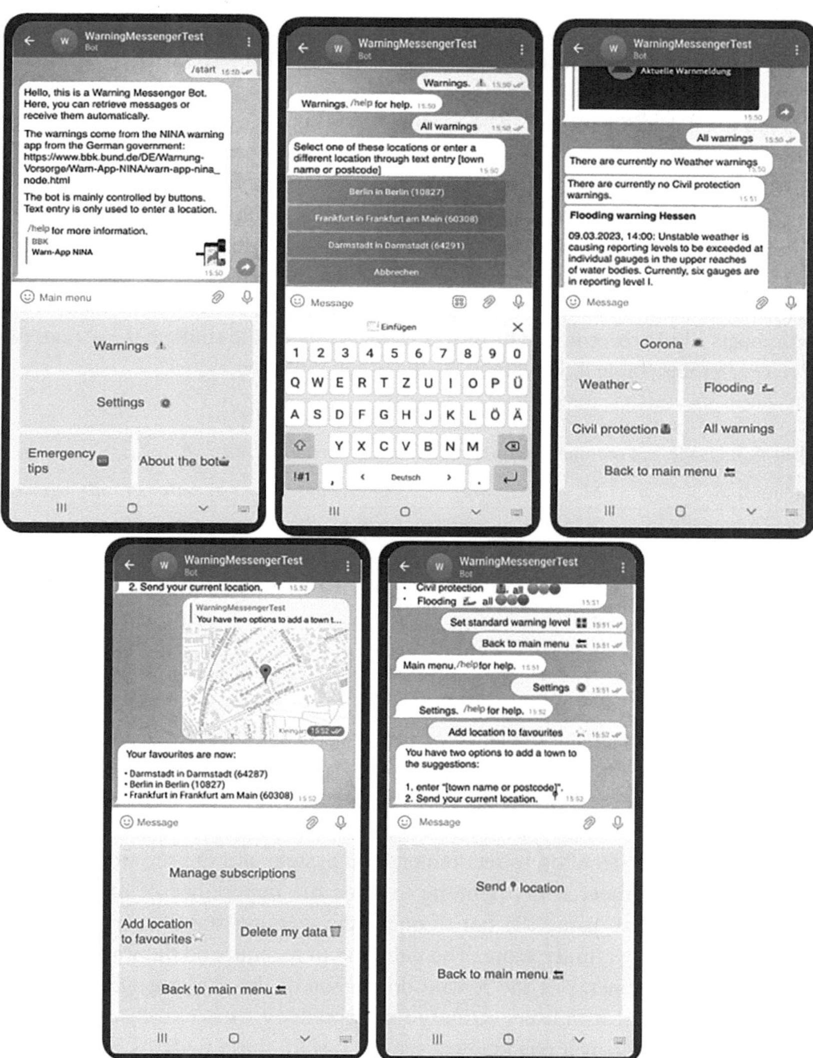

Fig. 4.2 Conversational interface design of a warning bot in a messaging app, presented in Chap. 14

Despite the challenges of a conversational interface in messaging apps, which makes it difficult to provide transparency about settings that were made by users, e.g. when personalising the warning types and locations, warning messaging channels received good usability scores. The System Usability rating, based on the 10-item SUS, is 76.88 ($M = 78.75$, $SD = 17.58$), indicating good usability (Sauro & Lewis, 2016). Forty percent of the study participants stated that they would likely adopt the messaging warning bot if it was available on their favourite messaging app . Primarily, participants would use the bot to receive warnings. Only three out of N=25 participants regarded querying information as more useful than the warnings, partly because they were already using a warning app. Around twenty percent of the participants would recommend the warning bot to others. This indicates a potentially large user group. Table 4.3 summarises the usage intention.

Table 4.3 Usage and recommendation intentions, reported in Chap. 14

Intention to	M	SD	Fully agree	Agree or fully agree
... use frequently	3.25	1.22	20%	44%
... use on favourite messenger				
... for subscriptions	3.83	1.37	40%	68%
... for inquiries	3.25	1.15	12%	44%
... recommend system	3.50	1.10	20%	48%

Looking at factors that influence these usage and recommendation intentions the results show usage intention is complex. The system usability score only correlated with the intention to recommend the system, not with the own usage intentions. The intention to use the system for its core functions, automatic warnings, particularly correlates with the intention to recommend the system and the reward dimension of the user engagement scale, pointing towards the importance of utility in this decision. In contrast, using the bot to query the warning status is related to the personal technology affinity score. The intention to recommend the warning bot to others, itself an elements of the reward dimension of the user engagement scale, positively correlated with users' affinity for technology, their usability rating, and all dimensions of the user engagement scale. Overall, these findings suggests that for personal usage intention, aspects related to reward, which is an indicator of the system's utility, are of central importance. For the intention to recommend a warning channel, a larger set of factors seems important. However, since the study sample was small (N=25), the results should be interpreted with caution and further studies should explore the relevance of these factors.

A main reason for adoption was ease of use, with participants stating that it was easy to set up the notifications and to join the warning channel. As expected, not needing to download a specific app is an advantage. Users would not use the warning channel when they did not perceive a need to be warned, as they felt safe or prefer to use internet searches or media.

Increasing Usefulness The studies have shown multiple ways of increasing the usefulness of warning apps from users' perspective, particularly by allowing users to get in touch with emergency services more easily through two-way communication modes, by offering more preparedness and first-aid advice and by adding features that allow users to more actively contribute to public safety.

Figure 4.3 shows the demand for features which help citizens contribute more actively, e.g. by helping find missing people, or by receiving first-aid instructions, which up to 70% of respondents judged as important.

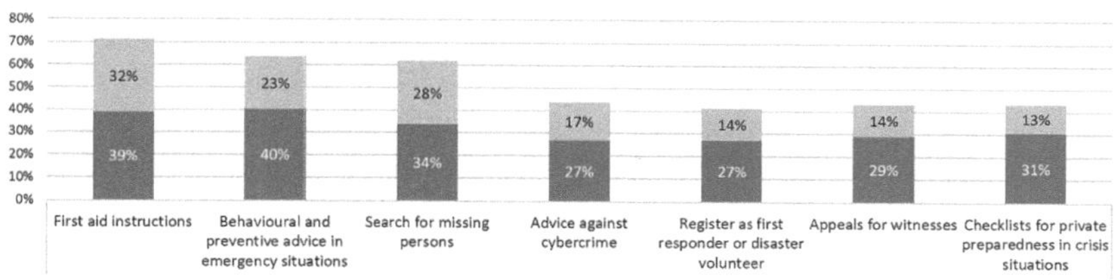

Fig. 4.3 Users' desired features and functions for citizen empowerment in emergencies, based on results presented in Chap. 8

Similarly, more functions for two-way communication would allow citizens to get in touch with emergency services, especially when they are in situations that do not warrant an emergency call. Fig. 4.4 shows that such communication features are highly demanded and are among the most desired features overall (see Chap. 8).

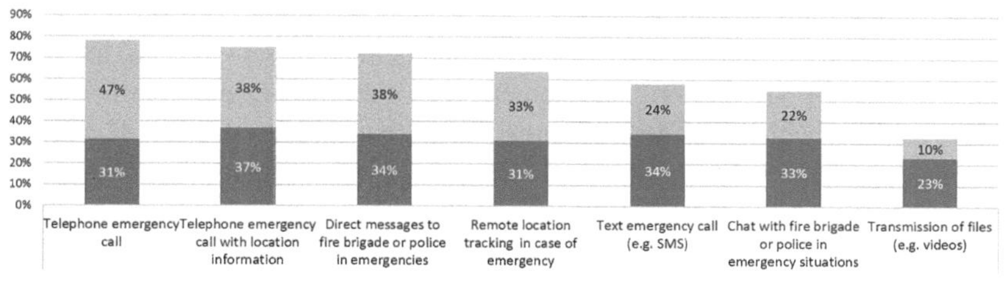

Fig. 4.4 Users' desired features and functions for citizen-agency communication, based on results presented in Chap. 8

In addition to citizens contacting emergency managers through a warning app, these could also be used to solicit information. An increasing number of people states that they would likely use a warning app to submit information to emergency managers in the future (61%).

Increase Engagement with Crisis Preparedness The studies presented in this dissertation have shown that despite a generally positive attitude towards warning apps, there is an adoption gap and further warning app functions are welcomed by users. To counter the low motivation of Germans to implement hazard preparedness and prevention measures (Appleby-Arnold & Brockdorff, 2018), motivational approaches for engaging citizens in public safety were explored by investigating the case of energy resilience. It showed that information provision can be coupled with nudges at low cost for both providing agencies and citizens to improve immediate decision-making, whereas persuasive technology design requires more investment but can lead to a reflective behaviour change (see Table 12.1 in Chap. 12).

To study the effect of these interventions in crisis preparedness, firstly, we explored the acceptance of proactive preparedness advice and nudging by current users and non-users of warning apps though a representative survey (see Chap. 8). The results show that people feel that information about preparedness measures should be integrated in a warning app ($M = 3.83$, $SD = .96$) rather than conveyed through a different channel ($M = 3.32$, $SD = 1.02$), supporting the relevance of the endeavour. Preventive measures should be event-related and sent along with acute warnings ($M = 3.93$, $SD = .93$), rather than unprompted by an imminent hazard ($M = 3.24$, $SD = 1.18$). Citizens are largely open towards nudging, with 52% regarding it as generally helpful (12% opposed, $M = 3.57$, $SD = 1.01$) and 47% stating that they would appreciate being nudged themselves (17% opposed, $M = 3.41$, $SD = 1.11$). More opposition (around 30%) is encountered when it comes to providing personal information to improve the preparedness messages, however around 35% of respondents were open to this ($M = 3.11$, $SD = 1.25$).

We found that current app users agreed that more that precaution advice should be delivered along with acute warnings ($M = 4.10$; $SD = 0.89$, non-users: $M = 3.87$; $SD = 0.93$) or for them to appear only as a menu item ($M = 3.93$; $SD = 0.88$, non-users: $M = 3.64$; $SD = 0.99$). Additionally, they were significantly more willing to provide information in order to improve the preparedness warnings ($M = 3.36$; $SD = 1.28$, non-users: $M = 3.03$; $SD = 1.23$) and the preparedness nudges ($M = 3.31$; $SD = 1.28$, non-users: $M = 2.99$; $SD = 1.24$).

We then explored the effect of nudging on implementing preparedness behaviours and on the evaluation of a warning app that proactively sends preparedness advice (see Chap. 13). The study comprised of two different nudging groups, who were shown interactive nudges along with some hazard prevention advice. We compared the reported behaviours of these groups with those of a control group, who also received prevention advice but who did not receive any interactive nudges. We analysed whether respondents in the different groups reported different intentions to comply with the recommendations directly after receiving them, whether they were more engaged in the app and whether they reported a different number of implemented behaviours after the study period.

To compare the groups, we conducted a robust mixed ANOVA involving a total of 59 participants, with 31 in the control group, 13 in the confrontational experimental group, and 15 in the social experimental group. The results indicated that there was no statistically significant main effect of the between-subject factor ($F(2,15.44)=1.26$, $p=.310$), indicating that the nudging groups showed no discernibly different intention to comply with preventive measures across the various emergency categories.

Looking at the number of preparedness measures taken after the study, a one-way ANOVA showed a statistically significant effect with a medium effect size of the nudge condition on reported preparedness measures ($F(2, 69) = 3.26$, $p < .05$, $\eta^2 = .08$). A Tukey post-hoc analysis revealed a significant difference ($p < .05$) in post-reported preparedness behaviour between the control group ($M = 1.51$, $SD = 1.47$) and the experimental group with confrontational nudges ($M = 2.73$, $SD = . 1.22$). There was no effect comparing the control group to the social influence nudge group ($M = 1.56$, $SD = 1.58$). The Tukey analysis was calculated with a 95% confidence interval of the difference between means from 0.03 to 2.41 points on a 0 to 12 scale. We can conclude that participants with a confrontational nudge reported significantly more implemented recommendations than participants without a nudge. Testing exploratively whether gender, age, experience with the warning app, experience with the emergency category, and relevance of the emergency category predict post-study reported implemented measures, we found no statistical significance ($F(9, 44) = 1.25$, $p = .289$). This indicates that the effect is not influenced by any of these variables. We also found that the interactive nudges did not affect how much time participants spent with the prototype or the number of interactions in the artefact. We calculated a robust mixed ANOVA to examine the effect of different nudging conditions on the time spent with the preparedness alerts for the different hazards ($N = 76$, $n = 43$ CG, $n = 15$ EG Confront, $n = 18$ EG Social. No statistically significant effects emerged, neither for the within-between interaction ($F(4, 14.18) = .56$, $p = .689$), nor for the main effect of the between factor ($F(2, 14.05) = .57$, $p = .578$), or the within factor ($F(2, 10.91) = 3.36$, $p = .072$). Overall, the experiment suggests that despite users reporting *no* increased motivation to implement preparedness measures, the confrontational nudge lead users to report a significantly greater number of implemented hazard prevention measures (Mean difference = 1.22 measures) (see the electronic supplementary material for the nudges shown and the list of reported measures).

To investigate how users evaluated the experience of receiving nudges and of receiving hazard prevention advice, we conducted a post-study survey. The study revealed that respondents generally valued preparedness advice but were not in favour of silent push notifications. Approximately 44% of participants expressed increased motivation to use a warning app if it included preventive information, while around 36% disagreed (M = 3.00, SD = 1.12). This motivation was significantly linked to learning something new from the warnings (τ_b = .29; p < .01), reevaluating one's behaviour (τ_b = .49; p <] .01), and implementing more safety measures with the app (τ_b = .27, p < .01). However, having felt disturbed by the messages (15%, M = 2.17, SD = 1.14) was associated with lower motivation to use a warning app (τ_b = −.34, p < .01). Surprisingly, feeling pushed towards a behaviour (11%, M = 2.04, SD = 1.03) did not significantly decrease the app usage motivation, despite a strong correlation between having felt disturbed and having felt pushed (τ_b = .42, p < .01).

Examining current warning app users, the main difference found was that they were *less* inclined to receive preparedness information in a manner similar to regular warnings in warning apps (τ_b = −.24, p < .05). This indicates that people with experience using a warning app want to be sure that all received push notifications that indicate an acute alert. However, considering the sample's demographic, with 77% of current warning app users being male, this discrepancy may be attributed to gender differences, as women tended to be more receptive to push notifications (τ_b = .24, p < .05) and the integration of preparedness information into warning apps (τ_b = .21, p < .05).

The majority (60%) reported learning something new, and a significant portion (61%) agreed that they should do more for their safety and security, prompting them to rethink their behaviour (see Table 4.4). Over a quarter (27%) claimed they took additional safety measures due to the app. The study found little variation in these responses based on the experimental groups, indicating that nudging warnings were perceived similarly to the plain warnings and that the nudging conditions did not have a significant impact on these evaluations. While most respondents (87%) welcomed preparedness information in a menu item in a warning app, the majority (57%) disapproved of using silent push notifications for warnings. However, 35% were open to these prompts in warning apps. When specified that these push notifications would have fewer permissions than acute warning alerts (e.g., a less forceful tone), agreement rose to 64%.

Table 4.4 Reflection of security behaviour and app experience after the study (5-point Likert scale), CG = control group, EG1= experimental group with confrontational nudge, EG2 = experimental group with social nudge, Rethought = Rethought behaviour, Increased = increased behaviour, Learned= learned something new, Disturbed = felt disturbed, WA = Motivation to use a warning app, Prev. app = Motivation to use a crisis prevention app, reported in Chap. 13

Group	N	Reflection of Security Behaviour and App Experience, M *(SD)*					
		Rethought	Increased	Learned	Disturbed	WA use	Prev. app
CG control	39	3.28 (1.15)	2.33 (1.25)	3.31 (1.13)	2.33 (1.20)	2.85 (1.16)	2.54 (1.27)
EG1 confront	14	3.43 (1.23)	2.86 (1.17)	3.71 (1.20)	2.14 (1.03)	3.20 (1.15)	3.01 (1.16)
EG2 social	17	3.71 (0.70)	2.82 (0.88)	3.35 (1.06)	1.82 (0.88)	3.12 (0.99)	2.53 (1.07)

Summary—RQ2: How can mobile crisis apps be designed and improved to enhance users' role in public safety

- **Increasing Usefulness:**
  - through extended functionalities in warning apps, including enhanced warning scenarios like policing information, search for missing persons, and cyber-security alerts.
  - through two-way communication features, including for sharing information with agencies and contacting them through the apps.
  - through advice, instructions and functions allowing users to make contributions and to provide information, e.g. in searching missing persons, by being better first-responders, registering as volunteers or receiving more advice on crisis preparedness and prevention of cyber incidents.
- **Choice Architecture:**
  - High acceptance of nudging, particularly by current users of warning apps, for disaster prevention.
  - Favourable nudging techniques: Rational-confrontational and social nudges with positive gain-oriented wording.
  - Positive impact of nudges on crisis preparedness behaviour and increased perceived knowledge.
- **Increase Ease of Use:**
  - Feasibility of implementing a warning channel in a messaging app for those hesitant to install a warning app.

- Large potential user group (40%) willing to adopt the messaging warning channel.
- Conversational interface design with limited user input and button guidance received positive usability scores.

Discussion 5

This chapter discusses findings' implications for mobile warning channel design (Sect. 5.1), the use of persuasive technology design (Sect. 5.2) and for policy (Sect. 5.3).

5.1 Implication for Design

The findings presented in this dissertation show that users generally feel that having and using warning apps is feasible. Even after the introduction of cell broadcast, many still state that they plan on using a warning app in the future. This suggests that investments in improving warning apps are justified.

From the user studies, a series of design implications can be drawn, which are summarised in Table 5.1. One of the main user wishes is for one standardised app, rather than having extremely specialised apps for different crises. This is in line with previous work that has pointed towards the combination of scenarios in multi-hazard apps (Dallo & Martí, 2021). The German warning apps are in line with this requirement in so far as they are compatible and receive civil protection, disaster and extreme weather warnings through the modular warning system. The rapid uptake of the regional warning app *hessenWARN* (see Chap. 8), which includes warnings by the police, supports the finding that people prefer even more crisis alerts to choose from, including warnings about fraud, cybercrime, school cancellations and missing person information. This is in line with a study investigating cybersecurity alerts and recommendations in Germany, which found that television and warning apps are preferred public channels for disseminating cybersecurity information (Kaufhold et al., 2022). This indicates that in this area, too, warning apps could be enhanced by working with cyber emergency response teams, particularly those

© The Author(s), under exclusive license to Springer Fachmedien Wiesbaden GmbH, part of Springer Nature 2025
J. Haunschild, *Enhancing Citizens' Role in Public Safety*, Technology, Peace and Security | Technologie, Frieden und Sicherheit,
https://doi.org/10.1007/978-3-658-46489-9_5

dedicated to public cybersecurity and public warnings (BSI, 2023; Kaufhold et al., 2021). While these currently offer newsletters and RSS feeds, the usability of the warnings for laypeople and the spread via channels, including warning apps, should be considered. Supporting the finding that users of warning apps also find direct contact with emergency services to be helpful, two-way communication features should be implemented for warning apps to take advantage of this affordance of a direct connection.

Table 5.1 Key design implications based on central findings

Empirical base, Chapter	Finding	Key Implications for Design
Representative survey, Chap. 7, Chap. 8	Over 50% of the population regard a range of further non-disaster scenarios as rather or very important	Integration of further safety and security scenarios, including crime, health, traffic warnings and first aid instructions
Representative survey, Chap. 8	Increasing majority (54%) would like to contact emergency services directly through a warning app and share information with them (61%)	Include two-way communication modes
Diary study, Chap. 10	Information gaps due to regionally and locally different regulation	Crisis regulation and pandemic rules should be communicated via warning apps
Conceptual investigation, Chap. 12	Behaviour change interventions differ with regard to resources, targeting decision or behaviour, and using reflective or automatic processes	For crisis communication, information provision can be paired with nudging or persuasive technology design
Experiment, Chap. 13	Acceptance of nudging for crisis preparedness, content that is perceived as informative increases motivation to use a warning app	Include interactive content and nudging to increase engagement with crisis preparedness advice
Experiment, Chap. 13	Push-notifications for preparedness content largely not accepted	Foster engagement via interactive content and optional reminders
Experiment, Chap. 13	Rational-confrontational nudges lead to an increase of implemented preparedness measures	Use statistic about costs and benefits to increase preparedness
User studies, Chap. 14	Personalising public warning subscriptions via conversational design yields good usability	Offer warning bot for a less motivated user group, use conversational interface design for better accessibility

The research has unearthed significant usage of the new regional warning app, *hessenWARN*, surpassing other warning applications in its region. In terms of design and functions, the app is very similar to one of the main warning apps, *KATWARN*, suggesting that usability is similar. For example, the apps have the same structure, similar menus and colouring, and both prominently display an identical warning map. However, the app has incorporated additional functions, particularly for policing as well as for traffic safety. The research in this dissertation suggests that these additional features are seen as an advantage that increases the utility of the app. Other aspects, such as the promotion or users' regional affiliations, might have contributed to the app's success compared to other apps and should be investigated in future research to further analyse factors of success. The finding of accelerated uptake of the app *hessenWARN* with its extended functions and users' wishes for additional functions also supports other research in HCI which has found that utility is particularly important for the continued use of warning apps (Tan et al., 2020c). The success of a warning app, that combines more warnings and crisis relevant information, is also in line with the finding of the Finnish app's popularity (Hauri et al., 2022). However, the Finish app, *112 Suomi*, goes even further by offering a large set of other crisis services, such as relevant phone numbers, first aid information or maps showing the location of defibrillators (Hauri et al., 2022). User needs may thus be better understood by considering more strongly the context of use (Jumisko-Pyykkö & Vainio, 2010), with the Finnish warning app being tailored to a broader emergency context. This approach recognises that warnings are important but are only only one element of the larger emergency context. This would mean a shift from "warning" to "public safety", including emergency management and risk communication. Drawing inspiration from the Finnish multi-purpose app (Hauri et al., 2022), this research could pave the way for innovative advancements in warning app design that open up the app to further emergency-related contexts and functions.

With regard to engaging users as co-producers of safety, the findings indicate that users are interested to collaborate as witnesses, to help in the search for missing persons, or to register as disaster volunteers (see Chap. 7). The majority sees these as important functions, even as a main function, with 5% and less strongly opposed. Similarly, supporting previous findings (Reuter & Kaufhold, 2018; Tan et al., 2017), two-way communication is a function that many users want and that is currently not supported by the German warning apps. In addition to further warning categories, users also want additional functions, including calling an emergency hotline via the app and sending the location, news tickers with updates and means of contacting fire departments or police via text in emergencies. Between 50% and 60% were also in favour of including preventive advice, links with further information and checklists (see Chap. 7).

Again, citizens' interest is not limited to disasters but extends to general emergencies, with a wish for first-aid instructions. In a study in 2021, around 50% of respondents' first aid training was more than 10 years ago and 65% of people stated that they would be interested in an app to help with first aid (ADAC e.V., 2021). This suggests that extending disaster apps by first aid functions could further improve the usefulness. Different regions have developed apps that alert registered first responders (Berliner Feuerwehr, 2023). Going a step further, a study indicates that emergency managers perceive that it would be helpful to activate first responders in case of a near-by emergency (Holzhüter et al., 2023); and their effectiveness has been shown in research (Naths et al., 2007). Including such a function in a warning app and making it available nationally could increase the reach of these first aid volunteers and increase the visibility of such a function.

Despite these wishes for more preparedness information, research also indicates that, particularly in Germany, the readiness to implement crisis preparedness measures is low (Appleby-Arnold & Brockdorff, 2018). The study of engaging users in contributing to the resilience of the energy network (Chap. 12) suggests that nudges and persuasive elements can be used to incur positive behaviour change, without changing the incentive structure. Due to their ubiquitousness, mobile phones are often used as mediums for conveying behaviour change interventions (McKay et al., 2018). This makes warning apps an ideal channel for integrating behaviour change mechanisms geared towards increasing crisis preparedness.

Individuals who have already been utilising a warning app tend to be more receptive to expanding its features, including providing information for crisis prevention and personalising advice and nudges (see Chap. 13). However, they exhibit more resistance to receiving push-notifications for preventive advice. This indicates that users predominantly view that push notifications should be reserved for severe alerts. This aligns with prior research emphasising the significance of utility, defined as the app's effectiveness in fulfilling its intended purpose (Tan et al., 2020c). It is thus important to ensure that users do not receive alerts that, in their view, distract from the main function of warnings. Instead, proactive preparedness alerts should be sent only to users who subscribe to the function or in the course of a specific event, such as a prevention day or a prevention week, thereby ensuring that users are aware and managing their expectations towards any push notifications. However, European studies indicate that citizens would like to receive preparedness information at least occasionally. The majority of individuals prefer to receive information about disaster preparedness for themselves and their families or friends at least once a year.

This preference varies from 62% in the Netherlands to 93% in Portugal and lies at 76% in Germany (and 40% biannually) (Appleby-Arnold & Brockdorff, 2018).

Finally, another design implication concerns the feasibility of using conversational interfaces, which would allow warning channels to be personalised for a new range of applications. In addition to reaching user who are unwilling to use a warning app, a conversational set-up could also be useful for people with limited accessibility, e.g. due to vision impairments. A study shows that conversational assistants provide a range of useful functions for blind people (Pradhan et al., 2018). With added natural language processing functions, the conversational design developed in the dissertation could be used to make personalised warnings accessible to further user groups who tend to prefer to be able to use mainstream channels rather than specifically designed ones (Gerber et al., 2018). However, moving towards accessibility would require extensive user studies and changes in the way that information and alert signals are presented (Sherman-Morris et al., 2020).

5.2   Persuasive System Design and Warning Apps

The experiment testing the effect of nudging in warning apps (see Chap. 13) has shown that the perceived value of preventive advice is contingent on whether it is perceived as interesting. This underscores the critical need for further investigation into the state of preparedness and preparedness knowledge in Germany, and the tailoring of advice to these specific states. Interestingly, the group exposed to a social nudge expressed the highest degree of reconsideration of their behaviour, while the rational-confrontational nudge tended to facilitate more learning. Importantly, stating that one had learned something new increased the motivation to adopt a warning app, which indicated that the primary focus, to ensure continued use, should be on informative preparedness information. The qualitative inquiries indicated that users preferred emotionally neutral, statistical information and a positive, gain-oriented wording. Further research should explore the specific information gaps and information needs of each countries' citizens, in order to create adequate recommendations.

In line with the particular correlation of rational risk appraisal and seeking of preparedness information found in Germany (Appleby-Arnold & Brockdorff, 2018), the rational-confrontational nudge was effective. Future research should investigate whether the social nudge is more effective in other countries, where preparedness is commonly more connected with worry—an emotional component (Appleby-Arnold & Brockdorff, 2018). To engage German users and motivate them to implement preparedness measures, providing statistical information about potential losses, and the costs and benefits of preventive measures are perceived as more acceptable. In addition, the experimental group that received these nudges, which contained more statistical information, in order to exemplify costs and benefits, lead users to state that they had learned new information about crisis preparedness. The nudges can thus serve as a positive example for crisis preparedness information. However, the nudges employed in this study may not yield the same effect in other countries, where an emotional approach may be more effective.

Despite the slight increase in implemented preparedness measures in the confrontational nudge group, this effect emerged even though users reported no change in their intention to implement measures. This aligns with research indicating a generally low intention to implement measures in Germany (Appleby-Arnold & Brockdorff, 2018). This effect of "disinclined actor" (Sheeran, 2002) indicates a reverse intention and behaviour gap and is puzzling, as it is very uncommon (Sheeran, 2002). Therefore, few explanations for this gap have been explored, and only the presence of outside pressure has been suggested (Polites & Karahanna, 2012). A potential explanation to be explored is personal norm, which describes a perceived moral obligation (Schwartz & Howard, 1984). In contrast to social norm, personal norm has been suggested to be relevant in influencing flood preparedness (Mol et al., 2021). However, personal norms were not specifically investigated in the study presented in this dissertation and its effect is unclear. Due to the high number of people disinclined to prepare in Germany (Appleby-Arnold & Brockdorff, 2018), participants may have perceived a social norm of non-preparedness, which might have made them state to not have been affected by the nudge, even while their personal norm may have been changing, resulting in the implementation of preparedness measures. Respondents in the social nudge group stated more strongly that they were reconsidering their behaviour, which might be indicative of a change in personal norms. However, the behaviour in this group had not changed by the end of the study. One possibility is that such changes are slower to occur. Future research should explore the effect of different nudges across a longer period of time. Due to the differently reported behaviours, future research investigating crisis preparedness should also consider asking about specific implemented actions that have been recommended.

Based on the results of the experiment exploring persuasive design, on the design workshops and user feedback, the set of suggestions can be made for choice architecture techniques that could increase the usage of warning apps and the engagement with crisis preparedness. Concerning the concrete design of prevention nudges, positive framing, emphasising the societal value of implementing preventive measures, as well as nudges presenting the costs and benefits of these measures, are more accepted by users. The design workshop uncovered a preference among users for positively worded and emotionally neutral nudges.

Considering persuasive design techniques beyond nudging, Münscher et al. (2016) describe three choice architecture categories and several techniques for each of them. Table 5.2 presents these and suggests techniques that can be implemented in warning apps. Looking at targeting the decision information (A), the experimentally tested nudges show that this type of intervention (A.2 and A.3) leads to an increase in implemented measures and a re-thinking of users' behaviour. At the same time, warnings and notifications should always be designed to reframe and translate crisis information for the public (A.1)—a process that was not done successfully with regard to the health regulations during the COVID-19 pandemic (see Chap. 10 and 11). Particularly with regard to measures that people tend to postpone (Eiser et al., 2012; Paton, 2006), despite judging them as important, such as taking first-aid courses, testing or installing smoke detectors or making an emergency plan with the family, the apps could do more with regard to how and when information about costs and benefits, social references or feedback are provided. The category of decision structure (B) targets the way that options, i.e., warnings and recommendations, are presented. While a reduced effort (B.1) and a no-action default (B.2) have been realised with cell broadcast, the same cannot be said for warning apps. As discussed in Sect. 5.3, changing the default with regard to pre-installing warning apps on smartphones is highly accepted. Similarly, push-notifications with relevant and interesting emergency and disaster-related content were accepted, provided there is an opt-out option, and these notifications can be easily differentiated from warnings. While the financial and physical effort of crisis preparedness cannot be changed, the installation of a warning channel itself can be more or less effortful and resource intensive, incurring maladaptive rewards (Fischer-Preßler et al., 2021). The implementation of a warning messenger has shown that affordances related to news consumption via messaging apps (Lou et al., 2021) can also be realised through a warning messaging channel. The feedback concerning the nudges and the design workshop indicated that preparedness advice may be best structured according to the simplicity of the measures. Further techniques can aim to change the consequences of options (B.4), e.g. by making positive actions visible to a community. However, such techniques would require a certain amount of networking features, reducing other design requirements such as privacy, anonymity and simplicity. While gamification and social networking elements could be included in a specific crisis preparedness app, it appears inadequate for the state-run warning apps. The two decision assistance techniques of providing reminders and facilitating commitments could be combined by offering a list of preparedness measures, from which users could select those relevant to them and make self-commitments to prepare. Optionally, push-notifications could be sent in user-specified intervals to remind them of their commitments. While the experiment has shown the effects of the nudges, the other techniques are based on the user statements gathered through the surveys, design workshops and evaluation, but their effectiveness has not been tested experimentally.

Table 5.2 Summary of suggested choice architecture categories, techniques and their suggested implementation for warning apps, column one and two based on Münscher et al. (2016)

Category	Technique	Implementation in Warning Apps
A. Decision information	A. 1 Translate information, *Includes: Reframe, simplify*	Warning app content, warning notifications
	A. 2 Make information visible, *Includes: Make own behaviour visible (feedback), make external information visible*	Confrontational nudge
	A. 3 Provide social reference point, *Includes: Refer to descriptive norm, refer to opinion leader*	Social nudge
B. Decision structure	B. 1 Change choice defaults, *Includes: Set no-action default, use prompted choice*	Pre-installing warning apps, opt-out preparedness reminders
	B. 2 Change option-related effort, *Includes: Increase/decrease physical/financial effort*	Offer channels with very low barriers to adoption
	B. 3 Change range or composition of options, *Includes: Change categories, change grouping of options*	Regrouping of preparedness information within the display, e.g. according to simplicity or effect
	B. 4 Change option consequences, *Includes: Connect decision to benefit/cost, change social consequences of the decision*	Out of scope
C. Decision assistance	C. 1 Provide reminders	Opt-in reminders, push notifications
	C. 2 Facilitate commitment, *Includes: Support self-commitment/public commitment*	Provide list for commitments to preparedness measures

5.3   **Policy Implications**

From the findings, several policy implications can be derived. The start of the COVID-19 pandemic revealed challenges with regard to crisis communication. Supporting previous research (Appleby-Arnold et al., 2019), users were willing to comply with regulations, primarily to protect others, expressing community values (see Chap. 10). However, due to the dynamic changes in regulation at the national, federal, and local level, citizens struggled to stay informed and receive reliable information. Chon and Park (2021) find that, in a different health crisis, high-risk and compliance motivated groups were using Facebook more than the other groups. This suggests that those seeking particularly to comply and to prepare were making more of an effort to find further information rather than relying mainly on TV news and on conversations with friends and family. The emergence of a new type of app, which aimed to portray current regulations related to the COVID-19 crisis, tried to fill this information gap. However, programmers struggled to keep the information updated based on legal texts, which is in line with Nikiforova (2021), who also showed that Germany's open data regarding COVID-19 had limitations regarding machine-readable information. The call for machine-readable information to improve governance and public digitalisation (Moon, 2020) is thus echoed here, as this type of information is particularly important in crises and a precondition for efficient crisis communication. In crisis communication, where awareness and compliance by the population are essential, measures should be taken to ensure that regulations and recommendations are made with crisis communication in mind, and with a plan for machine readable and understandable communication, to reduce cognitive load, combat misinformation and enable people from different cultural and language backgrounds to understand the information (Liu et al., 2017; Waidyanatha & Frommberger, 2022). This is especially true, since research indicates that related to wildfires, non-technical health related recommendations were easily recalled and widely followed, while technical health recommendations were not (Sugerman et al., 2012). Crisis communication during the COVID-19 pandemic was impeded by a lack of machine readable information about nation-wide, federal and local regulations. The dynamic changes made it difficult for citizens to stay informed about what was allowed and restricted in different locations, and the legal communication of the regulations made it difficult for state agencies and programmers to communicate these regulations via apps.

In line with research that has investigated the use of bot messaging apps for news consumption (Lou et al., 2021) and on the advantages of introducing "botplications" (Klopfenstein et al., 2017), the dissertation shows that a significant user group would use this type of technology, provided it is available via several widely used messaging apps. While warning apps have been pioneers in delivering "proactive" (Ayachi et al., 2016) services through location-based warnings, spreading warnings via messaging channels could reach a large user group and provide a more seamless interaction with the content (Lou et al., 2021). Governments should contemplate the addition of this channel to cater to a user group with moderate motivation for crisis information. By doing so, it can offer warnings that extend beyond the severity level of cell broadcast, while utilising a multi-purpose app that a substantial number of users have already installed. This approach effectively reduces barriers compared to mandating the installation of a dedicated warning app. The studies have shown that particularly younger people tend to use warning apps somewhat less. With news consumption shifting towards messaging apps (Newman et al., 2023), such a channel might be attractive to a younger demographic group, which should be explored by future research.

The design study suggests that for non-users, the expected performance is the most relevant factor that decreases the willingness to adopt a warning channel, even with good usability and low adoption barriers. For users of other warning channels, the status quo bias has a strong effect, which is, however, connected to the expected performance, with users stating that they expect the new channel to perform similarly. This is in line with with previous findings on e-government non-adoption which have found that a status quo bias has a strong effect: "The less need citizens feel for public services or the more convenient they find conventional ways of contacting administrations, the less they are inclined to use public e-services" (Distel, 2020, p. 13). Adapted to the use of warning apps, we also find that many respondents, even when indicating that the tool is useful and has high usability, state that they would continue using public media for warnings. Similar to another qualitative study on e-government, the findings suggest that a perceived lack of need or rare usage is one of the main barriers (Distel, 2018), along with the status quo (Polites & Karahanna, 2012). Further research is needed to explore and compare the affordances and (non-)adoption motivation of warning apps with messenger app-based warnings.

The proposed implications for warning app design would significantly change these apps and would also require emergency management organisations to accommodate and support such changes. Looking at the dimensions of crisis communication in public organisations (Olsson, 2014), the suggested changes would be in line with a shift of focus towards a more strategic and resilience-oriented communication style. Figure 5.1 shows the suggested contributions of warning apps for each communication style.

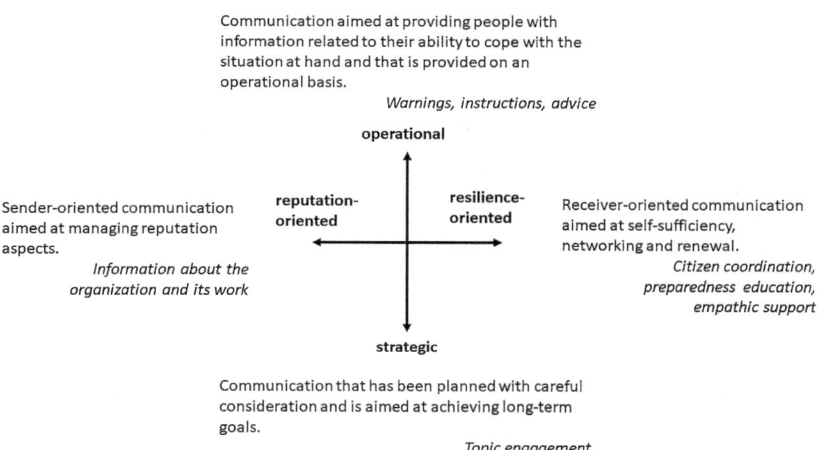

Figure 5.1 Dimensions of crisis communication adapted from (Olsson, 2014) extended by possible functions of warning app communications in *italics*

Looking at the limited functions, it seems that German warning apps are currently mainly used for operational communication, rather than for strategic purposes, which could include citizen empowerment or engagement. As warning messages often contain behaviour advice and can thus be classified as being somewhere on the resilience-oriented dimension (rather than portraying what the crisis management organisations are undertaking). However, with the lack of two-way communication modes or functions for citizen co-ordination and support, the warning apps represent a reputation-oriented communication strategy. While they do not, as such, portray the activities of crisis management organisations, they also do not "promote self-sufficiency, networking, and renewal" (Olsson, 2014, p. 114).

Olsson (2014) suggests that more resilience-oriented communication would include more empathetic engagement with citizens, their questions and concerns (Olsson, 2014). It could be argued that the affordances of other channels, especially social media, are more appropriate for this type of exchange. Research stresses the social media affordances of visibility, editability, persistence, association (Treem & Leonardi, 2013) or, from a more content-orientated perspective, persistence, replicability, scalability, and searchability (boyd, 2010). One analysis has looked at social media affordances in crisis management, mainly from the perspective of crisis managers (Eismann et al., 2021). The authors suggest that different affordances pertain to *internalising*, which help organisations draw in information, e.g. by monitoring the information steam, to *externalising* related to broadcasting and framing, and affordances for *collaboration* (Eismann et al., 2021). Based on Schlagwein et al. (2017), Eismann et al. (2021) show the "opening" functions of social media in crises. Firstly, they synthesise the ways that social media opens *resources*, from information to funds. Then they show opening *processes*, such as digital volunteerism. Finally, they present opening *effects* of social media, which are transparency, interactivity, responsiveness, empowerment and participation. Currently, due to the limited functions provided via most warning apps (Hauri et al., 2022), only very few of these affordances appear to also apply to warning apps.

However, studies on information dissemination during the COVID-19 pandemic also found that false information was pervasive and, because of this, some organisations stopped using some platforms during the pandemic (Tikka et al., 2023). Investigating ICT use by crisis managers, Tikka et al. (2023) found a lack of platforms for joint public communication, leading to a fragmentation of public knowledge. Both of these cases suggest that warning apps, which could provide more control of the message and discourse to crisis managers and which have already been providing a shared platform for different local and federal agencies, also have unique affordances for crisis managers.

Affordances of different media channels in crises have so far not been investigated from the user perspective. Looking at the domain of media consumption and e-government, it seems that affordances of mobile warnings might be similar to those of mobile news consumption (Lou et al., 2021) and include proactivity (Scholta & Lindgren, 2019; Sjöberg, 2018). The finding that warning app users also particularly prefer direct communication with emergency managers, and the finding that social media is regarded as, at times, an unreliable channel (Reuter, 2022), also indicates that trust and reliability are affordances of warning apps that could be used to further resilience-oriented and strategic, as well as operational crisis communication.

Operational functions are the core functions of warning apps, providing alert notifications, instructions and advice. For more strategic communication, warning apps may more broadly aim to engage citizens in the issue of crisis preparedness and management, e.g. through preparedness features. Chapter 13 shows that such a feature could not only increase preparedness, but also lead to an increased perceived usefulness of warning apps. However, as many use warning apps expecting to receive only severe warnings, preparedness features should be optional and users should be able to opt out of preparedness reminders. In addition, another strategic measure might be promoting alternative warning channels such as a messaging app channel to reach a larger group of people. Crisis managers should also consider adapting their systems to current trends, e.g. exploring what public warnings may look and sound like in smart home environments. Another strategic challenge to focus on might be using peer-to-peer networks as a resilient method for relaying messages in case of an internet outage. In the realm of disaster communication, the potential of employing peer-to-peer communication within messaging applications represents an intriguing avenue for exploration. This innovative approach leverages the widespread use of messaging apps to facilitate direct and immediate information exchange between individuals. Unlike traditional top-down communication models, peer-to-peer communication offers a decentralised framework, enabling users to share warnings—from agencies and their community—received via messaging apps. Chapter 14 demonstrates that using messaging apps for official warnings would be a viable option for a large user group. With technologies and trends changing quickly, and with the advantages that people perceive from warning apps compared to social media, it is feasible to continue developing complementary systems. In 2016, a report showed that messaging apps had 20% more active users than social media (Business Insider, 2016), and in many countries, messaging apps are more widely used for news consumption than social media platforms (Newman et al., 2023). Finally, to increase resilience-oriented communication, warning app messages, particularly in pre-crises phases, might focus more on the receivers, e.g. by acknowledging the challenges to behaviour change that users are facing when reflecting on crisis preparedness. As such, techniques to change behaviour, such as reminders, nudging or gamification may be used. Warning apps may include decision support systems for users to evaluate which preparedness measures are sensible for their situation. Finally, warning apps could also offer more features that foster neighbourhood support.

This suggests that in countries with state-oriented risk cultures, two strategies can be recommended. Firstly, research has shown the large influence of status-quo bias. For example, for organ donations, the default in the United States is that people are not assumed to be organ donors (opt-in), leading to only 17% of people donating organs. However, when asked to make a decision on the spot, 79% agreed to become organ donors (Johnson & Goldstein, 2004), showing a huge discrepancy. Similarly, in addition to people not being aware of the existence of warning apps, the status quo bias reduces the number of people who use warning apps. While cell broadcast was a positive example in that it is implemented as an opt-out system, with presidential alerts exempted from the opt-out option, which means that users receive warnings as the default (Bundesnetzagentur, 2022). Similarly, a warning app may be pre-installed on smartphones with the option to delete the app, which is highly accepted by users (Kaufhold, Haunschild, & Reuter 2020).

A second strategy might target the perceived ease of use of warning ICT. Particularly in state-oriented risk cultures, it appears that the intention to use a warning app is affected by the low perceived individual responsibility for disaster preparedness. This might negatively affect the perceived usefulness of crisis ICT. The limited expected usefulness could be counterbalanced by a favourably perceived ease of use, limiting the costs of adoption, e.g. by being very easy to access and set up. A study by Fischer-Preßler et al. (2021) shows that maladaptive rewards are a significant impediment to adoption and continuance of use of warning apps. Reviews of app markets indicate that for warning apps, such costs include battery draining, data and unnecessary and loud alerts at inadequate times (Tan et al., 2020b). Avoiding the costs by not using the apps results in maladaptive rewards. Therefore, agencies might choose channels that are used in daily life and engage users through these channels. The positive evaluation of the warning messenger bot indicates that this would increase the visibility of crisis information.

Finally, a strategy that was out of the scope of this dissertation is the promotion of warning apps. The regional app *HessenWARN* has been adopted more widely in that state than the more established national apps (see Chap. 8). While this success could be related to the additional policing information and the regional appeal, the introduction of the app might also have resulted in a greater awareness of its existence. Despite its more recent introduction, *HessenWARN* was known to 53% of the Hessian respondents, and thus better known than *NINA* (48%). However, the promotional efforts of the established and new app were not focused on in this dissertation and might be explored in future research.

Summary—Key Policy Implications

- **Change Default**: Pre-install warning app on smartphones with the option for uninstalling. The high acceptance for this strategy suggests that this can be done to change the default of warning app adoption.
- **Employ Warning Apps for Strategic Crisis Management**: Replace operational use of warning apps with usage for strategic and resilience-oriented communication and implement new desired features accordingly.
- **Implement Persuasive System Design**: Further explore user intentions to prepare and use persuasive design to help users close the intention-behaviour gap.
- **Engage More User Groups**: Provide interesting channels with low adoption barriers to target users with high affinity for technology interaction and with lower protection motivation.

5.4   Limitations and Future Work

The dissertation has several limitations which should be remedied by future research.

1. **National focus**: This dissertation focused on crisis communication in Germany. A culturally situated approach is warranted as countries differ in the way that they perceive crises and confront them (Cornia et al., 2016; Reuter, 2022). However, the focus on Europe perpetuates a bias towards investigating Western countries (McKay et al., 2022). A review of flood warnings found that 80% of articles focused on industrialised countries (Kuller et al., 2021), showing that this discrepancy also applies to the scientific crisis management discourse. With the advantage of low-bandwidth and requiring little internet data, mobile warnings via warning apps and messengers might be even more relevant in the Global South, where infrastructure investments impede the implementation of cell broadcast. More research should therefore investigate the use of mobile warnings in countries of the Global South. With regard to applying the findings to other contexts, the warning app requirements and user preferences identified in the dissertation mirror usability requirements found in other countries (Tan et al., 2020b) the dissertation shows that they are largely stable over time and despite crisis experiences. This indicates some level of independence of local context and experiences. However, due to Germany having been shown to be an outlier among other European states with regard to what drives a demand for crisis preparedness information and crisis preparedness intentions (Appleby-Arnold & Brockdorff, 2018), it seems likely that in other countries, other nudges and other persuasive techniques would have different effects (Caraban et al., 2019). Therefore, more motivational strategies and more countries should be explored in the future.
2. **Focus on pre-crisis phase**: In this dissertation, due to the dominance of warning apps among crisis apps, and due to the narrow way in which warning apps are currently used in most of Europe (Hauri et al., 2022), the focus has been on the pre-disaster and alert phase of crises. However, previous research has shown the many communication functions and activities that communication on social media offers *during* crises (e.g. Kaufhold, 2021; Reuter, 2022). Given the desired two-way communication functions by uses, future research should explore the possibilities and costs of integrating functions for involving citizens in crisis mitigation and fostering more extensive citizen engagement (Distel & Lindgren, 2019).

3. **Neglect of emergency management perspective**: The introduction for new technologies for crisis management and the extension of existing ones has implications for crisis management practice. For example, Reuter (2022) shows that the use of social media in crises has lead to a large section of the populations in Europe expecting crisis managers to respond to social media messages within one hour. In addition, looking at security organisations and social media, research has revealed a large set of practical, technological and organizational-cultural challenges to implementing social media strategies (Bullock et al., 2021; Crump, 2011; de Graaf & Meijer, 2019; Denef et al., 2013). Similarly, certain changes to warning apps would require an adjustment of emergency management practices, which are difficult to implement (Eismann et al., 2021). To investigate the possibilities and constraints of adapting warning apps, future research should take a multi-stakeholder perspective that includes emergency managers. For example, Stieglitz et al. (2022) have explored the use of social bots by emergency managers. Similarly, social bots may also be used to respond to a set of messages by citizens, reducing the burden on emergency management agencies.

4. **Resilient crisis communication**: Finally, a pressing challenge for crisis communication is resilience in case of infrastructure failure (Fischer et al., 2016). While cell broadcast technology stands out as a more robust communication method (Hauri et al., 2022), operating independently of the internet, it still relies on the functionality of cell towers, which can be impacted by crises. Nevertheless, there are promising advancements in enhancing internet resilience: Satellite-based internet, distinct from terrestrial internet infrastructures, shows potential for providing improved connectivity, particularly in rural areas, and offers redundancy in crisis situations (Möcker, 2021; Voelsen & Stiftung Wissenschaft und Politik, 2021). Furthermore, ongoing developments are focused on bolstering internet availability through device-to-device, mesh, or multi-hop communication, with such networks already undergoing exploration and testing for citizen coordination during crises (Haesler, Mogk, et al., 2021; Höchst & Baumgärtner, 2020). These advancements should be considered in the development of future warning applications.

Conclusion   6

With the pervasive use of smartphones, mobile warning systems have become an integral part of public warning systems, offering location-based warnings. Among the mobile warning systems, multi-purpose warning apps provide the greatest opportunities to adapt the interface to usage in crisis contexts. Additionally, they offer the greatest means of personalisation, allowing users to choose locations, events and severity levels that are relevant to their context. Increasing the usage of warning apps could lead to more people receiving timely warnings and an overall greater engagement of citizens in public safety. Little is known about the factors that influence the adoption of apps that focus on increasing individual and public safety.

In addition, Germany presents an interesting case with regard to engagement in crisis preparedness, as citizens base their risk perception on rational rather than emotional estimations, but do not adjust their preparedness intentions to their risk perceptions. With this pattern, Germany thus differs form other European countries (Appleby-Arnold & Brockdorff, 2018). A similar effect has been attested by studies investigating the role of risk cultures, which have identified Germany as a state-oriented risk culture in which citizens rely on state-agencies rather than taking individual responsibility (Cornia et al., 2016). A study of social media and warning app use suggests that this leads to a comparatively low use of social media in crises and scepticism towards the reliability of the information in Germany (Reuter et al., 2019). By contrast, warning apps are seen as a reliable source of information, and other studies have also shown that agency information is an important source in crises (Park et al., 2019). While crisis informatics has predominantly studied social media use in crises, this dissertation has set out to investigate user expectations for warning apps, answering the research question *RQ: How can warning apps be used to engage citizens in public safety and security.* To do this, a user-centred

J. Haunschild, *Enhancing Citizens' Role in Public Safety*, Technology, Peace and Security | Technologie, Frieden und Sicherheit, https://doi.org/10.1007/978-3-658-46489-9_6

approach is taken, in which a series of quantitative and qualitative empirical studies are conducted, as well as several design-focused studies.

Firstly, to explore how warning apps are currently used and on user expectations and preferences (RQ1), representative quantitative surveys, a repeated qualitative survey during the COVID-19 pandemic, and an analysis of user reviews about COVID-19 regulation apps were conducted. These analyses reveal the current usage patterns of warning apps across time and influences by a series of severe disasters. The outcomes signify that warning app users highly perceive the usefulness of such apps. It is also noteworthy that in addition to increased usage, citizens become even more ready to share information through warning apps. Overall, the findings strongly suggest that an enhancement of warning apps by including additional features and combining more warning types in the app is highly desired by users and prospective users (Chaps. 7 to 9).

The investigation of the COVID-19 health crisis and the information technologies and sources used during the first wave of the pandemic shows the challenge of communicating crisis information. Despite an overall satisfaction with the available information sources, significant gaps existed with regard to informing about federal, regional and local differences. These presented a great challenge for citizens, as well as for crisis managers seeking to communicate the frequent changes via apps (Chaps. 10 and 11).

Secondly, the dissertation has answered the question of how to increase user engagement with warning apps (RQ2). It has presented the findings of studies focusing on design interventions, investigating the impact of persuasive technology design and nudging on engagement and crisis preparedness adoption, as well as the effect of switching the warning channel to a messaging app channel. Looking at ways of engaging user in public safety through warning apps, adding interactive and persuasive functions can increase the motivation to use a warning app (Chaps. 12 and 13). The dissertation shows that nudging toward crisis preparedness via warning apps is both accepted and can lead to positive behaviour change. The chapters discuss the ways that the nudges should be designed and implemented in order to show these positive effects and increase the motivation to use a warning app.

Finally, the dissertation demonstrates that the delivery of public warnings could make use of "botplications", which are bots that take on the functions otherwise provided by bots. The dissertation indicates that bots are particularly useful for implementing the proactive delivery of warnings. It further shows that designing the setup of the bot through a conversational interface is a challenge that can be addressed by using design features to guide the process. The chapter discusses the provision of public warnings as an element of public services and discusses the added value of a particular user group (Chap. 14).

To summarise, the following key conclusions for developers and researchers with focus on crisis communication can be drawn from this dissertation:

1. *Relevance of warning apps:* Usage of warning apps has been increasing and more people state that they could use a warning app for a variety of functions in the future, suggesting that they remain an important channel of state-citizen crisis communication.
2. *User group:* Warning apps are particularly perceived as helpful by people who value receiving information directly from emergency managers, indicating distinct approaches for information gathering in crises.
3. *Usefulness:* Warning apps are perceived as the most useful online source of information by their users, suggesting that more attention should be paid to developing and promoting warning apps and other similar channels.
4. *Design more complex warning apps:* Users perceive that not only disaster and extreme weather events should be communicated via warning apps, but also more emergency and security related information, including crime-related warnings, crisis preparedness information, updates about pandemics and first aid instructions.
5. *More active role of citizens:* Citizens also highly value functions that allow them to be more actively involved, e.g. by helping find missing persons, registering as volunteers or providing information to emergency managers via apps.
6. *Include persuasive technology design:* Nudging towards crisis preparedness in the context of warning apps is largely accepted, especially by people already using a warning app, and can lead to improved crisis preparedness.
7. *Interactive and informative preparedness information:* Showing more preparedness information leads to an increased motivation to use a warning app, if the content is perceived as informative.
8. *Decrease adoption barriers:* Setting up personalised warning preferences for automatic warnings can be accomplished via conversational interfaces while maintaining good usability, suggesting that a broader range of warning channels would be feasible to increase the reach of warnings and accessibility, e.g. by visually impaired persons.

By synthesising and discussing the studies, the dissertation has contributed a series of empirical qualitative, quantitative and design-oriented studies that contribute to understanding the user perspective on warning apps, their role in the crisis information ecosystem and design interventions that can increase citizen engagement. From these studies, implications for design were derived to increase warning app adoption and their usefulness for citizens. Through these interdisciplinary studies, the dissertation advances scientific knowledge in crisis informatics and HCI and suggests technology design solution for citizen engagement in public safety.

Part II
Publications

Warning the Public: A Survey on Attitudes, Expectations and Use of Mobile Crisis Apps in Germany

7

Abstract

As part of information systems, the research field of crisis informatics increasingly investigates the potentials and limitations of mobile crisis apps, which constitute a relatively new public service for citizens and are specifically designed for the dissemination of disaster-related information and communication between authorities, organisations and citizens. While existing crisis apps, such as KATWARN or NINA in Germany, focus on preparatory information and warning functionality, there is a need for apps and research on police-related functionality, such as information on cybercrime, fraud offences, or search for missing persons. Based on a workshop with civil protection (N = 12) and police officers (N = 15), we designed a questionnaire and conducted a representative survey of German citizens (N = 1,219) on the past, current and future use, perceived helpfulness, deployment and behavioural preferences, configurability and most important functionality of mobile crisis apps. Our results indicate that in addition to emergency and weather warnings, crime and health-related warnings are also desired by many, as is the possibility for bidirectional communication. People also want one central app and are resistant to installing more than one crisis app. Furthermore, there are few significant differences between socio-economic groups.

Original Publication   Kaufhold, M.-A., Haunschild, J., & Reuter, C. (2020). Warning the Public: A Survey on Attitudes, Expectations and Use of Mobile Crisis Apps

Supplementary Information   The online version contains supplementary material available at https://doi.org/10.1007/978-3-658-46489-9_7.

in Germany. *Proceedings of the 28th European Conference on Information Systems (ECIS), 1–16. https://aisel.aisnet.org/ecis2020_rp/84*

7.1   Introduction

In the past two decades, information and communication technology (ICT) has become a relevant part of responding to crises and fostering public safety against bio-medical or chemical emergencies, large-scale traffic accidents, natural disasters and catastrophes, human-made at-tacks, terrorism and political uprisings (Olteanu et al., 2015; Palen & Anderson, 2016; Stieglitz, Mirbabaie, & Milde 2018). The corresponding multidisciplinary field of crisis informatics, which seeks to combine knowledge from computer science, information systems and social sciences, largely focuses on the utilisation of social media technologies, including artificial intelligence and social media analytics, applied to crisis management and response by emergency services (ES), affected citizens and digital volunteers (Alam et al., 2020; Kaufhold, Bayer, & Ruter 2020; Kaufhold, Rupp, Reuter, & Habdank 2020; Soden & Palen, 2018; Stieglitz, Mirbabaie, Fromm, & Melzer, 2018). However, an increasing body of research examines the development, evaluation, potentials and limitations of so-called crisis apps. These are understood as mobile smartphone apps specifically designed by public authorities for the distribution of disaster-related information and communication between au-thorities, organisations and citizens across different phases of the emergency management cycle (EMC) (Grinko et al., 2019; Helmerichs et al., 2018; Karl et al., 2015; Tan et al., 2017). Crisis apps nowadays complement existing multichannel warning systems, such as MoWaS (Modular Warn-ing System) in Germany, to increase urban resilience (Klafft, 2013; Weichselgartner et al., 2018). Despite Germany being frequently struck by severe floods and storms and recently experiencing terrorist attacks, such as the 2016 Munich shooting and 2016 Berlin truck attack, most German citizens have never experience a crisis and do not consider the risk very high (Guha-Sapir et al., 2004; Höppe, 2015; Mirbabaie et al., 2019). The lack of preparation enhances the potential damage inflicted by a crisis, which means locals' threat awareness should be raised and according measures for a higher crisis risk should be supported (BMI, 2009). Accordingly, existing crisis apps in Germany, such as KATWARN or NINA, which support different crisis types, focus on preparatory information and warning functionality, and provide different degrees of configurability (Kotthaus et al., 2016). Although existing representative studies for Germany examined the citizens' adoption and use of crisis apps, they did not include police-related functionality, such as information on cybercrime, fraud offences, or search for missing persons (Grinko et al., 2019; Reuter et al., 2019). Furthermore, as research indicates that only 16% of German citizens used crisis apps, ways to incentivise their use and continuance of use have to be explored (Reuter, Kaufhold, Leopold, & Knipp, 2017). In order to examine the use, perceived helpfulness, deployment and behavioural preferences, desired configurability and functionality of crisis apps, our paper integrates insights from both civil protection and police officers to answer the following research questions:

- RQ1: What is the citizens' past, current and expected future use and the perceived helpfulness of crisis apps in comparison to other media channels?
- RQ2: What are citizens' preferences on the deployment as well as the information and warning behaviour of crisis app?
- RQ3: What are citizens' demands on configurability, required and most important functionality of crisis apps?

The paper is structured as follows: After discussing related work on the characteristics and continuous use of crisis apps, outlining a research gap (Sect. 7.2), we introduce the overall method of our empirical study (Sect. 7.3). More specifically, based on a workshop with civil protection (N = 12) and police officers (N = 15) for the design of the questionnaire, we address these research questions by conducting a quantitative representative survey with German citizens (N = 1,219). We then present the descriptive and statistical results of our survey, also comparing some of them to findings of previously published work (Sect. 7.4). The subsequent section discusses our findings, the limitations of our study and concludes with potential future work (Sect. 7.5).

7.2   Related Work

Since September 11th, 2001, ICT have played an increasing role in fulfilling information demands in crises and helping concerned parties communicate with each other (Hughes et al., 2008; Reuter & Kaufhold, 2018). Facilitated by the wide distribution of smartphones and mobile apps, crisis and warning apps have been developed to supplement crisis communication and management efforts (Karl et al., 2015; Tan et al., 2017). Crisis apps refer to mobile smartphone apps which are designed for the purpose of disaster-related information and communication (Appleby-Arnold et al., 2019). Furthermore, the term warning apps is often used to refer to crisis apps which focus on warning functionality (Reuter, Kaufhold, Leopold, & Knipp 2017). These apps are typically used as a direct means of communication between emergency services and citizens and research shows that crisis apps serve a different audience than social media do (Kaufhold, Haunschild, & Reuter 2020), specifically a group that favours direct, privileged information.

7.2.1 Characteristics and Use of Crisis Apps

Among the most popular existing crisis apps in Germany are KATWARN and NINA, which deliver warning messages based on the users' GPS coordinates or user-defined locations and offer recommendations for actions as well as general tips (BBK, 2021; Fraunhofer FOKUS, 2018). Furthermore, affected people can inform contact points and tell them to what extent they are affected by the crisis or not. Groneberg et al. (2017) have so far conducted one of the broadest international crisis app comparison based on categories of information (e.g., push notifications, maps, news, organisational information), communication (e.g., social media integration, direct 112 emergency calls, contact directory, "I'm safe" notification) and preparation (e.g., emergency planning, behavioural tips, descriptions of dangers, trainings). They found that the most frequent crisis app functionalities in the information category are warnings, followed by maps and general information or news. The first aspect is also the most expected one, while (potential) users also often wish to receive behavioural advice and to help emergency services by providing on-site information (Reuter, Kaufhold, Leopold, & Knipp, 2017). However, communication and preparation functions are less widely spread in crisis apps (Groneberg et al., 2017). A review by Tan et al. (2017) revealed 57 crisis apps, whereof 16 had crowdsourcing, 13 collaboration, 13 alerting and information, nine collating, and six notification as primary purpose. In summary, 86% of the examined crisis apps contributed to the EMC's response phase, while only 26% had functionality contributing to each the mitigation, preparedness, and recovery phases.

One advantage of crisis apps lies in resilience against infrastructure breakdowns such as power outages, providing an additional channel for crisis communication, allowing for ubiquitous usage, utilising their battery life and providing recommendations for action even offline (Nestler, 2017). In a study of Markwart et al. (2019), people who received warning messages reached a safe spot faster than the control group, showing the apps' positive impact. Furthermore, crisis apps with formalised bidirectional communication functions enhance trust by giving users the impression that they are appreciated contributors (Appleby-Arnold et al., 2019). However, concerning their usage, low familiarity with such apps and their benefits is currently the greatest barrier: As found in a representative survey, only 16% of Europeans have been using crisis apps (Reuter, Kaufhold, Leopold, & Knipp, 2017). A following study by Reuter et al. (2019) also shows that risk culture is an important aspect for peoples' expectations in crises and that strong national differences exist in terms of crisis apps: while 28% of the Dutch population reported to have downloaded and used a crisis app, 16% in Germany and Italy, and only 7% in the UK had done so. Despite the findings that many people perceive a civic duty to stay informed and follow authorities' instructions (Appleby-Arnold et al., 2019), retention of such apps appear to be low. This may be due to the limited awareness of their availability, to their infrequent use, or the inadequacy of their design, which results in users' dissatisfaction and deletion of the app. By exploring peoples' wishes and needs, we can induce how an app should be designed to increase their continuous use.

7.2.2 Artefacts Mediating Citizen-Agency Interactions

Since many aspects of emergency interventions are a government service, the use of crisis apps can be regarded as an element of e-government. E-government means the use of ICT to improve government services and requires organisational change. Digitisation is changing the ways in which citizens and agencies interact (Lindgren et al., 2019) and crisis apps are changing crisis communication, data transfer and processing and visualisation of information (Tan et al., 2017). As Tan et al. (2017) show, about a third of crisis apps mainly support traditional one-way (one-to-one and one-to-many) interactions, regarding the public as a victim of disasters and as information receivers. About two thirds have followed the trend to (also) support the public's role as in-situ sensors and offsite volunteers, employing many-to-many interactions and/or many-to-one-to-many interactions. In this process, the skills required by citizens to access services or information and to participate can change. But agencies' tasks can also shift from case work towards educating people how to access digital resources (Pors, 2015) or move towards collating of content, alerting and informing, facilitating collaboration and crowdsourcing (Tan et al., 2017). In addition, while general-purpose apps are being used in crises because of peoples' familiarity with these apps, these may not be designed for the special requirements relevant in crises. In contrast, crisis apps are often less relevant to citizens' day-to-day life but designed to fulfil certain functions specifically in emergencies. While initiatives have already been taken to adjust emergency response to social media activity, for example by Virtual Operations Support Teams (VOST) (VOSG, 2019), this has so far been unexplored regarding crisis apps. This leaves the question open whether citizens' preferences in emergencies are best served by familiar general-purpose apps—requiring emergency services to adapt their organisation and processes to these tools—or by designing specific crisis apps, which can be more geared towards emergency services' structures and needs. So far, citizens' expectations have not been analysed with this perspective of required organisational adaption.

However, there is a risk that authority-centred apps will not be highly used by citizens. Tarute et al. (2017) find that continuous use of apps is mainly promoted by design and information quality. Information given by the app must be timely, relevant (Nikou & Mezei, 2013) and necessary (Tarute et al., 2017), while unnecessary information can lead to confusion and dissatisfaction. Other factors affecting user's satisfaction can be the ease of use (mental effort put into handling), privacy aspects, battery drainage (Olubusola, 2015) and ease of access and download (Kim et al., 2013). While there are many studies on retaining users on health and educational mobile apps, only few studies have explored factors that influence citizens' use of government-related apps. Their findings suggest that trust, performance expectancy and ease of use are the primary factors influencing whether or not such apps are engaged with (Sharma et al., 2018; Susanto et al., 2017). Research specifically on crisis apps shows that utility, dependability and output are significantly positive related to continuance, while graphics and input can negatively affect it (Tan et al., 2020c). Requirements and perceptions of IT and security are culturally influenced (Sturm & Nestler, 2018), therefore determining such requirements in a representative manner is crucial to designing usable systems. In addition, since crisis apps can help emergency services spread and receive important information, exploring what features are important to potential and current users is paramount to identifying which aspects would make such an app more useful to citizens and whether features could be added that avoid unnecessary information, yet provide value to users despite the infrequent nature of emergencies.

7.2.3   Research Gap

In crisis informatics, an increasing body of research examines the functionality, evaluation, potentials and limitations of crisis apps (Tan et al., 2017). Published representative surveys based on the Germany population investigated the citizens' adoption and use of mobile apps for civil protection (Grinko et al., 2019; Reuter et al., 2019). However, they did not provide insights on police-related functionality, which was requested by workshop participants (section 3). In addition, a risk perception paradox has been identified, which shows that a high perception of being at risk regarding natural hazards does not mean that precautions are being taken by citizens (Wachinger et al., 2013). Being reminded of previous experiences as well as trust in authorities' and experts' advice have been shown to increase citizen preparedness. Despite manifold potentials of crisis apps (Markwart et al., 2019), research indicates that only 16% of German citizens have used crisis apps, and ways to incentivise their use, desired configurability and functionality as well as continuance of use have to be explored (Reuter, Kaufhold, Leopold, & Knipp, 2017). In addition, while e-government is often regarded as the use of technology for delivering services to largely passive citizens, crisis research suggests that citizens are capable of participation and self-organisation (Kaufhold & Reuter, 2016; Wachinger et al., 2013) and that ICT may promote such active roles (Palen et al., 2010). In studying citizens' demands and preferences, we can deduce which organisational aspects may have to be changed to accommodate participation, different use patterns, to meet citizens' requirements and to be inclusive. Thus, our study seeks to contribute to the knowledge base by examining the citizens' past, current and expected future use and the perceived helpfulness of crisis apps in comparison to other media channels (RQ1), preferences on the deployment as well as the information and warning behaviour of crisis app (RQ2), and demands on configurability, required and most important functionality of crisis apps (RQ3). In addition to including questions that are particularly relevant to practitioners identified through a workshop, the analysis is based on a representative sample of the German population, which allows identifying usage patterns and the effect of socio-demographic factors and attitudes towards trust on the use and preferences towards crisis apps.

7.3   Empirical Study: Representative German Survey

7.3.1   Survey Design

On the 13th of February 2019, we conducted a workshop at a German central federal police agency that dealt with the use of mobile apps in crises, comprising civil protection (N = 12) and police (N = 15) officers. At the start we introduced the procedure for conducting a representative survey and the aim of this workshop to generate a questionnaire. Examples of closed and open-ended questions were introduced. After the presentation, the workshop comprised of three phases: In the reflection phase (10 minutes), based on their individual creativity, participants were instructed to note their ideas or questions on moderation tasks. In the presentation phase (20 minutes), participants presented their ideas and we subsequently arranged them thematically on a flip chart. The participants were encouraged to write down further ideas during the presentation phase. Finally, in the discussion phase (60 minutes), based on the group's collective creativity, participants discussed existing moderation cards, generated new ones and reflected upon their thematic grouping. In an integration phase, the workshop results were combined with a previously published survey (Grinko et al., 2019), whereof two questions (Q1, Q3) were re-integrated and extended to compare two datasets, resulting into seven distinct questions:

- **Helpfulness and Use of Crisis Apps.** Firstly, we designed two questions since officers were interested in the past, current and future use of crisis apps (Q1) and the perceived helpfulness of media channels in previous crises (Q2), which could impact the depth and breadth of their analysis and communication strategies.

- **Attitudes and Sharing Behaviour regarding Crisis Apps.** Secondly, we designed two additional questions since officers were interested in the citizens' attitude (Q3) and sharing behaviour (Q4) using crisis apps as a more detailed picture on motivations and fears would help them to feasibly adapt their communication strategy.

- **Configurability and Functionality of Crisis Apps.** Lastly, to further improve their analysis and communication strategies, emergency services were interested in citizens' expectations towards the configurability and functionality of crisis apps. Thus, we integrated three questions on the overall configurability (Q5), desired (Q6) and most important (Q7) functionality of crisis apps.

7.3.2   Data Collection and Analysis

After two rounds of feedback from the central federal police agency, the finalised questionnaire was self-hosted through LimeSurvey, then sent to the commercial and ISO-certified panel provider GapFish in May 2019, which ensures random sampling and representativeness according to age, gender, regional distribution and education. The survey covered the above dimensions which translated into seven closed and four socio-demographic questions about age (continuous 14–87 years), education (5-point ordinal scale "no school diploma" to "university degree"), urbanisation (4-point ordinal scale "less than 5.000 inhabitants" to "more than 100.000 inhabitants"), gender (3 categories, "male", "female" "other"[1], "no answer"). The definition of emergencies as unforeseeable events (such as epidemics, earthquakes, fires, big accidents, or floods) that impact several people and require immediate action to minimise negative consequences was given before the relevant questions. Information and warning apps were defined as smartphone applications that can provide information on how to behave before, during and after an emergency, and that can warn about imminent emergencies such as an attack, a bomb discovery, multiple vehicle collision, or storms. For all generated items, we followed guidelines for valid

---

[1] Since only one respondent identified as "other", the category could not be used for analysis due to the small group size.

item design, including phrasing positively, clearly, short, concisely, and understandably, limited to one statement per item and avoiding leading questions (Moosbrugger & Kelava, 2012). Though items should be related to the present (Mummendey & Grau, 2014), due to the infrequent nature of emergencies, we resorted to previous experiences, so potential effects of remembering should be taken in account in interpreting the results. Questions are either on a 5-point interval Likert scale (allowing for "no response") to evaluate the degree of agreement with a statement or the judgment of relevance of functions and design features. One categorical variable reveals the respondents' experience (16,5% were currently using an app, 26% had previous or current experience) and awareness of different warning and emergency apps. After elimination of incomplete answers, N = 1,219 reliable answers resulted representing the German population in age, gender, geography, urbanization and education (Statista, 2018), of whom 68% had been in an emergency (N = 827). An approximation of normal distribution of the data can be assumed due to the sample size (Leonhart, 2008).

We analyse the data using Chi Square tests, Cramer's V, Kendall's tau-b, ANOVA, Spearman's ρ and Pearson's r, depending on the scale of the dependent and independent variables. For the categorical variable "gender", we applied the t-test for independent samples, paying attention to the assumption of homogeneity of variance through a Levene test. We judge effect sizes of Pearson's r of | 0,10 | as a small, of | 0,3 | as a moderate, and of | 0,50 | as a strong correlation (Cohen, 1988). For the statistical analysis we use IBM SPSS Statistics Version 26. For each analysis, we chose the test that is most robust and allows for the most fine-grained scale. An exception is made when testing for non-linear correlations, for which data are recoded into categories to test group effects, such as binary categories for those under 25, over 45 and those over 60-years old. We test city size and age for collinearity and find no multicollinearity (VIF = 1). In Q1 and Q3, we furthermore compare the studies of 2017 (N = 1,069) and 2019 to see if and how crisis app use patterns and expectations have changed over recent time. The comparison data come from a 2017 survey conducted in Germany (Grinko et al., 2019).

7.4 Empirical Findings

7.4.1 Awareness, Use and Helpfulness of Crisis Apps

Awareness of crisis apps is currently not widespread in Germany (Fig. 7.1). The German Red Cross' (DRK) app is the only app that more than 50% of respondents were aware of, followed by Facebook's Safety Check, KATWARN (of the Fraunhofer Institute for Open Communication Systems) and NINA (of the Federal Office of Civil Protection and Disaster Assistance), which are known only to about a third of the population. Other apps are unknown to about 80% of respondents. But awareness does not equate to usage: Despite being well-known, the DRK's app is used by only 4% of the population, while the most actively used app, NINA, is used by 12%. Current or former usage varies between 16% (NINA) and only 2% (Safeture). Numbers on future use show a slightly better picture. Here, between 14% (DRK) and 4% (Cell Broadcast) of participants indicate that they would use a certain crisis app in the future. For most apps, the percentages of approximate future use are as high or even higher than those of current or former use. However, there is also a substantial group of respondents (10%–27%) who are aware of crisis apps, and despite never having used them are not planning on using them in the future.

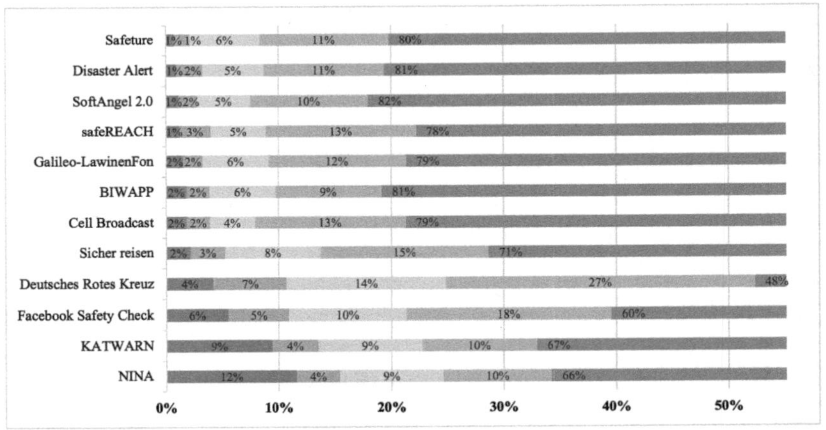

Fig. 7.1 Q1: Use of crisis apps (■ Yes, I am currently using it, ■ Yes, I have used it in the past (not anymore), ■ Yes, I want to use it in the future, ■ No, neither, ■ I do not know this app)

Looking across emergency apps, 16,5% of Germans currently use any app (19% when including Facebook's Safety Check) and 26% have current or previous experience with a crisis app. This leaves 9,4% of the population having tried out at least one app and discontinued use. Of those, 14% are now using the Facebook Safety Check feature. Looking at those who had used an app, its usefulness is judged positively (63%), more positively than other online sources, including social media. However, personal conversations and especially contact with emergency services surpass apps in their evaluations. An ANOVA shows that neither age (p = 0,093) nor gender (p = 0,083) have an effect on the evaluation of apps by those who had used them. However, gender is to a small degree significant for whether or not the app was used ($\chi^2(1)$ = 12,97, p < 0.001, ϕ = 0,14): Of those who had been in an emergency and had not used an app, 61% were female. Of those who had used the app, 46% were female, and 64% of females had used an app versus 77% of males. In contrast, age (Cramer's V = 0,11; p = 0,51) and city size (Cramer's V = 0,04; p = 0,85) do not influence the use of crisis apps in emergencies. While older age (over 65s, N = 108) does not decrease the chance of using an app, surprisingly youth (under 25s, N = 136) very slightly does ($\chi^2(1)$ = 8,175, p = 0.004, ϕ = −,104). Another small non-linear effect can be attributed to education (Cramer's V = 0,139, p = 0,006), with highest user rates of 75% and 77% among those without school diplomas and those with university degrees. Urbanisation is not a significant and relevant factor for the use of a crisis app: neither rural nor urban respondents have significantly more or less experience.

Comparing app use in 2017 with 2019 (Fig. 7.2), an increase in current use can be found across apps and the Facebook's safety feature, with the greatest increase for NINA. Especially NINA has enlarged its current userbase (4% vs. 12%). Furthermore, KATWARN (6% vs. 9%) and Facebook Safety Check (3% vs. 6%) can now report higher figures than two years ago, while Safeture remained at 1%, although it can report higher numbers on future use. A small increase in plans for using crisis apps can also be found for NINA and KATWARN (both 7% vs. 9%), while fewer people are planning on using Facebook's feature (13% vs 10%).

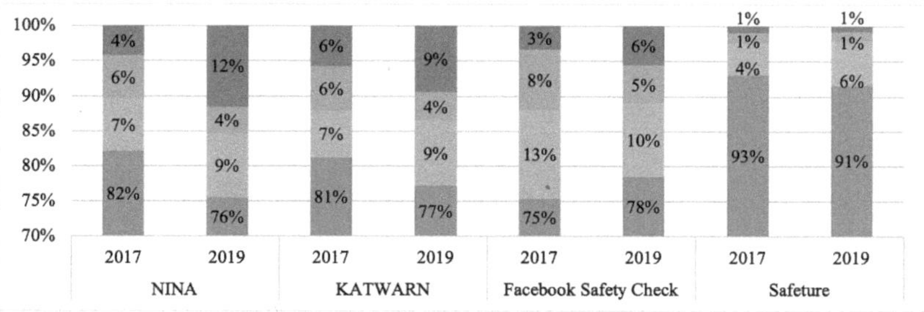

Fig. 7.2 Comparison of crisis apps use (■ Yes, I am currently using it, ▨ Yes, I have used it in the past (not anymore), ▤ Yes, I want to use it in the future, ■ Not sure / No, I haven't used it)

Regarding the judgement of helpfulness by those who had been in an emergency and used the source, crisis apps (N = 536) are well-rated (Fig. 7.3): 63% rate them as rather or very helpful, surpassing the judgements of social media and other online sources. However, radio, television, personal and telephone conversations and contact to rescue services are generally judged more positively. An analysis of preferences has identified distinct preference groups (Haunschild et al., 2020): Fans of crisis apps also particularly favour other online sources, contact with emergency services and local announcements. In contrast, those who favour newspapers do not judge apps positively, while they are a success for those who prefer direct contact with emergency services. However, those who like using social media in emergencies do not particularly appreciate apps. This suggests that all sources fulfil particular needs that differ between users. It also indicates that app users in emergencies are a group that particularly seeks out privileged information from emergency services, be it through an app, direct contact or websites. This groups does not find mainstream media particularly helpful. Other affected people may also be using their mobile phones in emergencies, but they may be fulfilling different needs by accessing social media. Then again, people who contact emergency services are also favourable towards apps, so this is a group that may be particularly interested in crisis apps (Table 7.1).

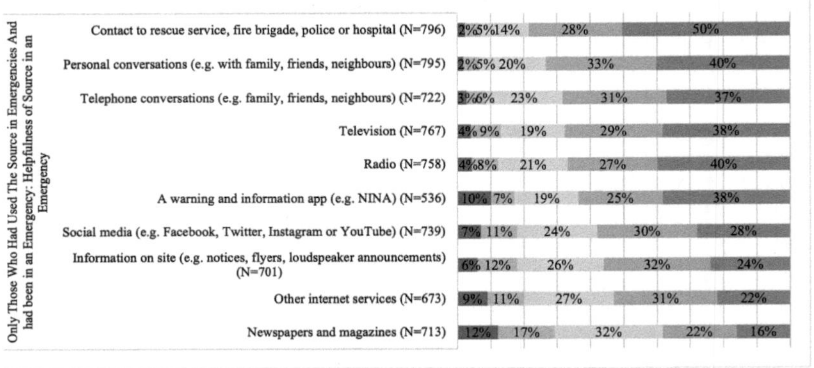

Fig. 7.3 Q2: Helpfulness of information sources (only those with emergency experience who used the source in an emergency) (■ not, ■ not very, ■ moderately, ■ quite, ■ very helpful)

Table 7.1 Insightful correlations of judgements of helpfulness of information sources, $p < 0.001$. SM= Social Media, ES= Emergency Services

Pearson's r	Crisis app	SM	Contact ES	News-paper	Local announce-ments	Per-sonal con-vers.	TV	Radio	Other internet
Crisis app	1	**0,274**	**0,393**	0,14	**0,390**	0,206	0,114	0,180	**0,424**
SM		1	0,206	0,19	**0,316**	0,188	0,296	**0,309**	**0,494**
Contact ES			1	0,156	**0,447**	**0,318**	0,26	0,297	0,225
Newspaper				1	0,281	0,2	**0,465**	**0,424**	0,237

7.4.2 Attitudes and Sharing Behaviour regarding Crisis Apps

Despite the low crisis app use, almost 70% agree that some form of crisis app implementation is desirable and reasonable. Strong agreement (65%) can be found regarding the wish for centralisation in one standardised app that is fed by several emergency services, which is mirrored by peoples' resistance to installing more than one crisis app (44%). Having an app preinstalled on phones upon purchase is also supported by 55%, while 19% are opposed. However, people mainly want to maintain control over being able to deactivate such a preinstalled app (45%). While a surprising number of people wish for an app to replace other existing information sources (41%), a large portion also oppose this idea (31%). Apps complementing other sources is widely supported (65%). Regarding people's design and implementation wishes for a crisis app, patterns are very similar (56% to 63% approval and 12% to 19% opposition). Most agree that such an app should only send messages in acute emergencies (63%), the status should always be visible on the lock screen, and messages should be sent each time an event is updated. In contrast, 50% are also in favour of only one message being sent for each event. Gender influences a few of these judgements, but only slightly: A t-test for independent samples shows that women are less prepared to have several crisis apps ($t(1119) = -3{,}204$, $p = 0.001$), and oppose apps replacing other sources more strongly ($t(1113) = -1{,}88$, $p = 0.06$), while they favour the idea of a pre-installed app more strongly than men ($t(1132) = 2{,}776$, $p = 0.006$). An ANOVA's R^2 shows that even where significant, age explains less than 2,5% of the observed variation and is therefore not relevantly connected to the judgements made. Support for crisis apps has not changed significantly since 2017 (Fig. 7.4). Although opposition against a standardised app has almost doubled (7% vs. 13%), the overall image shows a majority in favour (68% vs 65%). Support for pre-installed crisis apps has grown from 44% to 55% and resistance against pre-installed crisis apps has also dropped from 26% to 19%. However, while in 2017 the survey did not specify the origin of the pre-installed application, in 2019 it was specified as an app provided by the government, which may also play a role in the increase in support.

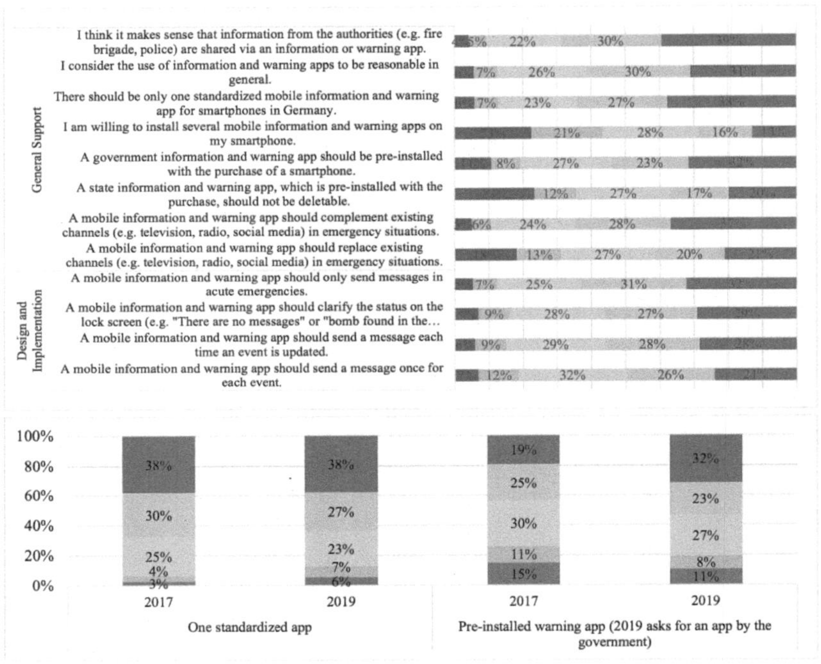

Fig. 7.4 Q3: Agreement to statements and comparison of crisis apps attitudes (■ strongly disagree, ▨ disagree, ▤ neutral, ▨ agree, ■ strongly agree)

Confidence in secure use of personal data by the police and fire brigades is relatively high (52%) and only 19% have no such confidence (Fig. 7.5). This is supported by the fact that more than 75% of participants would make image or video data available to the authorities and would also transmit live footage during a crisis. On the other hand, 32% indicate that they have no trust in a secure transmission of their location by an app of a private provider and only one third trusts rather than mistrusts software. Despite this mistrust, a majority is prepared to transmit their location automatically as part of an emergency call function or in emergency situations, while 64% would also like to be asked before each location transmission. A large group of smartphone users (42%) would be ready to contribute their phone as part of an ad-hoc network, potentially contributing to infrastructure resilience, which is a major concern regarding app use in crises. A test for Spearman's r shows significant correlations (all $p < 0.001$) between the statements about data sharing and distrust in the secure use of data. A large effect size is found between distrust towards police and fire departments and distrust towards private companies regarding GPS data transmission ($r = 0,473$), but only a slight negative connection with trust in software generally ($r = -0,107$). Another strong negative connection exists between preparedness to share photos and videos and distrust in ES ($r = -0,308$), while the effect is less strong regarding live transmission ($r = -0,185$). Distrust towards private companies' sharing of GPS data somewhat stops users from transmitting evidence ($r = -0,076$) and only slightly influences their reservations towards sharing GPS information outside of emergencies ($r = -0,226$) and their demands for being asked before transmitting information ($r = 0,204$). General trust in software enhances peoples' readiness to submit GPS data automatically ($r = 396$) and to share photos and videos with ES ($r = 0,196$), including live ($r = 0,223$). Gender plays a minor role for some judgements, with women slightly more prepared to share videos and photos (Cramer's $V = 0,108$; $p = 0.01$), live transmission ($V = 0,095$; $p = 0.037$) and they insist slightly more on being asked before transmitting GPS data ($V = 0,115$; $p = 0.005$). However, the connections are not strong and for many judgements, such as levels of trust, gender is not significant. Along with other studies

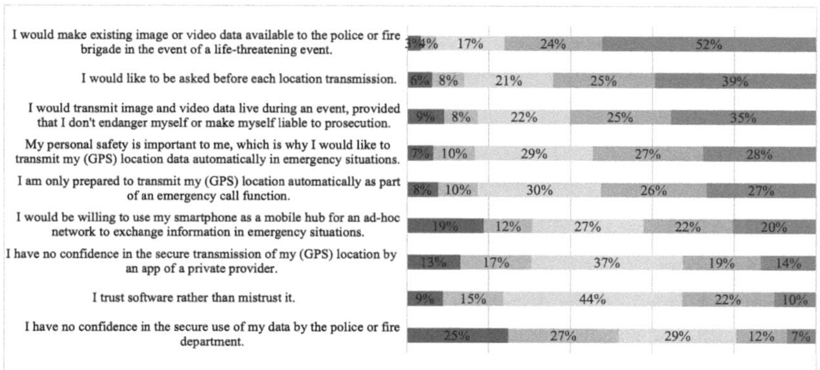

Fig. 7.5 Q4: Agreement regarding statements about willingness to submit multimedia or location data (■ strongly disagree, ■ disagree, ■ neutral, ■ agree, ■ strongly agree)

(Appleby-Arnold et al., 2019; Haunschild et al., 2020), our analysis finds that age is generally not a relevant factor in crisis communication. In our survey, age only correlates with increased trust in the secure transmission of GPS data to emergency services (Spearman's r = –0,125, p < 0,001).

7.4.3　Configurability and Functionality of Crisis Apps

All suggested design features receive wide support (Fig. 7.6). This ranges from 85% support for easy use and being self-explanatory to 61% supporting internationality and 59% multilingualism. More than 90% also consider a senior's mode, location-related information, operability for people with restrictions and the ability to turn on and off certain information types important. Age correlates (all p < 0.001) lightly with the wish for specific characteristics: Older people particularly value above all easy (r = 0.191) and inclusive usability (r = 0,185), followed by a seniors' mode (r = 0,172) and self-explanatory design (r = 0,171), location specific information (r = 0,12) and control over information types (r = 0,115).

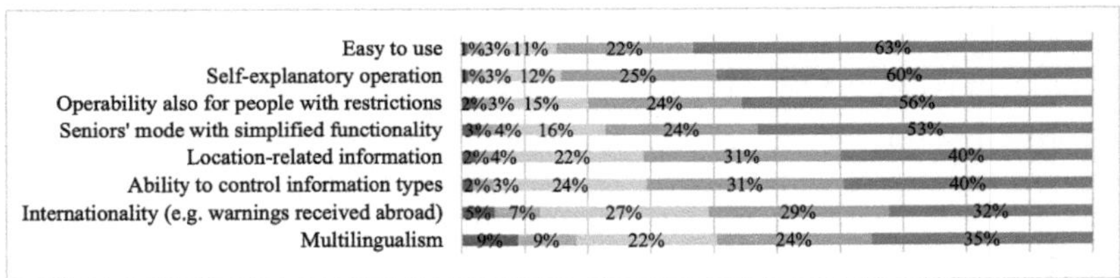

Fig. 7.6 Q5: Importance of characteristics (■ not, ▨ not very, ▨ moderately, ▨ quite, ■ very important)

Regarding emergency functions, both regular emergency calls and GPS enhanced calls are desired by over 75%, while direct messages are also a favoured feature. Over 60% support text emergency messages, a chat function and news tickers are also welcome. The large support for many warning functions shows the relevance of staying informed. While disaster warnings are the most popular feature, crime, traffic, health and weather-related warnings are also hugely supported. Warnings about speed gauges and school cancellation are regarded as less relevant and even encounter some resistance. Over 60% also wish for links to further information and instructions on how to prepare for and how to behave in an emergency. Almost 70% would also like to support the search for missing persons and over half would like to be able to register as volunteers and receive calls for eyewitnesses to come forward. This underlines the collaborative potential of such apps (Fig. 7.7).

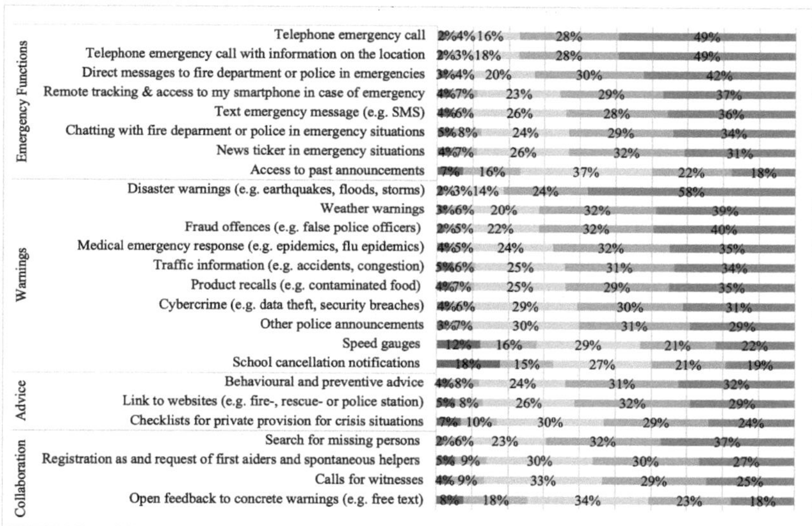

Fig. 7.7 Q6: Importance of functions (■ not, ■ not very, ■ moderately, ■ quite, ■ very important)

Here, gender plays a small role (Cramer's V between 0,1 and 0,15) for usability and accessibility features, but not for the other design aspects. Women also want to be slightly more involved in the search of missing persons ($V = 0,113$, $p = 0,006$), but for other collaborative features gender does not play a role. Women are also slightly more interested in some types of warnings (emergency warnings: $V = 0,133$, $p = 0,002$; police messages: $V = 0,121$, $p = 0,002$, school cancellations: $V = 0,102$, $p = 0,019$, cybercrime: $V = 0,105$, $p = 0,014$). They are also slightly more interested in direct messages to ES ($0,144$, $p < 0,001$), telephone ($V = 0,127$, $p = 0,001$) and GPS-enhanced emergency calls ($V = 0,137$, $p < 0,001$), and the access of their remote location ($V = 0,104$, $p = 0,017$). Analysing the preferences of those who had used an app but are not using it anymore, their preferences do not differ greatly from the others. In contrast to what could be expected, this groups' opinions are generally less extreme (with the exception of whether warnings, where both the groups wanting more and less warnings are bigger). Those who are not using apps anymore also prefer somewhat less advice, less access to past posts and less open feedback on disasters.

When asked about the three most important functions of crisis apps (Fig. 7.8), respondents show a clear prioritisation of disaster warnings, followed by GPS-enhanced and regular emergency calls. Weather and traffic warnings are also favoured. For 254 respondents, the possibility to assist in the search of missing persons is also a priority.

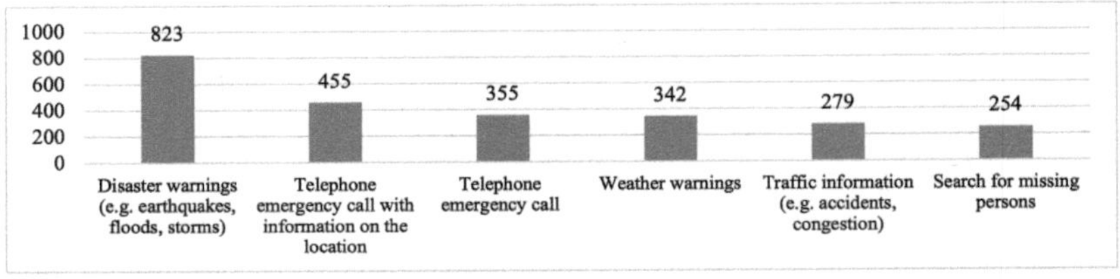

Fig. 7.8 Q7: Most important functions of a crisis app (most frequent mentions)

7.5 Discussion and Conclusion

In this paper, we examined the awareness, use, helpfulness, preferences, as well as favoured configurability and functionality of mobile crisis apps by German citizens. Our survey of a representative sample (N = 1,219) of the German population was informed by a workshop with civil protection and police officers and a previously published survey. However, as only two questions were re-integrated for comparative analysis and even extended, the paper contributes by the examination of five new questions. Our results provide insights into the following three research questions.

What is the citizens' past, current and expected future use and the perceived helpfulness of crisis apps in comparison to other media channels (RQ1)? We found that NINA and KATWARN are still the most used crisis apps in Germany, which slightly increased between 2017 and 2019. Almost two thirds of the participants who had been in an emergency rated crisis apps as quite or very helpful information source, which however is surpassed by radio, television, personal and telephone conversations and contact to rescue services. Especially those preferring direct contact to ES also like to use crisis apps. Where direct contact is putting a strain on ES or where citizens perceive that ES are too unavailable, promoting crisis apps can reassure particularly this group of citizens and save ES' resources that are under strain in emergencies. The use of crisis apps, which are a relatively new phenomenon, is on the rise and it is not crowded out by other channels or features, such as Facebook Safety Check. However, the analysis of use patterns and attitudes shows a discrepancy between the general support and perceived relevance of such apps and their actual use, which is currently at 16.5% including all relevant emergency apps in Germany. The analysis shows that almost 10% of the population had tested a crisis app at some point but is not currently using any, which calls for additional research on the experience, functionality, satisfaction and usefulness influencing the continuance of use (Fischer-Preßler et al., 2020).

What are citizens' preferences on the deployment as well as the information and warning behaviour of crisis app (RQ2)? In general, our participants had a positive attitude towards crisis apps but one of the biggest contentions for users is having to install more than one app, while great support exists for the concept of one single state-supported app, which may even come pre-installed (but deletable) on the smartphone. The finding that people who value apps also value direct privileged information from ES supports this wish for centralisation. This suggests that crisis apps are an area where people long for a trusted, centralised actor to combine information from different agencies. In addition, peoples' willingness to allow for apps to be preinstalled shows the positive potential that is expected from such apps and potentially also shows that citizens are aware of the risk perception biases identified in research (Wachinger et al., 2013). This would be in line with their high evaluation of using apps to receive advice. In crisis, they were furthermore willing to transmit their GPS location, image and video data to emergency services as long as they are not endangered. Allowing those measures may be related to the security-sensitive nature of crisis apps.

Investigating under which conditions and in which fields users desire a centralised app and perceive it being preinstalled as legitimate is an interesting question to explore in future research. However, while the availability of many options and the preinstallation of apps are desired, these should come paired with design features that allow to individually configure the types of emergencies that users receive warnings about, as well as the option to delete the app and to communicate with emergency services. While such communication requires resources on the part of ES, the potential for exchanging information, for preparing and guiding citizens before and during emergencies and for engaging them as volunteers and witnesses may outweigh these costs. It should be further explored which institutional changes are necessary to facilitate empowering citizens and enable the co-production of security by citizens and emergency services. One fruitful avenue may be to further explore how VOST or similar "trusted volunteers" could contribute to maintaining crisis apps (St. Denis et al, 2012). Further personal resources would allow the establishment of bidirectional communication, allowing citizens to report emergencies and situational updates as eyewitness or first responders and ES, for instance, to request additional information from the incident scene (Kaufhold et al., 2018).

What are citizens' demands on configurability, required and most important functionality of crisis apps (RQ3)? Besides ease of use and self-explanatory operation, citizens valued operability for people with restrictions, a simplified senior mode and a high degree of configurability in terms of content- and location-specific information. Here, gender and age to some degree influence preferences. Our results indicate that not only crisis-specific information, but also general, such as product recalls and traffic, and police-oriented information, such was cybercrime, fraud offences and search for missing people, are desired. Still, disaster warnings, telephone emergency calls (with information on the location) and weather warnings were mentioned as the most important functionalities of a crisis app. Since other studies have identified perceived benefit as a major factor shaping whether or not an app is installed, an app that unifies the functions of the available apps and the addition of further warning categories may be positive for increasing usage of crisis apps and providing more value despite the infrequent nature of crises. Especially supporting the search of missing persons is a priority for many. Peoples' interest in contacting ES through the app via direct messages or chat support other findings which have identified the need for bidirectional communication and the relevance of multi-functionality (Kaufhold et al., 2018). In accordance with other findings, we also find that age does not significantly influence peoples' willingness to use an emergency app or their preferences, and neither does urbanisation or education (Appleby-Arnold et al., 2019). Gender also plays only a minor role, which is supported by other findings (Reuter et al., 2019). Supporting previous findings about app functions, natural disaster and weather warnings are the most relevant aspects of crisis apps to users, but we also find that users are willing to receive advice for how to behave. We also confirm that people are willing to use apps to offer support in emergency situations. This indicates that despite Germans' predominantly state-oriented risk culture that emphasises the states' responsibility in emergencies (Dressel, 2015), there is an interest in taking on active roles as witnesses and volunteers (Kaufhold & Reuter, 2016).

For risk governance and communication, we can derive that the adaptation of a single authorised app and its preinstallation would be welcomed by most. Considering the high willingness to receive advice, administrators could use crisis apps to clarify the limits of state precautions and individual responsibilities and give advice to specific audiences, such as homeowners. This reaches a group that is different from the one using social media in emergencies (Haunschild et al., 2020). Designers should foster the centralisation of a crisis app that includes all branches of ES, such as fire and police departments. Such an app should enhance the users' safety but offer a high level of configurability to provide value despite the infrequent nature of crises, including health and (cyber-) crime related information, as well as varying user or risk group needs. Since users are also interested in participating as volunteers and witnesses, for instance to report incidents and seek missing people, a move towards bidirectional communication is suggested. Due to the willingness to receive advice, designers should also consider segmenting users according to risk-relevant information provided by users, in order to be able to transmit audience-specific advice and nudge vulnerable groups towards taking relevant action.

This paper is subject to limitations: The biggest challenge is that online surveys are biased towards people who engage online. Although the survey provides representative results according to age, gender, geography, urbanisation and education, it excludes those people who are most resistant to using technology. The categories for urbanisation used here should also be made more fine-grained in the future to further explore the potential differences in behaviour and needs in emergencies of particularly rural areas and very big cities. Furthermore, in multiple questions, a caveat results from capturing emergency services more generally, without differentiating between different types of organisations. Finally, the paper focuses on practical implications, with limited explanatory power. Therefore, future research should explore the role of crisis apps through a theoretical lens, i.e. using the framework of risk cultures, perception bias and nudging, the use of apps for community building sense-making in crises, trust towards authorities and target of blaming if emergency response is not successful (Dressel, 2015; Newman et al., 2019; Shklovski et al., 2008).

Perceptions and Use of Warning Apps—Do Crises Lead to Changes?

8

Abstract

Warning and emergency apps are an integral part of crisis informatics and particularly relevant in countries that currently do not have cell broadcast, such as Germany. Previous studies have shown that such apps are regarded as relevant, but only around 16% of German citizens used them in 2017 and 2019. With the COVID-19 pandemic and a devastating flash flood, Germany has recently experienced severe crisis-related losses. By comparing data from representative surveys from 2017, 2019 and 2021, this study investigates whether these events have changed the perceptions of warning apps and their usage patterns in Germany. The study shows that while multi-hazard emergency and warning apps have been easily surpassed in usage by COVID-19 contact tracing apps, the use of warning apps has also increased and the pandemic has added new desired features. While these have been little-used during the COVID-19 pandemic, especially non-users see smartphone messengers app channels as possible alternatives to warning apps. In addition, regional warning apps appear promising, possibly because they make choosing a warning app easier when there are several available on the market.

Original Publication Haunschild, J., Kaufhold, M.-A., & Reuter, C. (2022b). Perceptions and Use of Warning Apps—Did Recent Crises Lead to Changes in Germany? In M. Muhlhauser, C. Reuter, B. Pfleging, T. Kosch, A. Matviienko, K. Gerling, S. Mayer, W. Heuten, T. Doring, F. Muller, & M. Schmitz (Eds.),

Supplementary Information The online version contains supplementary material available at https://doi.org/10.1007/978-3-658-46489-9_8.

Proceedings of Mensch und Computer 2022 (pp. 25–40). Association for Computing Machinery. https://doi.org/10.1145/3543758.3543770

8.1   Introduction

The research field of crisis informatics investigates the potentials and limitations of mobile crisis apps, which are specifically designed for the dissemination of disaster-related information and communication between authorities, organisations and citizens (Reuter & Kaufhold, 2018). Especially in countries that are not currently using cell broadcast, mobile crisis apps are an integral part of timely warnings and meant to complement existing Modular Warning Systems (MoWaS). While some apps are used both in daily life and in emergencies, such as social media, messaging apps and news apps, other apps are specifically designed for emergencies (Tan et al., 2017). Their functions can include all phases of the emergency cycle, from preparedness, to response and recovery. Some crisis apps connect citizens with other citizens, especially for co-ordination and collective sense-making (Reuter & Kaufhold, 2018; Tan et al., 2017). Others primarily connect citizens and agencies, differing in the amount of communication that is one-way (often from agencies to citizens) or two-way, allowing citizens to communicate with agencies (Reuter & Kaufhold, 2018; Tan et al., 2017). Due to the infrequent occurrence of crises and the resulting infrequent use of the apps as well as due to the criticality of emergency apps, they have particular usability criteria related to reliability, salience of important information and trustworthiness (Tan et al., 2020a). A study in 2019 showed that while many Germans perceived warning apps as important and useful tools, only 17% were using a warning app. Since then, significant crises have occurred in Germany and potentially changed the attitudes and expectations towards warning apps. A flood in July 2021 had devastating effects, partly because of delayed or missing warnings (Fekete & Sandholz, 2021). The criticism regarding insufficient warnings was added to other failures identified during the national warning day (Tagesschau, 2020). Both events have lead to the decision to implement cell broadcast in Germany, following other European states (BMWK, 2021a,b).

In addition, the highly dynamic and global COVID-19 pandemic resulted in an "infodemic" with a lack of reliable information (Cinelli et al., 2020). Challenges included the dissemination of false information (Laato et al., 2020; Marchal & Au, 2020) and the effect of excessive information seeking, which can lead to distress, information avoidance and decreased compliance with recommendations (Park, 2019; Siebenhaar et al., 2020). A study investigating the information situation in Germany at the beginning of the pandemic showed that citizens were particularly challenged by regionally and locally differing regulations (Haunschild, Pauli, & Reuter, 2021). At the same time, they felt that it was their responsibility to stay informed, which could cause additional distress, because people who perceive information gathering as an obligatory task appear to be particularly prone to perceived information overload (Schmitt et al., 2018). Citizens expected authorities to provide easily accessible and understandable information (Haunschild, Pauli, & Reuter, 2021). In order to clearly present the temporally and locally varying regulations related to the pandemic, new apps emerged and warning apps in part started to include COVID-19 information (Haunschild & Reuter, 2021a). With these recent crises, this study investigates which crisis apps are currently used in Germany. In addition, it compares the results with data from 2017 and 2019, exploring which changes have occurred.Thus, our research questions are:

- RQ1: What was the usage and the expectations towards crisis information and warning apps in 2021 in Germany?
- RQ2: How have usage and expectations of crisis information and warning apps changed over time since 2017?

The study proceeds as follows: After presenting research findings in crisis app usage in general and during the COVID-19 pandemic in Sect. 8.2, we present how the data from German citizens was collected in a representative survey and how it is analysed in Sect. 8.3. Subsequently, in Sect. 8.4, we present the results, which are further discussed in Sect. 8.5. Finally, a conclusion is given in Sect. 8.5.3. In summary, the study shows that while multi-hazard emergency and warning apps have been easily surpassed in usage by COVID-19 tracking apps, the use of warning apps has also increased and the pandemic has added new desired features. The wish for integration of pandemic-related information is in line with a general preference for incorporating more topics, including police-related information. Messengers and regional emergency apps might be explored as alternatives to nation-wide warning apps.

8.2   Related Work

In the following, we present the state of the art of research investigating warning apps, their usage, continuance and discontinuance of use and the resulting research gap.

8.2.1 Warning App Usage

Cultural factors play an important role in whether people assume personal responsibility and prepare themselves in the event of a crisis (Cornia et al., 2016; Dressel, 2015). In research, several risk cultures are identified that differ in terms of the framing of disasters, the target of blame and the trust in authorities. For example, Germany is regarded as a state-oriented risk culture, which is characterised by high trust in authorities, which are assumed to be capable of handling and preventing disasters. In a state-oriented risk culture, citizens have little knowledge and confidence in their individual capability. In contrast, in an individualistic risk culture, it is perceived that individuals can actively shape the outcome in disasters, showing more knowledge and coping behaviours (Cornia et al., 2016; Dressel, 2015). Finally, in fatalistic risk cultures, trust in authorities as well as individual coping capabilities is low, leading to a perception that not much can be done to prevent or cope with a disaster (Cornia et al., 2016; Dressel, 2015). A country's risk culture also affects the way that information and communication technologies (ICT) are used (Reuter et al., 2019). Germans, with Germany being a state-oriented risk culture, rely more strongly on official communication channels and tend to distrust social media (Reuter et al., 2019).

Several studies confirm that users particularly demand a multi-hazard app that combines information about many important types of emergencies, instead of having to install several different apps (Dallo & Martí, 2021; Kaufhold, Haunschild, & Reuter, 2020). A previous study has shown that when it comes to warning app preferences, there are hardly any differences between age groups and gender (Kaufhold, Haunschild, & Reuter, 2020). Generally, warning apps should allow for transparent interactions, which includes salience of critical information, activation in emergencies and usability (Bonaretti & Fischer, 2021). In addition, they should allow for situational awareness, which consists of promptness and actionability (Bonaretti & Fischer, 2021). Finally, it is important that warning apps have "representational fidelity", meaning that they are a digital representation of the analogue emergency situation, which implies being exact, current, consistent, complete and relevant (Bonaretti & Fischer, 2021). While timeliness plays a role in all three dimensions, particularly representational fidelity is important for whether the app is trusted (Bonaretti & Fischer, 2021).

Other studies focus on the usability of disasters apps. A review study shows that for these apps it is particularly important to make critical information salient, to design the user interface with consideration of users' cognitive load in an emergency and to build trust bearing in mind the limited interactions with the app (Tan et al., 2020b). Multiple aspects concerning an app's usability are relevant, which has lead to specific usability guidelines for warning apps, which focus, in addition to common usability aspects, on app dependability, simplicity of design, minimal external links and audio output for critical situations (Tan et al., 2020a).

The use and intention to use a warning app can be conceptualised as a manifestation of protective behaviour motivation (Fischer-Preßler et al., 2021). This protection motivation is influenced by many aspects, including trust, social influence (whether friends and family use a warning app), risk and efficacy perceptions (Fischer-Preßler et al., 2021). Perceived vulnerability increases non-users' intention to start using a warning app (Fischer-Preßler et al., 2021). Therefore, particularly the experience of a severe flood may have changed the atmosphere in Germany in favour of warning apps.

Insight from the field of mobile enabled e-government suggests that users are insensitive to factors that prevent non-users from using a tool, such as concerns about behavioural control or perceived risks and costs (Susanto & Goodwin, 2013). Perceived trust and social norms (particularly whether a partner is using a warning app) influence the intention to start using a warning app (Fischer et al., 2019). According to a study by Fischer-Preßler et al. (2021), the construct of protection motivation is very good at explaining use intention for non-users, accounting for 69% of variance. In contrast, protection motivation only partly (45%) explains why users (dis-)continue using a warning app (Fischer-Preßler et al., 2021). Research indicated that overly complex warning apps are dropped: Users discontinue using apps that have complex user interface graphics or require too much user input (Tan et al., 2020c). Other factors positively influence the continuance of use, in particular the perception that the app performs its intended function, is reliable and error-free (Tan et al., 2020c). This is supported by a study that found response efficacy, mediated by trust, to be an important factor for continuance of use intention, in addition to social factors. For both users and non-users, hampering factors such as battery and memory use are important (Fischer-Preßler et al., 2021). Finally, users continue using warning apps that make critical information salient and easy to understand (Tan et al., 2020c).

8.2.2   COVID-19 Information and Contact Tracing Apps

The COVID-19 pandemic revealed some uncertainties about handling information in a dynamic and long-term crisis. In contrast to emergencies, which typically last only a short period of time, the pandemic, once it had started to spread, was on the one hand constant in most countries, but on the other hand also differing in severity, appearing in several waves of intense spreading, hospitalisations, intensive-care hospitalisations and death rates. Judgements therefore differed on whether and how information about the pandemic should be distributed through warning apps (Haunschild & Reuter, 2021a). The risk culture also appears to have influenced the perception of the information ecosystem at the beginning of the COVID-19 pandemic: Germans relied strongly on official media channels and press conferences and rather distrusted social media and private information exchanges (Haunschild, Pauli, & Reuter, 2021). In addition, research has found that during the COVID-19 pandemic exposure to information on social media was likely to result in information overload and thus information anxiety and avoidance (Soroya et al., 2021). Thus, many perceived a gap when it came to structured agency information about regulations (Haunschild, Pauli, & Reuter, 2021). This gap was partly closed by private sector initiative, which emerged to offer apps with information about COVID-19 regulations. In addition, the official warning app NINA also started including such information on regional regulations (Haunschild & Reuter, 2021a).

In response to the pandemic, many technological tools were developed to help cope with the crisis (Vargo et al., 2021). These included apps for identifying symptoms, tracking infections (Kumar et al., 2020) and coordinating volunteers (Haesler, Schmid, et al., 2021). Hundreds of apps were developed that related to the COVID-19 pandemic, with the greatest increase in spring 2020 (Tsinaraki et al., 2021). Around that time, most apps were dedicated to information and news and symptoms checking and many were provided by governments and involved health agencies, but many were also developed by private companies (Dieter et al., 2021).

In addition, during the COVID-19 pandemic, contact tracing apps emerged as a new type of app to help in the tracking of the spread of the virus and to warn people who had come in contact with an infected person (Ahmed et al., 2020; Min-Allah et al., 2021). Because these apps need access to precise location data as well as sensitive health information, their implementation was not without controversy. At the same time, it was clear that mobile contact tracing was only going to be successful if a large majority of the population was using it (Braithwaite et al., 2020). A large body of research thus investigated acceptance and worries related to tracing apps, privacy preserving features and ethical aspects of these apps (Altmann et al., 2020; Kitchin, 2020; Parker et al., 2020; Trang et al., 2020; von Wyl et al., 2021).

For example, a study analysing peoples' perceptions and news coverage of contact tracing apps in German-speaking countries during the pandemic's first wave has shown that there was a lot of distrust regarding contact tracing apps, framing them as governmental surveillance tools and thus a potential risk to individual freedom and privacy rights (Zimmermann, Fiske, et al., 2021). Unsurprisingly, user numbers were therefore relatively low. In Germany, for example, only 25% of the total population had downloaded the app and around 21% actively used it in 2020 (RKI, 2020). Privacy concerns pose the main hindering factor when it comes to user's willingness to use contact tracing apps. Motivating factors for using contact tracing apps are in turn a perceived benefit, the apps expected performance, social influence, and trust in government (Altmann et al., 2020; Zimmermann, Fiske, et al., 2021). A large scale qualitative study, analysing normative positions towards COVID-19 contact-tracing apps in nine European countries, found that positions are varied, ranging from clear opposition, over scepticism of feasibility and pondered deliberation, to resignation and distinct support (Lucivero et al., 2022). In addition, the functions of different apps are relevant for how they are judged, with contact tracing apps being viewed more positively than quarantine enforcement (Utz et al., 2021). Similarly to apps offering information on regulations, there were both official government and private contact tracing apps. In the context of Germany, Munzert et al. (Munzert et al., 2021a,b) have compared the two most popular COVID-19 warning apps, that is, the official "Corona-Warn-App" (Corona warning app) and the privately launched app "luca" at different times. Generally, they found that users of the official Corona warning app were more likely to comply with the COVID-19 rules and have greater trust in science, the government, and the German health care system. They also found that despite the advantages of the official app over private apps, scepticism about the app's privacy, security and efficiency prevailed among non-users.

8.2.3   Research Gap

Research suggests that experience with an emergency results in increased vulnerability perception and motivation to take precautions (Diekman et al., 2007). Germany has recently been affected by two large crises. First, the COVID-19 pandemic came to Germany in March 2021. The pandemic affected the whole population, leading to nation-wide restrictions. In the course of the pandemic, over 8,000 people died in the first three months with over 20,000 deaths in one month at its peak (RKI, 2022). Then, in July 2021, a flash flood incurred a large death toll of 182 killed people in Germany, partly due to a lack of evacuations and warnings (Löhe & Oswald, 2021). Both crises may have increased the perceived vulnerability and the perceived need to take preventive action and be informed in a timely manner, possibly leading to an increase in warning app use. At the same time, warning apps were criticised during both crises: The flood showed that even though the international weather service had issued a warning of a severe flood, in some regions the warnings were delayed, leading to missing warnings in the media and warning apps, and no actions were taken. In addition, cell broadcast is currently not available in Germany, meaning that only those actively using a warning app could receive the warning directly through the main warning channel. Such problems had already become clear in the evaluation of the first national warning day in September 2020 which was largely considered a failure (Tagesschau, 2020). While older systems such as sirens are no longer maintained in many places, the warning apps were over-burdened. Cell broadcast, which sends a text-based warning to all mobile phones in the relevant vicinity could be an alternative, as it requires less data and is universally available to all mobile phone users without having to install an app or own a smartphone. Due to the flood, a law was passed to implement cell broadcast in Germany as of 2022 (BMWK, 2021b).

During the first wave of the pandemic, German citizens on the one hand felt that sufficient information was available, but when searching for specific answers, e.g. about current laws in place to contain infections, which vary locally and regionally, they also often perceived that authorities should do more to provide reliable information (Haunschild, Pauli, & Reuter, 2021). While the warning app NINA (run by the Federal Office for Civil Protection and Disaster Assistance) included the first COVID-19 information around two months after the beginning of the pandemic in Germany, it only started including local regulations in December 2020, nine months after the pandemic's spread to Germany. Those apps that emerged to inform about pandemic-related regulations proved to be difficult to maintain, because the relevant information was often only available as a legal text or government announcement. Even though the apps informing about local regulations were highly appreciated, an analysis of user reviews showed that the lack of machine-readable information and personnel resources to update them lead to inconsistencies, uncertainty and diminishing trust in the provided information (Haunschild & Reuter, 2021a). Another new phenomenon are messenger channels maintained by official sources, such as the German Ministry of Health's WhatsApp and Telegram channel that informed about the pandemic, sought to counter false information and encourage vaccination. While the expansion of smartphone messengers and their development towards social media components has been noted (Newman et al., 2019), the use of messengers as a communication channel that could complement or replace warning apps has not been explored so far.

It remains unclear whether and how these different experiences in the last years have changed usage patterns and expectations of warning apps in Germany and whether these differ for current users and non-users of warning apps. This study addresses this gap by comparing data from October 2021 with data from 2019 and 2017. While previous studies have focused on representative investigations of social media in crises and a European comparison (Haunschild et al., 2020; Reuter et al., 2019), as well as on warning app design preferences (Kaufhold, Haunschild, & Reuter, 2020; Tan et al., 2020b) and user feedback (Fischer-Preßler et al., 2020), this study is the first one to investigate temporal changes in a replication study.

8.3   Method

To analyse the usage and expectations towards warning apps in 2021, we conducted a representative survey in Germany. We compare the data with representative survey data from 2017 (Grinko et al., 2019; Reuter, Kaufhold, Spielhofer, & Hahne, 2017) and 2019 (Kaufhold, Haunschild, & Reuter, 2020).

8.3.1   Survey Design

We designed the questionnaire to allow a comparison with previous surveys, but also to ask novel questions about the COVID-19 pandemic and changes in the crisis app market. First, in order to be able to analyse group differences and relevant independent variables and to ensure representativeness, we collected socio-demographic information on **age, gender, income, education** and **region (Q1–Q5)**. In order to evaluate whether people with emergency experiences have different attitudes and expectations than those without (Fischer-Preßler et al., 2021), the questionnaire asked about respondents' **experiences with acute emergency situations (Q6)**. We defined an emergency situation as "spontaneous and usually unforeseen event that affects several people and for which immediate action must be taken to minimise its negative impact. These include, but are not limited to: Epidemics, earthquakes, (large-scale) fires, major accidents (e.g., train, plane, pile-up), floods, severe storms, and other life-threatening emergency situations."

To compare warning apps with other sources and to evaluate their usefulness during the pandemic, we inquired about the **helpfulness of information sources**, including warning apps, in emergencies (**Q7a**). Because new apps and new information sources, such as ministry messenger channels, have emerged during the COVID-19 pandemic, another question explicitly asks about the helpfulness of sources and ICT during the COVID-19 pandemic (**Q7b**).

In order to compare whether **experience using an information and warning app** have changed over time, we asked respondents about their experiences with such apps: "By an information and warning app (e.g. NINA, KATWARN or Corona-Warn-App) we mean an application that is installed on a smartphone. On the one hand, this provides information on how to behave before, during and after an emergency situation (e.g. recommendations for action before, during and after a flood) and, on the other hand, also transmits warnings to the user about acute or imminent emergency situations (e.g. attack, discovery of a bomb, pileup or approaching storm front)." First, we asked whether respondents had ever downloaded an information and warning app, which type of app and which specific app (**Q8–Q10**). The question about the downloaded app types (Q9), which was asked in 2017, was revised for this survey to allow for a more nuanced insight, that now also includes whether a particular type of app is still being used, has been abandoned or whether its future use is planned. Similarly, in 2017 we only asked respondents to name the source that had been most helpful (Q7a), rather than judging the helpfulness of all sources, which allows for a more detailed comparison. Due to these differences, we exclude this question from the temporal comparison. Finally, the questionnaire contained questions about the **main aim of smartphone app use in emergencies (Q11)**, the **relevance of specific functions and features (Q12)** and **general preferences (Q13)**. Due to different emphases in the surveys, Q11, Q12 and Q13 were only asked in one of the past surveys, which still allows for insights into temporal changes and also indicates which attitudes have remained rather stable (see the online supplementary material for the questions).

8.3.2   Data Collection

To ensure a representative sample, we used the survey provider GapFish. GapFish is an ISO-certificated panel provider, guaranteeing panel quality, data quality, security, and survey quality through various (segmentation) measurements for each survey within their panel of 500,000 active participants. We submitted the questionnaire alongside the desired demographic distribution to GapFish, who then conducted the survey using their panel and afterwards provided the results via Excel and SPSS files. The survey was conducted in October 2021 and contained a total of 32 questions covering media and Information and Communication Technologies (ICT) use in crises and during the COVID-19 pandemic, of which those related to warning apps are analysed here. Only respondents who passed the quality assurance questions were considered, which resulted in N = 1,090 reliable answers. Details on the demographic variables and values are given in Table 8.1.

Table 8.1 Representative demographic variables and values

Variable	Values
Age	18–24 (8,9%), 25–34 (14,3%), 35–44 (15,0%), 45–54 (16,8%), 55–64 (18,3%), 65+ (26,6%)
Gender	Female (51,1%), male (48,7%), diverse (0,1%), not stated (0,1%)
Education	Lower secondary education (28,2%), middle or high school (55,9%), academic degree (16,0%)
State	BB (3,0%), BE (4,5%), BW (13,5%), BY (16,0%), HB (0,8%), HE (7,6%), HH (2,3%), MV (1,4%), NI (9,7%), NW (21,7%), RP (5,0%), SH (3,6%), SL (1,1%), SN (5,0%), ST (2,3%), TH (2,6%)
Income	Less than 1,500€ (23,7%), 1,500€ to 2,600€ (30,9%), 2,600€ to 4,500€ (29,0%), more than 4,500€ (16,4%)

8.3.3   Data Analysis

For all generated items, we followed guidelines for valid item design by phrasing the questions in a positive, clear, short, concise and understandable way, limiting them to one statement per item and avoiding leading questions (Moosbrugger & Kelava, 2012). Though items should be related to the present (Mummendey & Grau, 2014), due to the infrequent nature of emergencies, we resorted to previous experiences, so that answers relating to past events may be less reliable due to distortions resulting from remembering. Questions are on a 5-point interval Likert scale (allowing for "no response") to evaluate the degree of agreement with a statement, the judgement of relevance of functions and design features and the helpfulness of sources. An approximation of normal distribution can be assumed due to the sample size (Leonhart, 2008).

Depending on the scale of the dependent and independent variables, we use Cramer's V and Kendall's tau-b to analyse the data. For the categorical variable app use experience, we applied the t-test for independent samples, paying attention to the assumption of homogeneity of variance through a Levene test. For the statistical analysis we use IBM SPSS Statistics Version 27. We use the studies with data from 2017 (N = 1,069) and 2019 (N = 1,219) to see if and how crisis app use patterns and expectations have changed. The comparison data come from surveys conducted in Germany (Grinko et al., 2019; Kaufhold, Haunschild, & Reuter, 2020; Reuter, Kaufhold, Spielhofer, & Hahne, 2017). Because a previous study has shown that age and gender hardly have any significant influence on warning app preferences, we mainly focus in this study on differences between users and non-users of warning apps, which have been shown to be motivated by different aspects (Fischer-Preßler et al., 2021).

8.4   Findings

In the following, we present the findings concerning the use of warning apps and other crisis-related apps and how these have changed over time. We then present attitudes towards the apps and their functionalities.

8.4.1   The Use of Warning Apps in Emergencies

The survey shows that around one third of respondents (60.5%) stated that they had been in an acute emergency situation in the past, whereas 31% disagreed and 8.5% were unsure. As the definition of an emergency situation explicitly included pandemics, this shows that many participants had not perceived the pandemic as an acute emergency situation. Compared with the data from 2017 (Fig. 8.1), significantly more people (32%, compared with 6%) have ever downloaded an information or warning app (hereafter called "warning app"). Kendell's tau-b for ordinal variables shows that age is a significant but very small factor ($\tau b = 0.06$, $p = 0.038$), with fewer older people having ever downloaded a warning app.

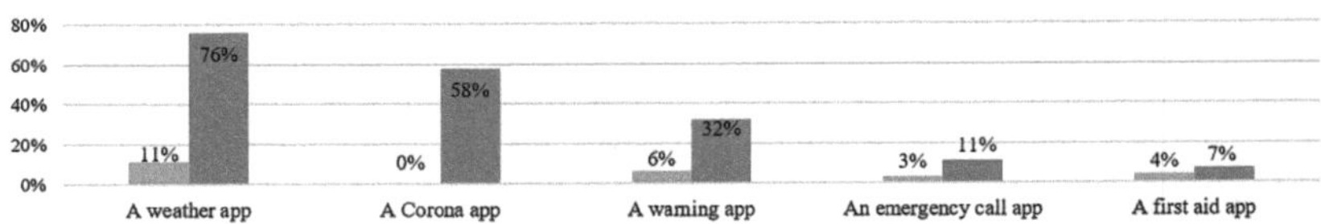

Fig. 8.1 Comparison of the usage of certain information and warning apps ▨ 2017 (left): Percentage having downloaded such an app, ▪ 2021 (right): Percentage currently using and having used such an app in the past

Regarding the types of apps used that are relevant for emergencies, the multipurpose weather apps are most widely used (Fig. 8.2). 71% currently have a weather app on their mobile phone, and only 20% have never used one or are unaware of such apps. The second most-widely used warning apps are single-hazard apps in the context of the COVID-19 pandemic, which 49% are currently using, 9% have stopped using, and 31% are not planning on using.

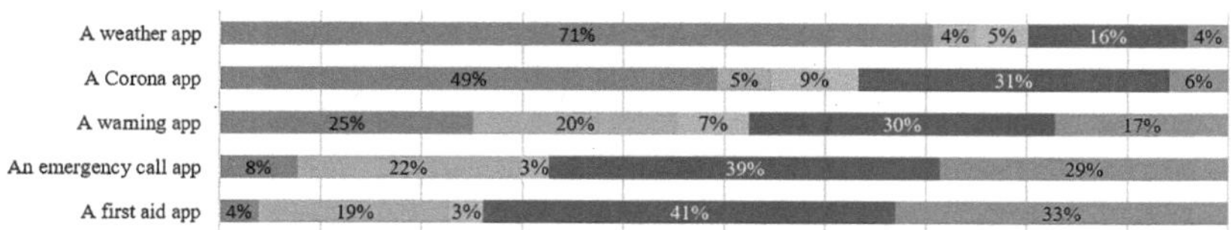

Fig. 8.2 Extent of current, past and future planned use and awareness of smartphone apps relevant for emergencies (left to right: ■ current, ▨ planned, ▨ past, ■ no use, ■ no awareness of such apps)

Warning apps for hazards other than specifically COVID-19 are currently used by 25% of respondents, a steep increase from 2019, when this number was just under 17%. Seven percent of respondents had abandoned the use of warning apps, while one third has no interest in using one. Remarkably, around 20% are planning on using a warning app, emergency call app or first aid app in the future. Only 17% are unaware of the existence of warning apps. In comparison, only 11% have experience in using an emergency call app and 7% in using a first aid app.

The analysis of the use of applications developed specifically for disasters shows that only a few apps are relevant (Fig. 8.3), which are the two COVID-19 contact tracing apps and the two nation-wide official multi-hazard warning apps *NINA* and *KATWARN*. These are followed by the federal ministry of foreign affair's app *SicherReisen*, which supports safe travelling abroad. A noteworthy newcomer is DarfIchDas (which translates to *"Am I Allowed"*), which is an app from the private sector that presents current regulations related to the COVID-19 pandemic. Another interesting case is the regional warning app *HessenWarn*, which is used by almost 17% of the residents of the federal state Hesse, surpassing the values of the national apps *NINA* (15%) and *KATWARN* (16%) in that region. The app is unknown to only 20% in that region. Further apps that were inquired about *(safeREACH, Galileo-LawinenFon, Safeture, CoroBuddy, Disaster Alert, EchoSOS, SoftAngel 2.0)* were all unknown to about 80% of the population, with only 1–2% current, planned and past use each. Looking at the most widely used warning apps *NINA* and *KATWARN*, we see that their usage has kept on increasing over time. Both of these apps are run by German agencies dedicated to crisis response. Users can choose warnings based on their GPS location or by defining regions that are relevant to them. In contrast to the growing usage of warning apps, the use of the Facebook tool for signalling that one is safe in emergencies has decreased in relevance (Fig. 8.4). The unequal growth rates between the different emergency-related apps indicate that the growing usage in the main warning apps represents a growing interest in these apps, rather than a general increase in app use.

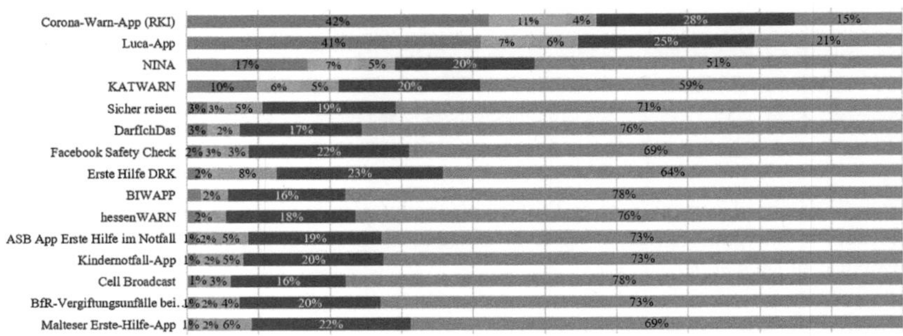

Fig. 8.3 Extent of current, past and future planned use and awareness of emergency warning and information apps (■ current, ▦ planned, ▨ past, ■ no use, ▦ no awareness of this app)

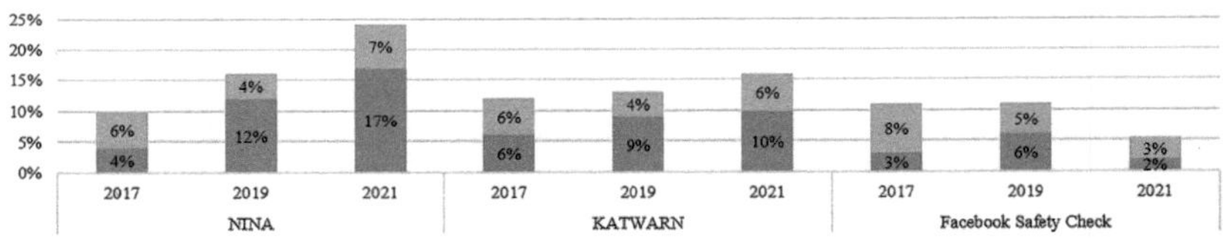

Fig. 8.4 Comparison of crisis apps use 2017, 2019 and 2021 (bottom: ■ current use at the time, top: ▦ planned future use)

8.4.2   Attitudes Towards Warning Apps and their Functionalities

One reason for adopting or ignoring warning apps could be their perceived usefulness. Therefore, we compare the perceived helpfulness of warning apps with other information sources and ICT. The analysis shows that channels and sources that had been helpful in emergencies had primarily been personal and telephone conversations, followed by direct contact with emergency services (ES) and the legacy media radio and TV (Fig. 8.5). Emergency apps were deemed similarly helpful as other internet sources and social media. Of those who have been in an emergency, 40% did not use a warning app. Looking at those who have been in an emergency and used warning apps, 69% stated that warning apps had been helpful or very helpful and only 27% disagreed (M = 3.55, SD = 1.4). There are no significant differences in the age groups (Kruskal Wallkis test: $p = 0.48$) and gender also does not play a role (Levene test: $p = 0.73$).

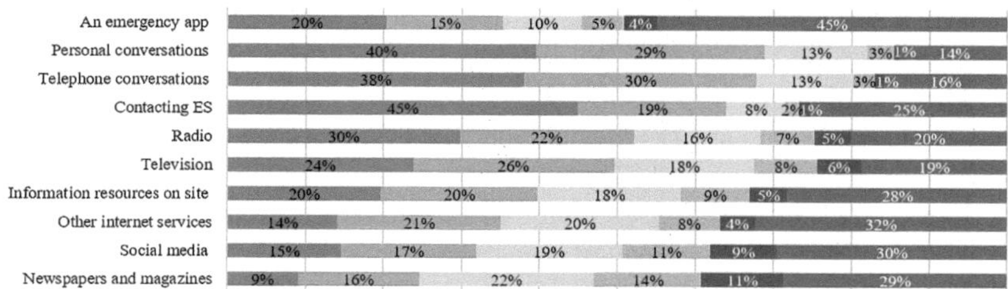

Fig. 8.5 Perceived helpfulness of sources and channels in past emergencies (■ very helpful, ▨ rather helpful, ░ moderately helpful, ▒ rather unhelpful, ■ very unhelpful, ■ did not use it)

Looking only at those respondents who had used the respective channel or source (Fig. 8.6), we see that emergency apps fare quite well, about as well as TV and radio, with over a third saying that they had been very helpful and about one quarter stating that they had been rather helpful. In 2019, the percentage of people who had found warning apps quite or very useful in an emergency was almost identical (63%) and surpassed by the same sources (Kaufhold, Haunschild, & Reuter, 2020). The only source that changes noticeably in 2021 is social media, falling from 58% to 46%.

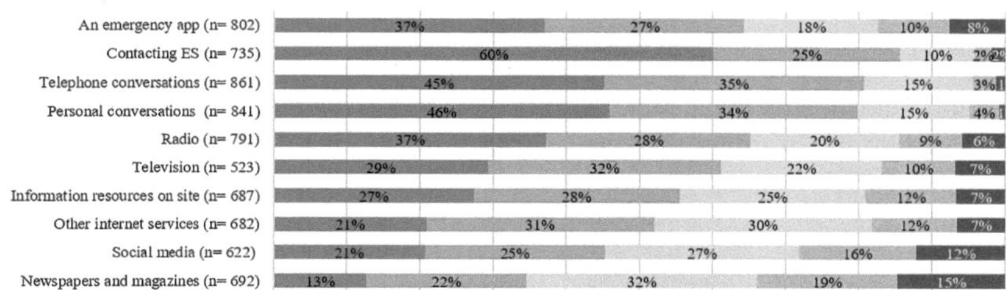

Fig. 8.6 *Only those who use the channel or source:* Perceived helpfulness of sources and channels in past emergencies (■ very helpful, ▨ rather helpful, ░ moderately helpful, ▒ rather unhelpful, ■ very unhelpful)

Looking at the most helpful sources of information in the recent health crisis, during the COVID-19 pandemic, reveals that similar to other crises, offline mainstream media sources were perceived as most helpful. Among the internet-based resources, health related and local city and county websites were perceived as most helpful. The contact tracing app is the most helpful app (helpful to 40% of all respondents), while general emergency apps are surpassed by apps dedicated to data about the pandemic. Although 40% of the respondents found contract tracing apps very or rather useful during the COVID-19 pandemic, apps for current information on COVID-19 (29%) or general emergency apps (18%) were mentioned less frequently as useful information sources (Fig. 8.7). Instead, television (55%), the website of the Robert Koch Institute (46%), radio (45%), and personal conversations (44%) were rated as the useful information sources by more respondents.

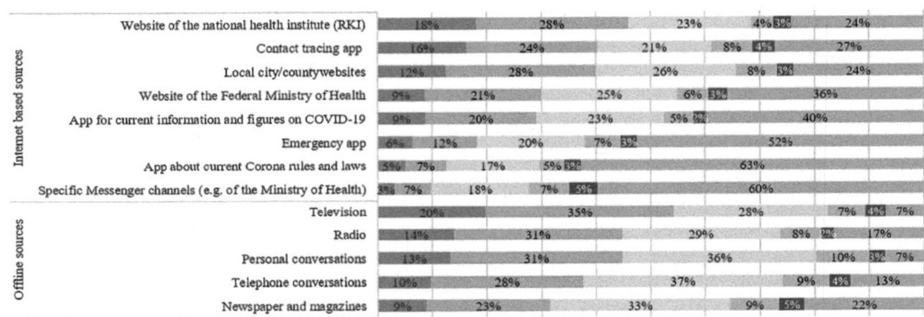

Fig. 8.7 Helpfulness of sources of information about the virus, associated restrictions and recommendations during the COVID-19 pandemic. (■ very helpful, ▦ rather helpful, ▒ moderately helpful, ▨ rather unhelpful, ■ very unhelpful, ▤ did not use it)

Compared with two years earlier, general attitudes towards information and warning apps remain largely the same in 2021 (Fig. 8.8). Even more people than before (73%) agree that warning apps are an important channel that should be used by agencies, complementing other information channels (79%).

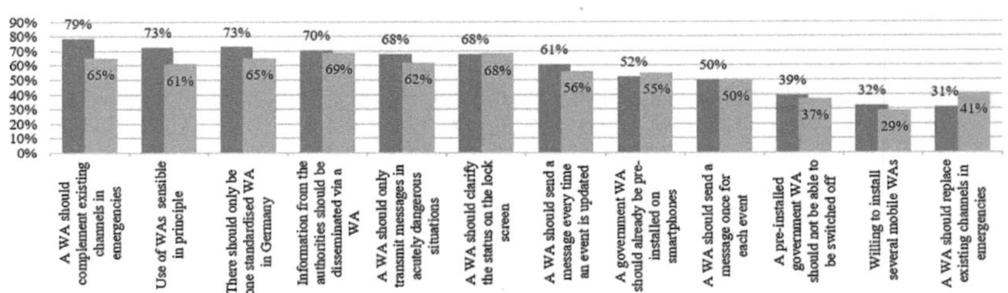

Fig. 8.8 Changes in general attitudes towards warning apps (WA) (sum of rather agree and totally agree, left: ■ 2021, right: ▦ 2019)

Fewer people are in favour of warning apps replacing other channels such as radio and TV for alerting, which could be a result of diminished trust in warning apps, possibly due to the failure of the warning day or during the flood, but it could also reflect the helpfulness of mainstream media during the COVID-19 pandemic. The attitude concerning the replacing or complementing of other channels is the only attitude that has changed by more than a few percent points. When it comes to the number of warning apps, 73% of the population, slightly more than in 2017 (68%) and 2019 (65%), feel that there should be only one emergency app. Although slightly decreasing, a slim majority still feels that a warning app could even be pre-installed upon purchase of a smartphone (2017: 44%, 2019: 56%, 2021: 52%). The comparison across all three survey instances shows that there is no clear trend, but only minor shifts (Kaufhold, Haunschild, & Reuter, 2020).

The main purposes of using a smartphone app in an emergency is receiving an alert (over 75%, see Fig. 8.9). However, for more than half of respondents, emergency-related advice, receiving information and sharing information are also important features. Indeed, only 3% state that the only purpose would be the alert. For almost 60%, both alert and advice are main purposes, indicating that when using cell broadcast, adding advice should also be considered. For just under half of the respondents, contacting emergency services through the app and coordination with other citizens are also important. All aspects of warning and emergency apps have increased in relevance since 2017, indicating a growing interest in these apps. Interestingly, the areas that have gained the most interest concern those that are related to contacting emergency services. Over half of the population can see themselves using an app instead of the emergency hotline, with only 22% opposed. In 2017, 42% still rejected this idea. Similarly, 20% more people envision themselves using an app in the future to share information with emergency services. Only 14% oppose this idea in 2021, whereas this number was 33% in 2017. This indicates a greater openness towards sharing data and contacting emergency services even from those people who are not currently using a warning app. There are no stark differences regarding these aspects between current users and non-users of warning apps.

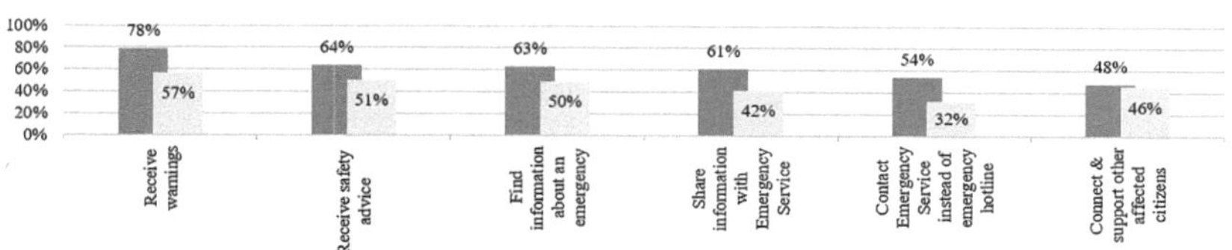

Fig. 8.9 Main purposes for using a smartphone app in an emergency in the future (sum of very and quite likely, left: ■ 2021, right: 2017)

8.4.3   Messengers as Additions to Warning Apps

Research has shown for a while that messenger apps increasingly include functionalities similar to social networks, such as communication in large anonymous groups (Newman et al., 2019). Our findings indicate that German citizens favour the use of multi-purpose apps, including information and warnings regarding natural hazards, crimes, and pandemics (Fig. 8.10). Although disaster warnings were still perceived as the most relevant functionality (86%), respondents also valued opportunities for bidirectional communication, e.g., integrated telephone emergency calls (78%) or sending direct messages to emergency services (72%). Besides natural disasters, more than half of the respondents also welcomed the integration of crime and police related information, such as search for missing persons (62%), fraud offences (57%), or warnings about cybercrime (53%). With regard to pandemics, most respondents liked to received information on medical hazard prevention (58%) and 45% also incidence level updates. Updates about pandemic-related rule (58%) emerged as one of the most relevant warning categories. Compared with attitudes in 2019, there are almost no changes with regard to the other items, only the wish for direct contact with emergency services has increased (by 9%) (Kaufhold, Haunschild, & Reuter, 2020). Most features are thus regarded as helpful and respondents are in favour of integrating more functions. Only few items are controversial, especially whether to include school cancellation and speed measuring notifications. This stresses the importance of allowing for personalisation in the apps.

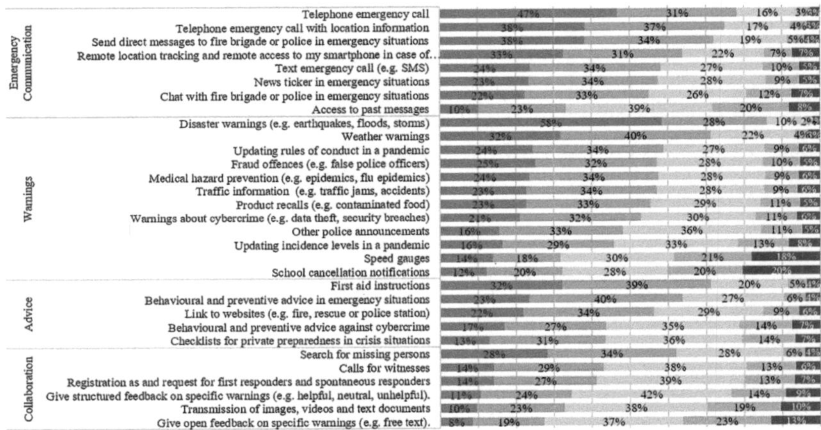

Fig. 8.10 Importance of warning and emergency app functions 2021 (■ very, ▓ rather, ░ moderately, ▒ rather not, ■ not at all important)

In addition, the pandemic with its long-term effects and dynamic need for regulation and policy change has opened up new demands for information (Haunschild, Pauli, & Reuter, 2021; Haunschild & Reuter, 2021a). We therefore asked participants whether a) rules and regulations and b) relevant incidence levels should be included in warning apps. The results show that for a large majority these are important aspects that should indeed be included (Fig. 8.11). This applies to the inclusion of currently applicable rules (73% agree, 12% disagree; of warning app users even 82% agree and only 6% opposed) and for updates of risk levels (70% in favour, 10% opposed; of warning app users even 77% agree and only 9% opposed). This shows that pandemic-related updates are an important new field for warning apps. Another prospective change in the field may be the use of messenger channels. One third agrees that they would rather receive warnings through an official messenger channel, while another third disagrees. However, of the current non-users of warning apps 36% strongly or rather prefer such a channel (versus 27% of current users). This represents a small significant difference between the groups (t(1002) = 3.0; p = 0.003; d = 0.21).

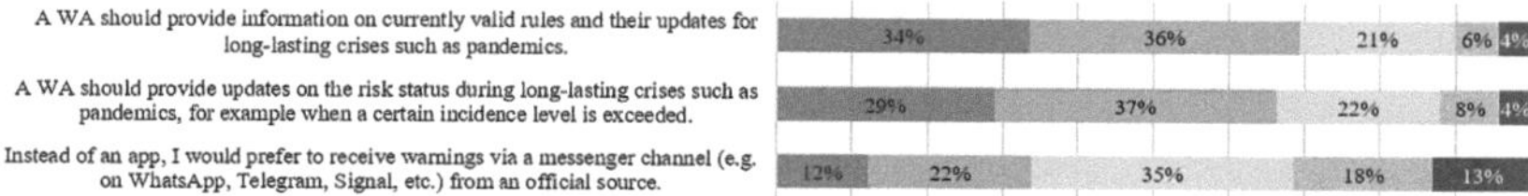

Fig. 8.11 Importance of COVID-19-related information in warning apps and messengers as alternatives to warning apps (■ very, ■ rather, ▪ moderately, ▪ rather not, ■ not at all important)

8.5   **Discussion and Conclusion**

In this study, we have collected representative data about the use of warning apps and attitudes towards them for Germany. With the experience of the COVID-19 pandemic, a failed warning day and a flood in 2021, we investigate whether there are any changes in the warning app landscape by comparing these data with data from 2019 and 2017. In addition, we investigate aspects that concern the inclusion of pandemic-related information into warning apps and the increased significance of messenger apps as a new trend that might impact warning apps. The study contributes these main findings:

- Between 2017 and 2021, the use of warning apps has continually increased.
- For those people who have used a warning app in an emergency, it has been more helpful than other internet services or social media.
- Multi-purpose warning apps have been overtaken by specific COVID-19 related apps for contact tracing.
- The regional app hessenWARN has overtaken the national apps in the relevant region.
- The inclusion of COVID-19 related information about infections data and regulations are strongly desired features, especially by users of warning apps.
- Most attitudes and preferences have remained largely stable.
- People are now more open towards using warning apps to contact emergency services in emergencies.
- Over a third of non-users of warning apps would prefer a warning messenger app channel to warning apps.

8.5.1 The Usage and Expectations Towards Warning Apps Between 2017 and 2021

Answering RQ1, we find that 25% of Germany were using a warning app in 2021. This number has risen from 17%. However, weather apps dominate, indicating the prevalence of apps designed for everyday use. Emergency apps continue to be evaluated by their users as more helpful than other online sources, but less helpful than direct contact with emergency services, personal exchanges or mainstream media. During the COVID-19 pandemic, other online sources have been perceived as more helpful, supporting previous findings that the apps failed to provide important information that instead had to be gathered through other channels, often leading to insecurity about the quality of information (Haunschild, Pauli, & Reuter, 2021, Haunschild & Reuter, 2021a).

Around every fifth person states that they plan on using a warning app in the future. Asking about the planned use of specific emergency apps, these numbers are significantly smaller, suggesting that one aspect preventing citizens from using warning apps may be lacking information about which app to chose. This would be in line with previous research that suggests that users strongly prefer having only one app for all related functions and hazards (Dallo & Martí, 2021; Kaufhold, Haunschild, & Reuter, 2020). Emergency managers may thus focus their public relations efforts more on informing about these apps and any differences, but also emphasise that the main functions of alerting are guaranteed in all major emergency apps.

At first glance, it seems that the new regional warning app hessenWARN, which was launched in 2020, is insignificant with only 2% current users. However, it has a large user group among the residents of that federal state, with 17% of Hessian respondents currently using the app. Indeed, it appears that adding another app has increased usage in that state, with 10% more Hessians using a warning app than the national average. It appears that the attention has even increased the awareness and usage of the national version of the app, KATWARN, which is also used by 6% more Hessians and is known to 60% of Hessians (compared to 40% national average). HessenWARN is compatible with NINA and KATWARN and also reports all warnings that are transmitted through these apps. The app can thus also send warning related to regions outside Hesse. While the design is similar to KATWARN, hessen-WARN the app has put into practice this study's and previous findings (Kaufhold, Haunschild, & Reuter, 2020) in that it has integrated police-related matters, including warnings about cyber fraud, missing persons and product recalls, while offering the option to personalise the app so that it only sends the warning types that the individual user deems relevant. While other warning apps can also be set up to only convey regionally relevant warnings based on GPS and selecting personally relevant regions, the explicit local connection may increase users' trust in the relevance of the notifications and make it easier to choose an app. However, whether it is the increased public relations effort connected to the launch of a new app, the inclusion of more warning types or the fact that users feel "at home" using a regional app, remains an open question.

Looking at how the usage and expectations towards warning apps have changed (RQ2), we see that usage of the main warning apps has increased, while other emergency related apps have remained unused and less-known. Respondents report an increased relevance of all warning app functions that increases their likelihood of using such an app in the future. A noticeable increase can be found in the readiness to use a warning app to share information with emergency services and contact them in case of an emergency. When it comes to the relevance of functions and design preferences, there are only small changes. However, functions related to health crises, including updates of rules related to the pandemic, are now regarded as some of the most relevant warning types (see Sect. 8.5.2). Due to the increased usage of the main warning apps, which is not paralleled by other emergency-related apps, it appears that, despite the criticism that these apps and the warnings system as a whole has received in Germany, the recent past has motivated rather than demotivated the use of warning apps. However, looking at the number of people who can see themselves using a warning app to receive alters (78%) and for other functions, at the number of people who are planning on using a warning app in the future (20%), as well as considering the helpfulness that users of warning apps attest these tools, there seems to be a considerable adoption gap. Some of this can be explained by the lack of awareness of these apps. Looking at NINA, half of the population is unaware of the app. Of those who know it, only 5% have abandoned the app and 20% have no interest in using it. If this is similar among the 50% who are unaware of the app, another 25% of the population might be interested in using the app when made aware of it. In this study, we reaffirm users' strong desire to have a large number of issues, including police-related notifications and significant COVID-19-related changes, included in the same app (Dallo & Martí, 2021; Kaufhold, Haunschild, & Reuter, 2020). This insight is something that we see implemented in hessenWARN, which is widely used where it is known. We can thus derive one technical and one socio-political implication: Rather than fearing that warning apps' notifications will be perceived as not warranting a warning, designers of warning apps should include a large set of warning types and leave it to users to personalise the app to fit their needs. Secondly, a lot more should be done to increase awareness of these apps. Because users want only one app that covers all warnings, such PR efforts should stress the compatibility of all the main warning apps and clearly show that using any of the main apps, users will receive largely the same warnings. This might make it easier for people to choose. The public discussion of warning apps should also consider attitudes that have to do with Germany being a state-centric risk culture (Reuter et al., 2019), which makes people less inclined to feel personally responsible for their own safety. Appealing to citizens' cooperation and communicating that safety depends both on emergency services *and* citizens might increase citizens' willingness to take the step and install a warning app (Dressel, 2015; Fischer et al., 2019). Since about half of the population agrees that a warning app may be pre-installed on smartphones upon purchase, this could also be a avenue that could be discussed.

It is unclear, whether the insights from this study can be applied to other countries. In 2017, a European study compared the use of warning apps across Germany, Italy, the Netherlands and the United Kingdom (Reuter et al., 2019). According to the study, warning apps were only used by 17% in total, but in larger proportions in the Netherlands (28%). The most downloaded apps were weather apps, used by 42% in total, mostly by Germans (69%), and warning or alert apps, used by 42% in total, with a visibly higher proportion in the Netherlands (53%). For future use, the most likely aims mentioned were receiving emergency warnings, mentioned by 57% in total, with the UK (39%) having the least interest. Receiving information about an emergency was mentioned in the countries by between 40% to 65% (54% on average across all countries). Advice about how to stay safe was similarly envisioned as a reason for using a warning app in the future, by 50% on average, with Italians in the lead (61%). The average numbers across the surveyed European countries were similar to those of Germany in 2017, suggesting that changes seen here may be similar across Europe. However, looking at individual states, differences also emerged that can be explained by differences in risk cultures (Reuter et al., 2019). For example, respondents in the United Kingdom judged themselves to be significantly less likely to use an emergency app for any of the purposes in the future. Some countries, such as the UK, are thus less likely to follow a trajectory similar to Germany.

8.5.2   On the Relation of COVID-19 Contact Tracing and General Warning Apps

Comparing the COVID-19 contact tracing apps with warning and crisis apps reveals stark differences. Warning apps have existed for a long time and users have always stressed the importance of combining all hazards into one app (Dallo & Martí, 2021; Kaufhold, Haunschild, & Reuter, 2020). It could thus be expected that similarly to other crises, integration of COVID-19 information in a warning app would be the best option. Clearly, this is not the case for contact tracing apps, which perform very specific crisis related tasks that differ significantly from those of warning apps. In the COVID-19 pandemic, the main challenge was not to raise awareness of the danger and the measures to limit the risks (even though keeping track of local regulation was a challenge from citizens' perspective (Haunschild, Pauli, & Reuter, 2021), but to warn people who may have become infected because they have been in the vicinity of infected people. This required access to sensitive data and high trust in the app-providing organisation.

Due to privacy concerns, the German version of the app uses Bluetooth instead of GPS, contains features that hide the identity of people who have been tested positively for the virus and is managed by the national health institute RKI (Munzert et al., 2021b, p. 9–10). As this sensitive type of information is not needed to alert users of warning apps, a separation of apps was feasible in this case. The high number of users of the COVID-19 apps could be explained by the public attention that it received and the social pressure that was generated to use them, because it was expected that it would only work if many people used it. In addition, *luca* and later on also the *Corona-Warn-App* made it more convenient to check in at locations, thus offering advantages beyond tracking one's COVID-19 risk. While public attention and social pressure were likely factors to increase the usage of the COVID-19 apps, the warning apps also received a lot of attention since 2020, albeit mainly centred around the apps' failures.

The data show that one new important field for warning apps are pandemic-related updates, for which a strong demand exists, especially among current users of these apps. The relevance of apps showing pandemic-related data and of websites with local information indicates that in the case of the COVID-19 pandemic, warning apps have failed to integrate pandemic-related data in time. However, previous research also shows that maintaining information about locally specific and dynamically changing regulations is currently a manual and resource-intensive task (Haunschild & Reuter, 2021a). The huge demand for this type of information shows that investing resources and establishing collaborations that enable relevant agencies to submit this type of information in a machine readable format would be worthwhile. While more controversial, a third of respondents also perceives that messenger apps could be good alternatives to warning apps. Another interesting aspect is the decrease of the Facebook safety check feature in relevance. Since it is a feature that is not prominent in warning apps, it is more likely to be attributed to the decreasing relevance of social media for interpersonal communication and the increased relevance of smartphone messenger apps.

8.5.3   Limitations and Future Work

This research is subject to limitations and potentials for future work. The online survey's results might be biased due to possible self-selection of volunteering individuals. Our findings are based on individuals' answers and not observation of actual behaviour. However, citizens' perceptions were our focus and, as such, the study provides valuable results, not at least with respect to potential further implementation of crisis and warning apps. Due to the high interest in COVID-19 related information within warning apps and the interest in messengers in the context of crises and emergencies, these aspects should be explored in future crisis informatics research. Due to the conversational nature of messenger channels, this would require investigating their usability and how conversational agents could be used to personalise the information on such a channel (Stieglitz et al., 2022; Tan et al., 2020a).

With the decision to implement cell broadcast in Germany (BMWK, 2021b), future research could look into how warning apps and cell broadcast interact. Past research indicates that in the Netherlands, where cell broadcast has been used since 2012, warning app use is relatively low (Reuter et al., 2019). Cell broadcast might thus make warning apps obsolete for alerting in acute emergencies and even offer the advantage of being resilient to internet outages and available for older mobile phones and thus more citizens. On the other hand, due to the text-based nature of cell broadcast notifications, they are limited in the media formats that can be transmitted and their only way of differentiating between users is based on their location. Warning apps might therefore still have a lot to offer when it comes to reliable information, including about non-life threatening incidents, with regard to prevention and multi-media offers. For example, warning apps allow users to set e.g. thresholds for incidence numbers that are relevant for their protective behaviour. In addition, areas other than one's current location may be of interest, e.g. the place of work, where one's parents live etc., which would not be covered by cell broadcast notifications. By allowing to set individual preferences, users can stay informed and receive reliable safety and security information and receive acute or non-acute information as per their preferences, optimising the app's utility and representational fidelity to fit their perception (Fischer-Preßler et al., 2020; Tan et al., 2020c).

Because research on risk cultures shows that trust in authorities is an important factor, future research should investigate whether the past experiences in Germany have resulted in a decline in the typically quite high trust in agencies in Germany and whether this also leads to limited trust in the tools provided by these agencies, or rather in an increased perceived need to take individual measures, e.g. by downloading a warning app. These open questions apply not only to safety threats experienced during the flood and the pandemic, but also to the security threat resulting from the the war in Ukraine 2022, which might change the perception of preparedness as a value and increase the urgency of the question what role ICT can play for civil defence.

Finally, we have seen that the regional app hessenWARN has been successfully established in the federal state that initiated its development and gives the app its name, leading to more warning app users in Hesse compared to the national average. Future work should explore the reasons for that increased usage by focusing on the adoption process by hessenWARN users. These insights might be used to also increase warning app adoption in other regions, leading to an overall increase in crisis preparedness.

Use and Perception of Social Media in Emergencies and Comparison with Warning Apps

9

Abstract

Crisis informatics has examined the use, potentials and weaknesses of social media in emergencies across different events (e.g., man-made, natural or hybrid), countries and heterogeneous participants (e.g., citizens or emergency services) for almost two decades. While most research analyses specific cases, few studies have focused on citizens' perceptions of different social media platforms in emergencies using a representative sample. Basing our questionnaire on a workshop with police officers, we present the results of a representative study on citizens' perception of social media in emergencies that we conducted in Germany. Our study suggests that when it comes to emergencies, socio-demographic differences are largely insignificant and no clear preferences for emergency services' social media strategies exist. Due to the widespread searching behaviour on some platforms, emergency services can reach a wide audience by turning to certain channels but should account for groups with distinct preferences.

Original Publication Haunschild, J., Kaufhold, M.-A., & Reuter, C. (2020). Sticking with Landlines? Citizens' and Police Social Media Use and Expectation During Emergencies. *Proceedings of the International Conference on Wirtschaftsinformatik (WI) (Best Paper Social Impact Award)*, 1–16. https://doi.org/10.30844/wi_2020_o2-haunschild

Supplementary Information The online version contains supplementary material available at https://doi.org/10.1007/978-3-658-46489-9_9.

9.1   Introduction

Social media interactions have become a relevant part of responding to crises including natural disasters, human-made attacks and political uprisings. They are used by volunteers and professional emergency services (ES) to coordinate action, express solidarity and exchange information (Reuter & Kaufhold, 2018). However, social media use by state organisations is also controversial. Criticism concerns unequal access to digital infrastructure and resulting biases in resource distribution (Crawford & Finn, 2015). Indeed, we have a limited understanding of who does and does not use social media in emergencies, which platforms are used and with what intentions. In addition, we do not yet know what people's expectations are regarding ES' use of social media generally and in crises. The reactions of ES to social media and their use of it also varies greatly across countries, departments, social media platforms and different types of ES, employing different communication styles (Hughes et al., 2014). Although some works try to formulate social media guidelines of practical use for ES and citizens (Kaufhold et al., 2019) or to generalise strategic communication in social media (Lwin et al., 2018), there is still uncertainty among ES on how to best use social media in different cases and for different audiences, leading to the following research questions:

- **RQ1**: Which patterns of social media use, perceptions and expectations are prevalent in the German population regarding their use in emergency situations? Do different socio-demographic groups show different use and expectation patterns?
- **RQ2**: Which implications and guidelines, also considering the change of social media use over time, can be deduced for the design of social media engagement strategies for effective analysis of data and communications for emergency services?

To answer these questions, we conducted a workshop with participants (N = 15) from a German central federal police agency to explore which aspects are relevant to them. From this we developed a questionnaire that in addition also asked questions similar to some we had previously asked in smaller samples in 2015 and 2016 (Reuter, Kaufhold, Spielhofer, & Hahne, 2017; Reuter & Spielhofer, 2017). We then conducted a representative national German survey (N = 1,219), asking about social media behaviour, use and evaluation of different platforms, shared content, advantages and disadvantages of its use in emergencies, expectations towards ES' social media behaviour and socio-demographic data (age, gender, education, federal state residency, urbanisation of residence). In this paper, we first outline current research and research gaps (Sect. 9.2). We then portray our data gathering and insights through workshop discussions (Sect. 9.3.1) and describe the survey and used methods of analysis (Sect. 9.3.2). We then analyse the survey, additionally comparing it to results from 2015 and 2016 (Sect. 9.3.3). Lastly, we discuss our findings of both approaches (Sect. 9.4) and draw conclusions for ES' social media use in emergencies (Sect. 9.5).

9.2   Literature Review

Crisis informatics examines the use of information and communication technologies (ICT) in crises, often focusing on social media (Reuter & Kaufhold, 2018; Soden & Palen, 2018). With few exceptions (Kim et al., 2017; Olteanu et al., 2015), insights into ES' activities are fragmented and limited to specific locations (de Graaf & Meijer, 2019) and specific crisis events such as earthquakes, hurricanes, riots, shootings or terrorist attacks (Hughes et al., 2014; Ross et al., 2018; Tagliacozzo, 2018). Indeed, more progress has been made to understand and conceptualise the role of citizens, such as citizen-to-citizen and citizen-to-authority communication (Brynielsson et al., 2018), and the public, which uses social media for self-coordination and collaborative ICT in crises, as opposed to the roles of state authorities, both in the real and virtual realm (Reuter & Kaufhold, 2018). Yet how the general population views the use of social media in crisis situations, its use by emergency managers and reasons for not engaging with modern technology in crises (Reuter et al., 2016) remain understudied.

9.2.1   Barriers and Potentials of Social Media Use and Communication

Research identifies different strategies and tactics for social media use by emergency managers (Kaufhold, Bayer, & Reuter, 2020; Wukich, 2015) such as information dissemination, data monitoring/analysis and conversations/coordinated action. However, analyses of social media engagement more broadly find that ES are using new tools in a way that is very similar to traditional media: mainly broadcasting information, with little back-and-forth engagement and building on pre-existing communication styles (Meijer & Thaens, 2013). Considering the relevance of informal behaviour and improvisation (Ley et al., 2012), it is unclear how the potential informality, speed and transparency of social media will influence the perception of ES' work and efficiency. In this vein, (Meijer & Torenvlied, 2016) identifies risks that are associated with the new decentralised communication styles of agencies on social media, among them the risk of missing important information as well as the risk of damaging one's reputation online because of its use by untrained officers.

Despite a lack of skills and tailored tools for analysing social media data, ES across Europe appear to see a need to get active on social media (Kaufhold, Rupp, Reuter, & Habdank, 2020). A survey with ES staff (N = 761) outlines ES' intention to significantly increase social media use for sharing emergency-related information, establishing bidirectional communication and improving situational awareness based on situation updates, multimedia files and public mood (Reuter et al., 2016). While the previous survey largely reflected fire departments' points of view, the aims of police departments include avoiding other actors filling that space, framing the online conversation and preventing users from taking unguided policing actions (e.g., publishing missing person reports), reaching a wide audience, nurturing trust and understanding online deviance (Denef et al., 2012; Williams et al., 2018). Instructive is an analysis of public information officers (PIOs), the public relations component of the US National Incident Management System, which finds that PIOs' roles are changing considerably (Hughes & Palen, 2012): Most PIOs use social media to ease their work by directly communicating with affected populations and directing them to online help resources. In this way, they are empowered through social media. However, PIOs also lose control over information because private individuals are taking part in reporting. This has also negative consequences for their relationship with traditional media, which has been sized down and relies more on private sources as well as "citizen journalism" (Stuart & Thorsen, 2009). Social media thus offer more independence from traditional media and control over delivering messages, while at the same time ES lose control over information to citizens who publish information on their matters online. Insights into citizens' attitudes and behavioural traits are thus important for refining social media communication strategies (Mirbabaie et al., 2019).

9.2.2   Research gap

While such mainly qualitative works have made great contributions to understanding the varied ways in which social media are used in different crises (Reuter & Kaufhold, 2018), they often focus on a single platform, especially Twitter. Research that provides insights into general attitudes and usage patterns across various platforms, their change over time and how they are influenced by sociodemographic characteristics, is still limited. Currently only a few works deal with some of these aspects (Graham et al., 2015; Plotnick & Hiltz, 2016). There are some quantitative surveys, however, they are not based on representative samples (Reuter et al., 2016). A representative survey on social media use in Germany was conducted in the EmerGent (Greenlaw et al., 2014) project but it is based on one point of enquiry, not allowing a trend analysis and missing important details about social media platforms and citizens' preferences (Reuter, Kaufhold, Spielhofer, & Hahne, 2017). To this end, our study is motivated by a workshop with practitioners and contributes generalisable findings from a representative survey of the German population with 1,219 participants. It covers all relevant social media platforms, privacy and security attitudes, and socio-demographic data to test assumptions about user groups and usage patterns on specific platforms. Apart from enabling generalisations, such research sheds light on those people who do not turn to social media in crises and who are often overlooked by studies of use patterns.

9.3   Empirical Study: Representative Survey of the German Population

9.3.1   Survey Design

On February 13, 2019 we conducted a workshop at a German central federal police agency which focused on the use of social media in emergencies, comprised police officers (N = 15) and lasted 1.5 hours. At the start we introduced the procedure for conducting a representative survey and the aim of this workshop to generate a questionnaire with a short presentation. Examples of closed and open-ended questions were introduced. After the presentation, the workshop was based on three phases: In the **reflection phase** (10 minutes), based on their individual creativity, participants were instructed to note their ideas or questions on moderation tasks. In the **presentation phase** (20 minutes), participants presented their ideas and we subsequently arranged them thematically on a flip chart. The participants were encouraged to write down further ideas during the presentation phase. Finally, in the **discussion phase** (60 minutes), based on the group's collective creativity, participants discussed existing moderation cards, generated new ones and reflected upon their thematic grouping. Through open coding the three dimensions of *use*, *perception* and *expectation* emerged from the data, which were subsequently also reflected in the survey design.

- **Citizen's *Use* of Social Media in Everyday Life and Emergencies.** Police officers were interested in both how citizens use social media in emergencies and which platforms are used, as this could impact the depth and breadth of their analysis and communication strategies. Thus, we designed three questions to address use of media in emergencies (Q1), evaluation of information sources (Q2), as well as types of content shared in crises (Q3).

- *Perceived* **Benefits and Barriers of Social Media in Emergencies.** Officers were interested in the perceived benefits (e.g., easy, fast and detailed emergency information) of and barriers (e.g., data privacy, fake news and rumors) to using social media. A more detailed picture on motivations and fears would help them to feasibly adapt their communication strategy. Thus, we designed two questions to ask for benefits of (Q4) and barriers to (Q5) social media use in emergencies.
- **Citizens'** *Expectations* **Towards the Use of Social Media by Emergency Services.** Lastly, to further improve their analysis and communication strategies, police officers were interested in citizens' expectations towards social media use by ES, especially in terms of their monitoring behaviour, the desired number of social media channels and language style. All interests are reflected in our last survey question (Q6).

In an **integration phase** the workshop results were combined with a published questionnaire (Reuter, Kaufhold, Spielhofer, & Hahne, 2017; Reuter & Spielhofer, 2017), whereof some existing questions were extended by new items (Q1, Q5, Q6) for additional insights (in conjunction with Q2 which was added as a new question) and some were re-integrated (Q3, Q4) to allow a temporal comparison of two datasets on how use, perceptions and expectations of citizens changed. While the previously published questionnaires were developed in the EmerGent (Greenlaw et al., 2014) project whose consortium comprises fire departments and rescue services, the workshop with police officers allowed us to enrich the questionnaire with their perspective. Then, the first draft of a questionnaire was sent to the central federal police agency for two rounds of feedback.

9.3.2 Data Collection and Analysis

The finalised questionnaire was self-hosted through LimeSurvey, then sent to the commercial and ISO-certificated panel provider GapFish in May 2019. The **survey** covered the dimensions of use, perception and expectation towards social media in emergencies, which was translated into six closed questions (see Sect. 9.3.1). Definitions were given before relevant items. Emergencies were defined as unforeseeable events (such as epidemics, earthquakes, fires, big accidents or floods) that impact several people and require immediate action to minimise negative consequences. Social media were defined along with the German dictionary as social networks through which users can connect with each other, communicate, have exchanges and generate content. For all generated items we followed guidelines for valid item design, including phrasing positively, clearly, short, concisely and understandably, limited to one statement per item and avoiding leading questions (Moosbrugger & Kelava, 2012). Though items should be related to the present (Mummendey & Grau, 2014) due to the infrequent nature of emergencies, we resorted to previous experiences, so potential effects of remembering should be taken into account in interpreting the results. Questions are either on a 5-point interval Likert scale or categorical. With regard to **participants**, the sample was adapted to represent the German population in age, gender, geography, urbanisation and education (Statista, 2018). These criteria ensure that we can infer the German usage patterns with minimal biases, avoiding the selection biases inherent in surveys, depending on where participants are recruited, typically favouring groups with more time.

For our **data analysis** we eliminated incomplete answers, resulting in N = 1,219 reliable answers. Due to the sample size an approximation of normal distribution of the data can be assumed (Leonhart, 2008). Depending on the scale of the dependent and independent variables, we used Chi-Square tests, Phi (ϕ), Cramer's V, Kendall's tau-b, ANOVA, and Pearson's correlation coefficient (r) to examine the relationships and associations within our data. For the categorical variable "gender" we applied the t-test for independent samples, paying attention to the assumption of homogeneity of variance through a Levene test. We judge effect sizes of Pearson's r of |0.10| as a small, of |0.30| as a moderate, and of |0.50| as a strong correlation (Cohen, 1988). For the statistical analysis we use IBM SPSS Statistics 26. For each analysis we chose the test that is most robust and allows for the most fine-grained scale. An exception is made when testing for non-linear correlations, for which data are recoded into categories to test group effects, such as binary categories for those under 25, over 45 and those over 60-years old. In addition, we recode the categorical variables that capture the use types of various social media platforms in emergencies into ordinal variables to represent social media engagement. This we derived from answers to whether and how people used social media in emergencies (*not used, just to search, just to share, to search and share*) and interpret it such that searching is a less intense engagement than sharing, while doing both signifies most intense engagement. To perform tests of this assumption and group effects, we also code social media use as a bivariate variable that delineates any type of use of a platform ("1") from no use of that platform ("0"), and various combined user or non-user groups. We test city size and age for collinearity and find no multicollinearity (VIF = 1).

We furthermore compare the studies of 2015 (N = 1,034), 2016 (N = 1,069) and 2019 to see if and how social media behaviour and expectations have changed over recent time. The comparison data come from a 2015 snowball survey conducted in several countries and a 2016 representative survey conducted in Germany (Reuter,

Kaufhold, Spielhofer, & Hahne, 2017; Reuter & Spielhofer, 2017). Since social media is a fast-developing field, these three years of difference between surveys can yield interesting insights. However, in part items varied slightly between these surveys. Thus, newer and more fine-grained questions on the use of social media platforms of the most recent study were merged to capture the same dimensions found in the other studies concerning any use of social media in emergencies. In addition, what respondents think about when reporting on "social media" might also have changed over time. While we explicitly include WhatsApp as a social media platform in this survey, in 2015 and 2016 this platform may not have appeared to all respondents as such (Reuter, Kaufhold, Spielhofer, & Hahne, 2017; Reuter & Spielhofer, 2017). Thus, we exclude WhatsApp in the survey comparison.

9.3.3   Results

Widespread Use but Limited Helpfulness of Social Media Platforms (Q1 and Q2). Almost two thirds of respondents (63%) seek out information by reading social media messages in emergencies (see Fig. 9.1). About half of all respondents made posts on social media sites, equally aiming to obtain information, to share information, or both. Two thirds of respondents used their smartphone in an emergency, 13% used it only to broadcast, 27% only to share information and another 27% to both share and search. When it comes to specific platforms, people have mainly used WhatsApp, which is also the platform through which most people share information (55%). In its use, WhatsApp is followed by Facebook, YouTube and Instagram. Twitter, Snapchat and Periscope were used by less than 25%. YouTube is particularly popular for searching information. On all platforms, it is uncommon to only post information and to not use it also retrieve information. In Germany, WhatsApp

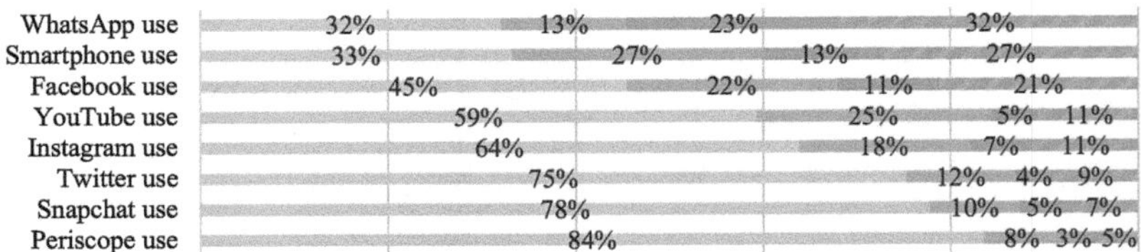

Fig. 9.1 Q1: Please indicate the extent to which you have used the following social media in an emergency situation affecting you (█ not used, █ to search information, █ to share information, █ to search and share)

and Facebook are the only platforms on which widespread sharing behaviour is taking place in emergencies. 26% use none of the social media platforms we asked about, 24% use no other platforms apart from Facebook and WhatsApp, 10% use exclusively WhatsApp, and 73% use a combination of social media that includes Facebook or WhatsApp.

Engagement on all platforms correlates significantly ($p < 0.001$) with all other platforms. Kendall's tau-b for ordinal variables shows effect sizes of between 0.33 and 0.74. However, the correlation between WhatsApp and other social media platforms is the smallest (ranging between 0.14 (Periscope) and 0.4 (Facebook)), followed by Facebook (0.33 to 0.44). In contrast, use of less common social media platforms (YouTube, Twitter, Instagram, Periscope and Snapchat) correlates strongly with effect sizes between 0.5 to 0.74. This indicates that while Facebook and WhatsApp are more frequently used independently from other platforms, high engagement on the more peripheral social media channels is strongly linked to engagement on more platforms.

Analysing the effect of city size on social media engagement (see Sect. 9.3.2) on the different platforms, we find no significant connection between the two variables. To control for non-linear effects in different age groups, we analyse groups of 1) under 25-year-olds, 2) over 45-year-olds and 3) over 60-year-olds separately as binary variables with a Mann-Whitney-U-test for two independent samples and ordinal dependent variables. Surprisingly, the results show that when it comes to activity in emergencies, there are no significant differences related to age groups. To ensure that this finding does not result from the recoding of categorical social media use to ordinal social media engagement, we again recode social media use as a binary variable for any use on a particular platform in an emergency. We then perform logistic regressions over the binary variables of social media platform use and the continuous variable age and find no significant influence.

People's strong use of WhatsApp is in line with respondents' opinions on the usefulness of information sources (see Fig. 9.2), where contact with professional ES or personal contacts such as family and friends are regarded as the most helpful sources (with both deemed quite or very useful by 70% of respondents). Social media get the same positive evaluation only from 50%.

Looking only at those who did use emergency service contact or social media respectively in an emergency, positive evaluation lies even at 80% with ES and 55% with social media. For those 64% that had been in an emergency situation before (N = 827) to judge the helpfulness of sources, there are moderate to strong positive correlations between all sources (for all p < 0.001, see Table 9.1). We can distinguish between different groups: First, social media users do not necessarily support all things digital in emergencies: Despite a strong connection with

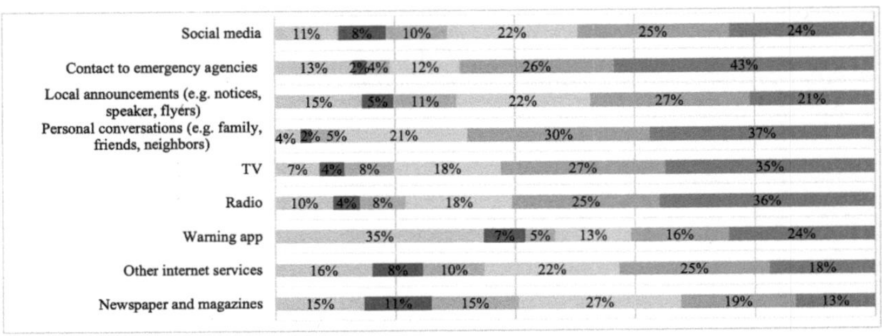

Fig. 9.2 Q2: Please indicate how helpful you perceived the following sources of information in an emergency situation affecting you (not used; ■ not, not very, moderately, quite, ■ very helpful)

other internet sources, warning apps lag behind in this group. They are moderate supporters of TV, radio and local announcements. However, they appreciate newspapers, personal conversations and telephone conversations somewhat less. This indicates that social media use is indeed not equivalent to using technology for different types of technologically enhanced exchanges with ES, friends or family. In contrast, those who appreciate contact with ES appear to also favour more direct forms of exchange through personal conversations and authoritative information through local announcements and warning apps. A third group whose members prefer more traditional sources such as newspapers, TV and radio can be identified.

Table 9.1 Insightful correlations of helpfulness of information sources, $p < 0.001$. SM = Social Media, ES = Emergency Services

Pearson's r	SM	Contact ES	News-paper	Local announcem.	Personal conver-sations	TV	Radio	Warning app	Other internet
SM	1	0.206	**0.19**	**0.316**	**0.188**	0.296	**0.309**	0.27	**0.494**
Contact ES		1	**0.156**	**0.447**	**0.318**	0.26	0.297	**0.393**	0.225
Newspaper			1	0.281	0.2	**0.465**	**0.424**	**0.14**	0.237

Only for the evaluation of social media we find no significant influence of gender (p = 0.097). Despite this, females evaluate all other sources more positively. Though gender is significant here, effect sized are very small to negligible (e.g., helpfulness of contact with ES: Cramer's V = 0.107, p = 0.003). Gender is thus not relevant for the perception of helpfulness of social media in emergencies and its influence on the evaluation of other sources is negligible.

Increase in Sharing of Content (Q1 and Q3). Comparing those who are concerned with their privacy and data safety with those less concerned in regard to their sharing behaviour on social media in emergencies, using logistic regression we find no significant differences in their sharing of types of information. Looking at the data of 2016 and 2019 (see Fig. 9.3), there is a noticeable decrease of citizens who never shared information in emergencies (56% to 37%). Especially the number of citizens who searched and shared information increased (19% to 36%), which could be affected by the recent increase of man-made disasters, such as the 2016 Munich Shooting and the 2016 Berlin truck attack on a Christmas market, that were extensively covered across German media.

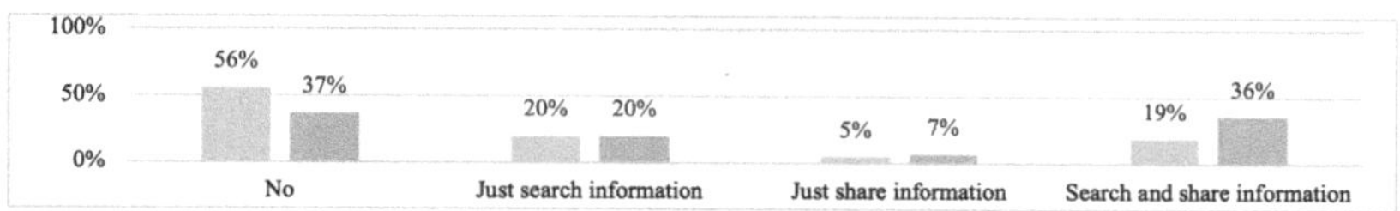

Fig. 9.3 Q1: Please indicate the extent to which you have used social media in an emergency situation affecting you (■   2016 vs. ■   2019) (For 2019: all platforms merged except WhatsApp)

Participants shared on average two types of content, 85% having shared between one and three types. About one in four people shared photos, videos or testimonies as eyewitnesses (see Fig. 9.4). Among those who became active as eyewitnesses, a significant but slight negative correlation with age appears ($\tau b = -0.15$, $p < 0.001$), but none with any particular social media platform or city size. Of those 68% that had been in an emergency, weather information and road/traffic warnings were most often shared, followed by measures taken for one's safety, one's location, feelings and emotions about the event, while 18% did not share any information. Gender plays a significant but only small role: Women are more likely to share feelings on social media ($X(2) = 24.907$, $p < 0.001$; $\phi = 0.174$). In contrast, men were more likely to share their location ($X(2) = 10.36$, $p = 0.006$; $\phi = 0.112$), videos ($X(2) = 16.23$, $p < 0.001$; $\phi = 0.14$) and witness accounts ($X(2) = 6.928$, $p < 0.001$; $\phi = 0.088$). Similarly small significant differences can be found in relation to age: People under the age of 25 are more likely to share feelings on social media ($X(1) = 31.097$, $p < 0.001$; $\phi = 0.194$) as well as videos ($X(1) = 4.612$, $p = 0.032$; $\phi = 0.075$). In addition, they are less likely to not have shared any type of content ($X(2) = 10.012$, p = 0.002; $\phi = -0.11$). In contrast, those over 45 years old were less likely to share most types of content or any content at all ($X(2) = 13.222$, p < 0.001; $\phi = 0.127$).

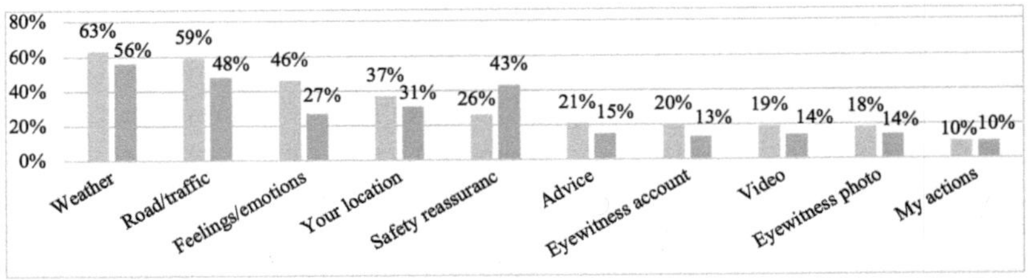

Fig. 9.4 Q3: Please indicate the types of information you have shared in social media in emergency situations affecting you (█ 2016 vs. █ 2019)

Despite the overall increase of shared information from 2016 to 2019, the amount of different information types decreased except for safety reassurance (26% to 43%) making it the third most shared information type after weather (56%) and road/traffic conditions (48%). The increase for safety reassurance may be linked to the use of Facebook Safety Check, which appears to have become more widely used during the above-mentioned events. Furthermore, there is a significant drop of citizens sharing feelings or emotional content on social media (46% vs. 27%). Further research should investigate this phenomenon, which may be caused by a habituation effect due to the increasing number of large-scale emergencies that are heavily covered by media.

Preference for Emergency Hotline Despite Speed of Social Media (Q4, Q5). Regarding advantages of social media in contrast to traditional sources, respondents particularly value the speed with which information is available (61%) and its easy accessibility (60%). On the other hand, less than 30% regard social media as more accurate, trustworthy and reliable than traditional media. A significant majority (71%) states that it is better to call an emergency number. Many are also concerned with rumors (64%) and false news (56%), as well as an excess of information (53%). Half of the respondents also regard data security and privacy as a disadvantage (50%) as well as the possibility that social media might be inaccessible in an emergency (54%). While social media is granted with providing accessible and speedy information, only a minority of between 8% and 20% disagrees with the limitations of and concerns about social media, painting a picture of social media users who appear unenthusiastic about relying on social media in emergencies.

Analysing judgements about social media advantages and disadvantages shows that all advantages correlate significantly and moderately to strongly (p = 0.001, 0.32 < r < 0.83) among themselves, as do disadvantages (p = 0.001, 0.27 < r < 0.83). Especially accuracy, trustworthiness and reliability occur together (r = 0.8 – 0.83), while trustworthiness and speed of availability show the smallest correlation (r = 0.27). Regarding the disadvantages, people who prefer to call an emergency hotline are also highly concerned with rumors on social media (r = 0.83), and security and privacy concerns occur especially in conjunction with concerns over false news (r = 0.71). The data also shows that respondents differentiate between rumors and fake news (r = 0.57). How can the attitude towards calling emergency numbers be explained? Again, there is no correlation between age or gender and a favourable opinion of emergency hotlines. Nor do we find significant correlations between people's favouring of phone calls and their expectations towards ES' social media use, such as their capacity to monitor the platforms or their expected response time. A t-test for independent samples, comparing those with and without emergency experience, shows no significant differences. Thus, emergency experience has no effect on social media perceptions.

Looking at our data sets from 2015 and 2019 (see Fig. 9.5), the perceived accessibility (54% to 60%), richness (29% to 36%), reliability (22% to 24%) and accuracy (13% to 27%) of information increased noticeably. Although the perceived faster availability of information dropped (76% to 61%) together with accessibility, these characteristics are still most dominant in comparison to traditional media. Despite

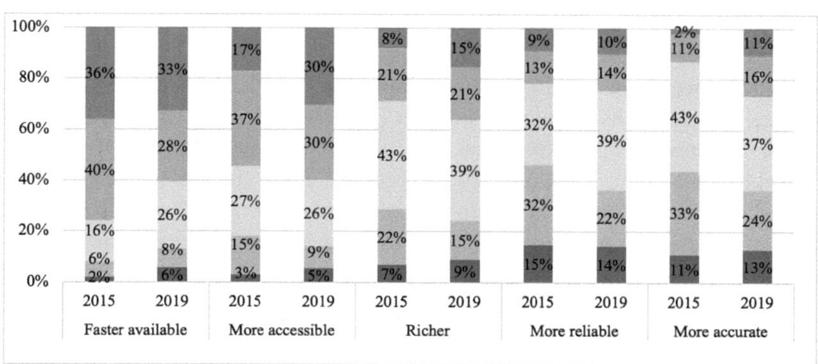

Fig. 9.5 Q4: Which advantages do you assume for social media compared to other information sources (e.g., television, radio or traditional website) (■ strongly disagree, ■ disagree, ■ neutral, ■ agree, ■ strongly agree)?

the increase in richness, reliability and accuracy, social media is still perceived worse in these characteristics compared to traditional media (see Fig. 9.6). Compared to 2016, social media were perceived as less disadvantageous: the survey participants showed less skepticism with regard to false rumors (74% to 64%), data privacy (62% vs. 50%), malfunction in emergencies (60% to 54%) and missing reliability (65% to 38%). Still, except for reliability, at least half of participants were skeptical with regard to dangers emanating from social media. Interestingly, while the preference of emergency calls over social media was balanced in 2016, in this survey 71% preferred an emergency hotline over using social media in emergencies.

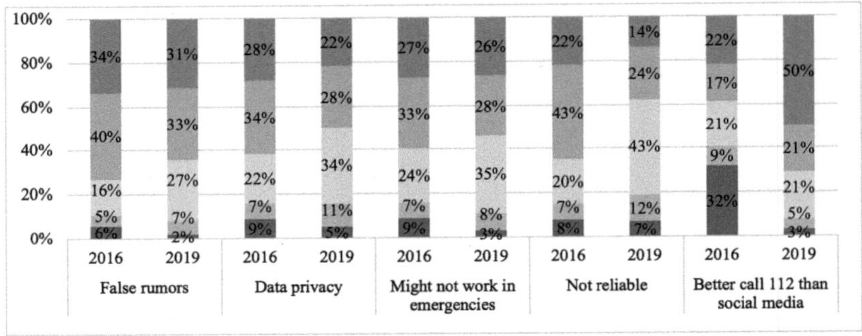

Fig. 9.6 Q5: Which disadvantages do you assume when using social media during an emergency situation (■ strongly disagree, ■ disagree, ■ neutral, ■ agree, ■ strongly agree)?

High Expectations, High Understanding (Q6). With regard to ES' activities on social media, many respondents do not have clear expectations (see Fig. 9.7). On the one hand, people expect that ES should continuously monitor their social media channels (58%) and almost 40% expect that a post (including one asking for help) should be responded to within one hour, although this idea is also rejected by a large portion of respondents (27%). 46% expect that in emergencies ES will be too busy to analyse social media. While 40% regard the current extent of ES' social media presence as sufficient, 47% wish to find news about ES' day-to-day activities on social media. In our survey, a tendency exists to favour a variety of regionally and thematically differentiated social media channels. 44% of respondents would welcome direct messages (e.g., through social media chats) by ES in case of emergencies, while 22% were against such direct contact.

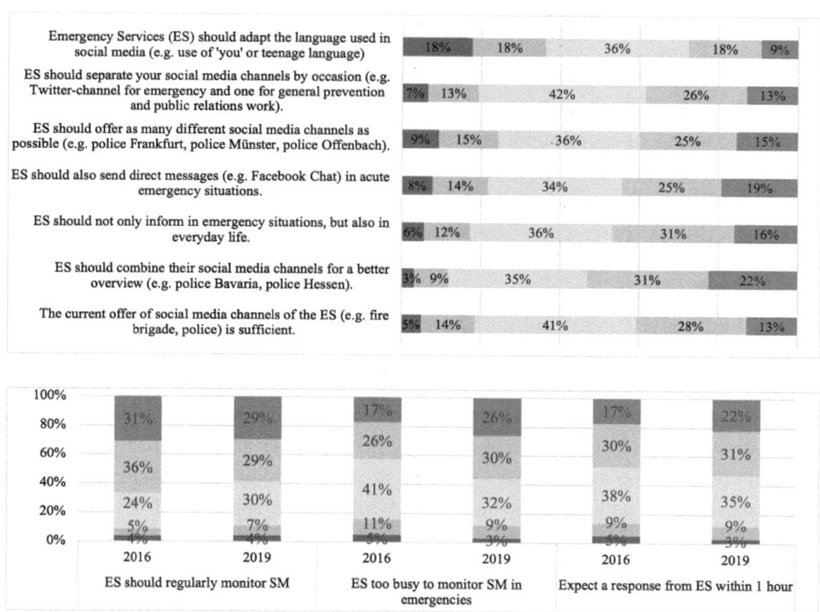

Fig. 9.7 Q6: Emergency Services (e.g., fire brigades, ambulance services or police) also use social media in emergency situations (■ strongly disagree, ▦ disagree, ░ neutral, ▒ agree, ▓ strongly agree)

Relatively strong opposition is found with regards to the question of whether to use more colloquial language on social media, with only 27% in favour and 36% rejecting the proposal. Of those firmly in opposition, only about half used Facebook in emergencies and only less than 20% used Twitter. The Mann-Whitney-U-test and an analysis of variance (ANOVA) show that neither age, gender nor education play a role. Comparing the datasets of 2016 and 2019, there is a decrease in the perception that ES should monitor social media (67% to 58%) and participants increasingly think that ES are too busy to monitor social media during emergencies (43% to 56%). While both of these developments seem to point in the same direction, the expectation that ES should respond within one hour to social media posts increased (47% to 53%).

9.4   Discussion

9.4.1   Contribution and Policy Implications

With regard to use of social media in emergencies (**RQ1**), one of the unique contributions of this paper is the differentiation of platforms, which shows that WhatsApp and Facebook dominate and are together used by over 70% of respondents. Only on these platforms sharing is widespread, while YouTube is popular for searching information. The predominant use of social media for searching information fits with many ES' strategy of using social media to broadcast information and supports previous findings (Reuter, Kaufhold, Spielhofer, & Hahne, 2017). Despite Twitter's importance for many ES for fast updates, YouTube has so far been more relevant to people in emergencies. This suggests that for a majority virtually "returning" (Hughes et al., 2008) may be more relevant than factual updates. People's sharing behaviour on WhatsApp and Facebook may be explained by the still more private nature of social media's chat tools (Newman et al., 2019) and the growth in personal safety assurance. A new finding is also that while the perceptions of helpfulness of sources all positively correlate some clusters can be found, showing that consumers of traditional information sources and those seeking personal and authorities' information are less likely to respond well to social media information. This may mean that ES' spreading of information through social media may not have to be prioritised at this point, or that it is not fulfilling users' demands: Those favouring social media evaluated contact with ES as comparatively less helpful; instead, they value other online content, likely generated by peers and online journalists. In contrast, those who find contact with ES helpful do not find social media very helpful in emergencies.

Our survey supports findings on age and gender as partly relevant (Reuter et al., 2019) but mainly taking on only a small to negligible role, while education and urbanisation are not central for social media use in emergencies. While the study design is biased towards people who negotiate the online realm, the finding may also suggest that in cases of emergencies, people diversify their strategies: Those who normally use social media less may explore new channels for seeking out information and for reaching out, while people may also be drawn to less frequently used but familiar ways of communication that are perceived as reliable, such as placing an emergency call. Despite the frequent use of social media, traditional emergency phone calls find strong support among respondents. Those favouring such calls are particularly skeptical of the trustworthiness, accuracy and reliability of social media. While those who are concerned with data security and privacy do not show different sharing and use patterns in emergencies, they are strongly concerned with the existence of fake news and whether social media channels are reliable in crises.

Regarding the adoption of casual language in social media channels, strong resistance is voiced. Of those concerned, however, a large proportion did not use social media in emergencies, while those who are active on Facebook appear to be largely indifferent. Turning to wishes regarding social media accounts run by ES, there are no strong preferences to be found overall and no big differences between those having used social media in emergencies. This suggests that no clear preferences have emerged within the German population and a variety of strategies can be successful. This supports the findings that a diverse set of social media strategies can be relatively successful within ES (Wukich, 2015).

With a view to recommendations for ES' social media strategies (**RQ2**), we can derive preliminary suggestions: Most respondents see advantages in social media and many have used it in situations of uncertainty. However, people are also aware of the downsides of social media and its uncertain reliability in crises. This suggests that the use of hybrid strategies (Hughes & Palen, 2012; Meijer & Torenvlied, 2016), incorporating elements of social media while maintaining traditional channels of communication, is recommendable. Although it is likely to increase burdens on ES, our survey also finds that respondents are understanding those challenges particularly in stressful situations, suggesting that people may be prepared to use offline modes of communication in case social media remains unanswered. However, this flexibility may also be required due to social media not fulfilling all expectations. This would have to be explored further by future qualitative research. It also appears that social media communication in emergencies is not and perhaps does not need to be tailored to specific socio-demographic groups, since their experiences and evaluations are similar. Since there are so far no clear preferences for how to implement social media strategies, this should be ever more closely surveyed among the population, as opinions may only form while making experiences with different modes.

Changes over time indicate an overall increase of people who have been using social media in emergencies, especially those who both searched for and actively shared information despite decreasing diversity of different content types. While the increase of safety assurance may be connected to the use of Facebook Safety Check, the noticeable decrease of emotional content requires further research to identify its cause. We observed an increase of perceived social media potentials but citizens still rate the quality of other media, except for accessibility and speed, higher and the 2015 survey is based on a snowball sample which limits the comparability of results. In terms of disadvantages, there is a noticeable decrease, but most barriers are still perceived as a problem by more than 50% of participants. Further inquiries are necessary to identify long-term trends. Despite the fact that a larger number of participants expect ES to be too busy to monitor social media, interestingly, more of them demanded an answer on their posting within one hour, indicating a widening gap between perceived reality and desired state.

The comparison of this survey with data from 2015 and 2016 also shows the strong fluctuation of opinions. Considering the strong changes also seen in the development of social media channels, e.g., the emergence of chat features in Facebook, as well as the emergence of large groups on WhatsApp, continuing surveillance of changes appears necessary to adjust strategies in due time. It seems likely that this will challenge more rigid styles of organisational communication. For now, a focus on Facebook and WhatsApp as the dominant communication channels appears to reach a vast group of people. They are also the best sources to attain information through citizens' content. For broadcasting information, ES may consider including YouTube. Since mobile phones are generally wide-spread, further exploration of warning apps appears to be a promising avenue (Kaufhold et al., 2018).

9.4.2 Limitations and Future Work

This paper is subject to limitations: The biggest challenge is that online surveys are biased towards people who engage online. Although the survey provides representative results according to the criteria of age, gender, geography, urbanisation and education, it excludes those people who are most resistant to using technology. A normalising effect may also be incurred by remembering events in the past. In addition, it is unclear when the emergencies took place that respondents were referring to, which may influence the channels used at that point in time. Since several social media platforms consist of components that can be used as a public forum or private messaging tools, it is not completely clear whether "posting" or "reading" on social media exclusively refers to the public creation and recreation of content, or also to private communication. In future research it seems fruitful to differentiate between such more private use of social media directed at acquaintances or towards bigger circles of strangers and public institutions, especially as behaviour might be relevant for communication about emergencies and with ES. Further research should thus explore how people interact with technologically enhanced means of communication in the private and public realms, as well as regarding how, when and why they turn to both the public realm and state institutions.

Regarding urbanisation truly rural areas and very big cities that might have specific needs in emergencies may not be sufficiently visible, since our categories start at 5,000 inhabitants and end at 100,000. More refined categories could test the lack of differences found in this study between people living in highly urbanised or rural areas and their particular challenges. A caveat results from not differentiating between different types of ES. Furthermore, respondents' lack of decisiveness regarding their preferences for ES' social media strategies may be a result of limited experience with different online communication modes and will likely yield more opinionated results in the future. Finally, the paper focuses on practical and policy implications, lacking an underlying theory guiding the questionnaire design and an explanatory framework. Therefore, future qualitative research should explore social media use, perceptions and expectations from a theoretical lens, i.e., using the framework of risk cultures, which differentiates the framing of incidents, trust towards authorities and targets of blaming if emergency response is not successful (Dressel, 2015; Newman et al., 2019).

9.5   Conclusion

In this paper, we examined the use, perceptions and expectations of German citizens towards social media in emergencies. Our survey of a representative sample (N = 1,219) of the German population was informed by a workshop with a German central federal police agency. We found a widespread use of WhatsApp, Facebook and YouTube in emergencies, which increased in the last three years. Despite social media being perceived as faster and more accessible, citizens appraise traditional media as more accurate and reliable information sources, which is supplemented by the fear of false rumors and issues of data privacy in social media. Furthermore, many citizens expect emergency services to monitor social media, establish bidirectional communication, publish information not only in emergency situations but also in everyday life, and provide both regional and federal social media presences. Finally, those valuing social media in emergencies also favour other online sources, but not necessarily apps, contact with ES or personal conversations.

Citizens' Perceived Information Responsibilities and Information Challenges During the COVID-19 Pandemic

10

Abstract

In crises, citizens show changes in their information behaviour, which is mediated by trust in sources, personal relations, online and offline news outlets and information and communication technologies such as apps and social media. Through a repeated one-week survey with closed and open questions of German citizens during the beginning of the COVID-19 pandemic, this study examines citizens' perceptions of information responsibilities, their satisfaction with the fulfillment of these responsibilities and their wishes for improving the information flow. The study shows that the dynamism of the crisis and the federally varying strategies burden citizens who perceive an obligation to stay informed, but view agencies as responsible for making information readily available. The study contributes a deeper understanding of citizens' needs in crises and discusses implications for design of communication tools for dynamic situations that reduce information overload while fulfilling citizens' desire to stay informed.

Original Publication Haunschild, J., Pauli, S., & Reuter, C. (2021). Citizens' Perceived Information Responsibilities and Information Challenges During the COVID-19 Pandemic. *Proceedings of the Conference on Information Technology for Social Good*, 151–156. https://doi.org/10.1145/3462203.3475886

Supplementary Information The online version contains supplementary material available at https://doi.org/10.1007/978-3-658-46489-9_10.

10.1 Introduction

Information and communication technologies (ICT) are essential tools for sense-making, social interactions and information gathering in crises, including during the COVID-19 pandemic. Cultural values surrounding risk, community cohesion and trust in hierarchical structures in crises (Appleby-Arnold et al., 2019, 2020, Reuter et al., 2019) have shown to influence the use of warning apps and Social Media (SM). With the COVID-19 pandemic being a protracted crisis, it is unclear whether it leads to similar information behaviour as other crises or emergencies. At the same time, "the relatively concentrated media world has given way to a new ecology of diffuse information sources" (Robinson et al., 2019) in which citizens are increasingly both senders and receivers of information, using SM and increasingly messenger apps (Newman et al., 2019) to share, receive and search for information in crises (Haunschild et al., 2020). News agencies but also state agencies, decision-makers and experts are increasingly offering content online, through a variety of ICT, from websites, apps, SM platforms to specific crisis tools such as warning apps (Reuter, Kaufhold, Leopold, & Knipp, 2017). This diversity of information and sources can lead to information overload, which is likely to occur when people perceive a high need to stay informed (Schmitt et al., 2018) and which might foster news avoidance (Park, 2019). Yet, information overload is typically only investigated in the context of SM. Therefore, in this study we address the question *how citizens perceive the information availability in the COVID-19 pandemic*. We conducted a one-week study with 47 German participants consisting of daily open surveys, complemented by two questionnaires. After the introduction, we discuss the state of the art on crisis and COVID-19 communication (Sect. 10.2). In Sect. 10.3 we describe the survey, followed by the analysis (Sect. 10.4), before we discuss the results and implications for design in Sect. 10.5 and conclude in Sect. 10.6.

10.2 Related Work

Insights into the COVID-19 pandemic show the dynamics of news consumption: First announcements of infections in the USA lead to increased Google queries about COVID-19, followed by a decreasing trend back to baseline (Bento et al., 2020). Germans informed themselves several times per day during the first wave through various sources, whereas such activity later decreased (Viehmann et al., 2022), despite perceiving the threat of COVID-19 to be very high (Bendau et al., 2021). A study of Flemish Android users found that during the first wave, mobile phone use duration increased by 15% (Ohme et al., 2020). Especially the use of mobile news apps and SM was spurred by and remained high after the first announcement of measures, whereas the use of mobile web browsing remained largely as before (Ohme et al., 2020). This indicates a persisting increase of information strategies as opposed to less structures information seeking through web browsers and an increase in social outreach as soon as public measures were announced. Being affected by public measures changed information behaviour more drastically than the first case of a pandemic-related death in the country (Ohme et al., 2020).

Previous research shows the role of messengers and SM in crises: In Germany, 73% of citizens used either WhatsApp or Facebook during an emergency, while 25% had used YouTube as an information source (Haunschild et al., 2020). While SM is often investigated for crisis communication (Reuter & Kaufhold, 2018), messengers are on the rise, with around 50% using WhatsApp and around 25% using Facebook Messenger for news consumption in some countries (Newman et al., 2019). Still, legacy technologies are important during crises (Dailey et al., 2016), with television being the preferred information source (Petrun Sayers et al., 2021). Yet, during public health crises, online sources such as websites and SM gain in importance compared to traditional media (Park & Avery, 2018).

Online behaviour in crises typically falls into one of six categories: Helping, being anxious, returning, supporting, mourning and exploiting (Hughes et al., 2008). Some research has focused on the role of peer-to-peer communication, revealing citizens as contributors to situational awareness and as self-organising helpers (Palen et al., 2010, Reuter & Kaufhold, 2018) and in a dual role of sender-receiver (Shi et al., 2018). Models also show different communicative actions in crises, from information selection to information acquisition and information transmission in the process of problem-solving and such information behaviour depends on the trust in the information sources (Austin et al., 2012, Fischer et al., 2019).

Other research has revealed that especially local emergency responders and actors perceived as acting with expertise and without a political agenda are particularly trusted in crises (Appleby-Arnold et al., 2020). Due to this trust and the reliability of their information, communication from authorities to citizens, e.g. through warning apps is an important aspect of crisis informatics. However, a study shows that in Germany only 16,5% were using any warning app in 2019, despite high recognition of their relevance (Kaufhold, Haunschild, & Reuter, 2020). The national context and risk culture also has an impact on societies' behaviour and risk related expectations in crises. Three specific ideal types of risk cultures have been identified: state-oriented, individual-oriented and fatalistic risk culture, which differ regarding trust in authorities and blaming (Cornia et al., 2016). Germany is regarded as a "state-oriented risk culture", characterised by a believe that disasters can be prevented by public authorities, who are highly trusted and responsible for citizens' safety (Cornia et al., 2016). The risk culture also affects SM in crises, with Germans more skeptical of citizen-generated content (Reuter et al., 2019). Previous findings suggest that for other crises high risk perception leads to an increased usage of warning apps, as well as compliance with their behaviour recommendations (Fischer et al., 2019), while other studies have found this effect to be low for COVID-19 (Rahn et al., 2021) and depending more on the believe in effectiveness of precautions (Clark et al., 2020). Investigating the relationship between trust in authorities and the use of SM and warning apps, Appleby-Arnold et al. (2020) found that citizens perceived a responsibility to trust hierarchical orders and were likely to cooperate with authorities in crises despite possible individual distrust based on negative personal experiences with security agencies.

Looking at the state of research, we can identify a lack of qualitative research that elicits citizens' perspectives and their information motivations during the COVID-19 pandemic, regarding all ICT artefacts. While qualitative and ethnographic research has investigated information needs, such as in reaction to disasters (Hughes et al., 2008) and regarding specific artefacts such as warning apps and SM (Appleby-Arnold et al., 2020), it is currently unclear which information responsibilities are perceived by citizens and how this may affect information behaviour, information overload and coping strategies. We therefore designed a study to answer the research questions:

RQ1: Which actors are perceived as responsible for citizens' being informed about COVID-19 related information and which are perceived as reliable sources?

RQ2: What are citizens' unaddressed information needs, their information wishes and implications for designing communication in the context of a dynamic pandemic?

10.3   Method: Repeated One-Week Survey

10.3.1 Recruitment and Sample

The study was conducted via SoSci Survey from 14th-20th of April 2020, during the first wave of COVID-19 infections in Germany while many restrictions were in place, such as shop and school closures or social distancing. Due to the federally varying regulations, the study focuses on the federal state Hesse to ensure exposure to a similar context. Hesse is located in the centre of Germany, features both rural areas and urban centres and is rather representative of (western) Germany in terms of size, population density, unemployment and income. Potential participants were selected by circulating the link to the study, offering a 20 Euro voucher for participation. After review of a selection questionnaire, demographic gaps were filled by recruiting in Facebook groups related to Hesse. The selection characteristics consisted of participants' age, gender, judgement of oneself belonging to a COVID-19 risk group and urbanisation. Signing up required some level of computer literacy. Due to the focus on interactions between the digital and analogue realms, we consciously chose participants who would have some manner of online engagement and included diverse participants from different educational backgrounds, living with and without children in the household and with and without proximity to other persons considered at high risk regarding the pandemic (see the electronic supplementary material A.3.1 for respondent details). The survey started with N = 47 participants (60% female, 40% male) and ended with N = 24 on day 7, providing 133 instances of diary entries (72 on workdays, 61 on the weekend). A similar socio-economic distribution was maintained throughout the study.

10.3.2 Survey Design

Each day for one week participants were invited to an online survey according to their preferences via email, SMS or Telegram. The item and question design built on previous research on crisis communication and crisis informatics, while giving options for additional media types and artefacts and posing open questions to elicit the respondents' perspectives and experiences (see the electronic supplementary material A.3.1 for an example question and the coding scheme). The quantitative questions either asked for the most accurate description or offered responses on a 5-point Likert scale. The survey consisted of a predominantly quantitative pre- and post-questionnaire (socio-demographics, risk and obligation perceptions) on days 1 and 7, whereas days 2, 3, 5 and 6 posed identical open questions concerning ICT use and information behaviour and sources used that day, offering prompt questions to support participants' ideas of aspects that they may describe (see Tab. 10.1). Additional questions explored information overflow and evaluations of information sources and kept the journaling interesting. Misinformation addressed at day 4 is not discussed in this paper.

Table 10.1 Survey questions by days (only questions that were included in the analysis, which also excludes questions from day 4)

Day(s)	Quantitative closed question; qualitative open question
Day 1	*What do you think is your primary obligation in this crisis? To whom do you feel obliged?*
Days 2–6	*What is your current experience*
	… with personal or technical communication with friends, family and acquaintances regarding COVID-19?
	… with communication with larger groups (e.g. in messenger, in social media etc.) regarding COVID-19?
	… with communicating with authorities regarding COVID-19?
	… with experts regarding COVID-19?
	… with different information channels regarding COVID-19?
	What else do you have in mind about communication and distribution of information regarding COVID-19?
Day 6	How do you perceive the information content on different channels, platforms or media?
	How do you deal with information overload?
	How did you react to information overflow or doubts about the quality
	of the information?
	Did you adapt media consumption or your communication behaviour?
	With regard to which aspects of the pandemic do you feel insufficiently or
	unreliably informed?
	How do you deal with the lack of information?
	Where do you look for it?
	Is there any information that you cannot find despite your research?
Day 7	What have been your primary motivations for gathering information about the COVID-19 pandemic?
	What amount of information did you get from authorities such as the federal government, police, ministry of health on new COVID-19 measures, recommendations and laws through which channel?
	Which amount of agency information related to COVID-19 regulations and recommendations did you receive via these channels?
	How much important information about COVID-19 did you obtain through these circles/media/channels?
	Was there any information and types of communication that particularly
	helped you to behave positively regarding the pandemic?
	If so, can you please describe it for us?
	What kind of information was that?
	Where did this information come from/In what circle and through what channels
	did you have this exchange?

10.3.3 Coding

For coding and analysis, we used the R 4.0.0, package RQDA (version 0.3–1) for Qualitative Data Analysis. After a first round of independent open coding (Strauss & Corbin, 1998) by two researchers, we abductively constructed similar coding schemes (Timmermans & Tavory, 2012). Codes were derived from thorough reading of answers to open-ended questions in the first coding phase as well as from code categories informed by crisis communication theories. The artefacts and sources relevant for crisis communication were derived from communication theories and previous crisis informatics research on citizens' and agencies' use of SM in crises (Haunschild et al., 2020, Reuter & Kaufhold, 2018). The codes particularly describing information content, communication, challenges, and coping strategies emerged. With the coding system reaffirmed through comparison and discussion, seven participants were randomly selected and coded by two researchers to test intercoder reliability using Cohen's Kappa k. After a training round, we reached a very good value of $k = 0.81$ for intercoder reliability ($k > .80$; (Altman, 1991, Landis & Koch, 1977)).

10.4   Results

10.4.1 Motivations, Responsibilities and Trusted Actors

Protecting friends and family was the primary reason for staying informed, followed by knowing the current regulations, evaluating adequate personal behaviour and being up to date with the Hessian and German situation. Less important were aspects related to political discussions, donating or volunteering and following social and economic developments. Participants felt responsible to comply with measures to reduce the risk of infecting themselves and others, in particular people at high risk, their family, but also to support health staff. In contrast, few felt responsible for sharing information which they received. Participants stated particularly that they wanted to be informed about current measures, which was often considered easy. In contrast, some participants saw primarily agencies as responsible; *"I continue to see the agencies as having an obligation to better communicate the ever-changing regulations. Communication must be better"* (P33). Such agency communication tool place mainly through TV or radio, and through often localised news from specific federal news channels and municipalities' websites. Many participants described frustration with the lack of direct agency information. *"Authorities should provide standardised information on rules via certified, simple channels. And this before any other media do it!!"* (P42), describing a channel similar to a state-run information app, even though this participant was using the local warning app HessenWARN. Warning apps were also criticised for being too slow in comparison to public news media: *"Since I had already read the news, I felt well informed and didn't read the information in the warning app"* (P4). Overall, the use of warning apps or agency messaging channels was low, while agency websites were often visited. Agency websites provided some information about regulation and recommendation to 83% of the participants, while only 30% indicated receiving any information through agency apps and messengers (see the electronic supplementary material A.3.2 for the figures). Press conferences by politicians were frequently mentioned and deliberately sought out by many respondents, including via live streams, leading to positive emotions; *"Today I feel very well informed, because today the press conference with Angela Merkel took place, which was broadcast and I watched it on TV"* (P25). Information from both agency channels (75% reliable or very reliable) and warning apps (44%) was deemed somewhat less reliable than TV or radio news. SM and private conversations through messengers were deemed less reliable (19% and 3% reliable or very reliable). In comparison to social media, messenger apps are often neglected, whereas around 20% of the respondents indicated that they had obtained much or the majority of the important information through messengers. Only 9% obtained much important information through agency channels and over 30% none (see Fig. 10.1). Citizens-to-citizen communication provided the majority of information to only 13%, while only 4% received no information through word-of-mouth.

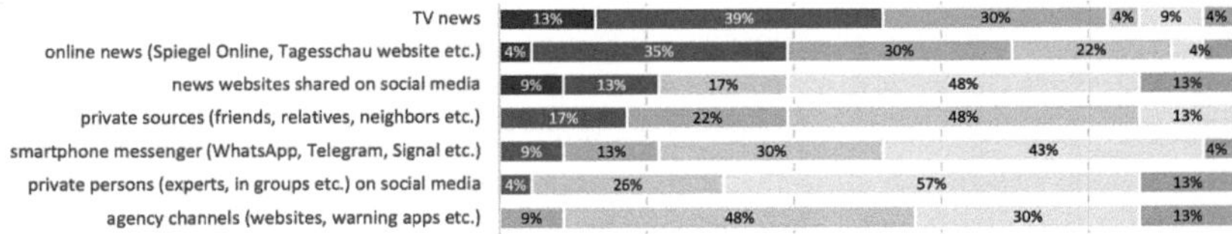

Figure 10.1 "How much important information about COVID-19 did you obtain through the circles/media/channels listed here?" (left to right: ■ all, ■ majority, ▨ a lot, ▨ some, ▨ none, ▨ not used/answered), N = 23

Descriptions of helpful information included statistics, TV reports, local news through local radio and local news agencies' SM presences or municipal agency websites, expert podcasts and press conferences, while discussions and deliberation with friends were helpful for sense-making (e.g., *"With friends over Skype meetings we discussed how to evaluate the measures and how to deal with them"* (P42)). News apps were frequently mentioned, *"especially for a daily update and later to search for news about the new rules of the prime ministers"* (P1). They were often seen as successful filters that displayed news *"directly on the start page without me having to search for it"* (P4).

10.4.2 (Dis)satisfactions, Unaddressed Needs, Suggestions for Improvement

Some respondents reported challenges related to information overload; *"It always oscillates between a very large need for extensive information and a need to receive no news at all"* (P47). One person connected this overflow with digital media, blaming themselves: *"I am increasingly annoyed by digital media, as they provide too much information and steal a lot of time (but this is more my personal problem and less due to the inability of the media)"* (P35). Filtering and reduced media consumption were used as strategies for countering overload, e.g. *"I only watch/listen to serious news (Tagesschau, heute, hr-info)"* (P8). Many reported not reading app or push messages or taking breaks from information seeking, e.g., on the weekend or when the weather was good or avoiding channels such as Facebook that some perceived as toxic. One participant explained changes in media consumption, limiting themselves to watching the news and redirecting personal conversations to other topics. Fake news was often mentioned as a challenge, which was associated with SM and group chats, especially among family members. People associated this with unquestioned sharing of information and thus stated that they prefer not to share information: *"information is typically falsified by forwarding"* (P7), whereas one person took on the responsibility to correct false information as a moderator in a Facebook group. The participants were split on whether there was a lack of information. A large proportion of the sample felt that the information was sufficient (12 out of 32, e.g. *"I feel sufficiently informed about all aspects at the moment"* (P35)), while many missed a clear overview of regulations, e.g. *"We [...] concluded that nobody knows exactly what is allowed and what is not. Everyone has heard something different or remembered something different"* (P7).

This insecurity was often related to location-specific regulation and differences between federal states, particularly for people commuting between two states. Complaints regarded the lack of central information, e.g., the wearing of face masks: *"I had to search for the cities that interest me. I was annoyed that there is no overview [about what applies in which city] on the Hessian site"* (P41). This was an issue even for people who knew of agency channels: *"E.g., Katwarn is not used for its purpose. Looking for information oneself can be misleading. In this tense situation, this leads to even more confusion and speculation"* (P23). One of the dominant wishes was for a tool that gives a quick overview, also over different regulations in federal states and neighbouring countries: *"It would be nice to have an official portal (e.g., app), where all important information is collected, and which points to other pages or articles"* (P7). Information should be centralised, coming *"[…] through ONE government platform. NOT through various different platforms of individual authorities, ministries, federal states etc."* (P42). Other participants had identified an agency channel, the Telegram COVID-19 channel of the Federal Ministry of Health, but perceived that it was not well-known. Another wish was for personalised information, such as regular reports about the own and neighbouring municipality, as well as age-specific or profession-specific recommendations, e.g., for teachers. Typically, reasoning and debate about regulation was not expected, rather there were complaints about reporting *"about things that are still being decided […] and thus confusing whether it is already in effect or only in planning"* (P6).

10.5   Discussion and Implications for Design

In the participants' responses we can identify coping behaviour, with participants limiting their information seeking at certain times, supporting findings that information overload leads to retreat not just on SM (Park, 2019) but also in general. While community values such as protecting others were the driving forces for staying informed about current measures and adequate personal behaviour, participants did not feel responsible for sharing information with others and even avoided it, fearing the spread of false information. This appears to partly contradict findings in the US where participants intended to forwards trusted health information (Chon & Park, 2021). It may be explained by the German state-oriented risk culture, which implies that Germans are skeptical of user-generated content and of information shared through messaging channels (Reuter et al., 2019). The high level of trust in authorities, another aspect of state-oriented risk cultures, is further shown by very positive evaluations of press conferences and the wish for more structured agency information. At the same time, about half of the participants received no information through agency channels. Live tickers were mentioned in connection with information overload, while news apps were mainly positively described. However, many found it difficult to identify the outcomes amidst the debate that is shown in news apps. Warning apps, which can reduce information overload through regional settings, reliable information and updates of important information through push notifications (Dallo & Martí, 2021; Reuter, Kaufhold, Leopold, & Knipp, 2017; Tan et al., 2017), did not present themselves as adequate solutions, because they were slow and did not offer the information that was sought. However, since then the German warning app NINA has been extended to include local regulation related to COVID-19 (Bundesregierung, 2021). Since information overload can be worsened by push notifications (Schmitt et al., 2018), users' preferences regarding immediate push notifications, regular updates and "pull" information in this context, which is neither an immediate emergency, nor everyday-life, should be further investigated.

Looking at information gaps, challenges relate to local, federal and sometimes national differences in regulation for specific topics. Specific COVID-19 information apps such as *CoroBuddy* and *DarfIchDas* are increasingly offering an overview of current regulations sorted by theme, such as face masks, travel, commerce or religious communities, partly offering the option to mark topics as favourites. All the information is offered on a "pull" basis, requiring users to look up changes, which are updated manually by the app providers. At this point, the apps primarily show federal regulation, neglecting local regulations, e.g. about childcare facilities. While the apps display incidence numbers of several locations, they do not allow an overview of differences, e.g. for commuters. Setting nuanced preferences to avoid information overload and achieve information satisfaction is another challenge that should be further investigated. Some messenger apps, which are wide-spread and relevant also in daily life, offer new options for news consumption through news bots (Lou et al., 2021) and increasing development options for bots (Klopfenstein et al., 2017) that could send updates on previously specified topics at specified times. Bots such as the *D64 Covidbot* (https://covidbot.d-64.org/) already exist, but they are currently limited to machine-readable and readily available statistical information.

In addition, reliability and receiving only up-to-date information was a challenge. It should thus be investigated how this affects the COVID-19 information apps, where the state-run app NINA, the pro bono app DarfIchDas and the volunteered app CoroBuddy may be compared. When involving local authorities, it should also be investigated how the back-end may be designed to enable and support authorities in entering locally specific information. Natural language processing and machine learning, which is already investigated for news bots (Jones & Jones, 2019), could be used to process laws and regulations and support agencies in providing the information.

10.6   Conclusion and Limitations

Through a qualitative repeated survey this study investigated how German citizens perceive and evaluate their own and agencies' information responsibility in the COVID-19 pandemic. Previous studies indicate that cultural values, including trust in agencies and community cohesion, influence information behaviour in crises (Appleby-Arnold et al., 2019, 2020). At the same time, information demands change in crises, during the COVID-19 pandemic particularly around times when new measures are announced (Ohme et al., 2020). The perceived need to stay informed for daily life has, however, also been found to increase information overload, which is worsened by certain ICT (Schmitt et al., 2018). The key findings of this study are that

- Germans perceive themselves as responsible and motivated to stay informed in order to behave adequately and protect their family and friends;
- they do not feel responsible for sharing information with others;
- citizens perceived a shared information responsibilities with agencies, who were regarded as responsible for preparing information in a concise manner;
- they perceived information seeking as strenuous, indicating information overload and as a consequence consciously limited their news consumption or use of specific tools;
- they were skeptical towards social media, perceived news media as very reliable, and used press conferences to decrease insecurity;
- a large group who found it difficult to identify the outcomes and current measures for personally relevant topics and regions;
- a gap exists for ICT that portray reliable, fast information that depicts personally relevant local, federal and to a smaller degree national differences in regulation.

As changing risks and regulations are highly relevant to daily life, attaining this information appears to be particularly stressful and prone to causing information overload (Schmitt et al., 2018). We therefore suggest more research on tools that cover the space of dynamically changing regulation that do not qualify as emergencies—and may therefore not warrant warnings in crisis apps –, but that require the cooperation of citizens, can have severe consequences for them and vary locally and nationally. An information tool could be helpful to achieve this, which should consider demands that citizens express in the context of warning apps (Dallo & Martí, 2021; Kaufhold, Haunschild, & Reuter, 2020; Tan et al., 2017), which similarly provide information in crises, e.g. for highly relevant information as push notifications. Providing information, a feeling of "being well-informed" while reducing information overload is an open challenge in the protracted and dynamic crisis. As agencies were regarded as both co-responsible and reliable, but their channels were not widely used, their potential role in providing accessible overviews of up-to-date local regulations should be explored more. Participants were selected to cover the broad Hessian population, which also reflects the different characteristics of the German population. While the study is limited by a small, non-representative sample, due to its qualitative nature, it has revealed information challenges and needs for dynamically and locally varying crises that are likely shared by many. These challenges that are likely also relevant in other dynamic situations with changing regulation.

COVID-19 Regulation and Warning Apps: User Expectations, Developer Challenges, and Insights from App Store Reviews 11

Abstract

During a dynamic and protracted crisis such as the COVID-19 pandemic, citizens are continuously challenged with making decisions under uncertainty. In addition to evaluating the risk of their behaviours to themselves and others, citizens also have to consider the most current regulation, which often varies federally and locally and by incidence numbers. Few tools help to stay informed about the current rules. The state-run German multi-hazard warning app *NINA* incorporated a feature for COVID-19, while two apps, *DarfIchDas* and *CoroBuddy*, focus only on COVID-19 regulation and are privately run. To investigate users' expectations, perceived advantages, and gaps as well as the developers' challenges, we analyse recent app store reviews of the apps and developers' replies. We show that the warning app and the COVID-19 regulation apps are evaluated on different terms, that the correctness and portrayal of complex rules are the main challenges and that developers and editors are underusing users' potential for crowdsourcing.

Original Publication Haunschild, J., & Reuter, C. (2021a). Bridging from Crisis to Everyday Life – An Analysis of User Reviews of the Warning App NINA and the COVID-19 Information Apps CoroBuddy and DarfIchDas. *Companion Publication of the 2021 Conference on Computer Supported Cooperative Work and Social Computing*, 72–78. https://doi.org/10.1145/3462204.3481745

Supplementary Information The online version contains supplementary material available at https://doi.org/10.1007/978-3-658-46489-9_11.

11.1   Introduction and Related Work

During a dynamic and protracted crisis such as the COVID-19 pandemic, citizens are continuously challenged with making decisions under uncertainty. In addition to evaluating the risk of their behaviours to themselves and others, citizens also have to consider the most current regulation, which often varies federally and locally and by incidence numbers. Few tools help to stay informed about the current rules. The state-run German multi-hazard warning app *NINA* incorporated a feature for COVID-19, while two apps, *DarfIchDas* and *CoroBuddy*, focus only on COVID-19 regulation and are privately run. To investigate users' expectations, perceived advantages, and gaps as well as the developers' challenges, we analyse recent app store reviews of the apps and developers' replies. We show that the warning app and the COVID-19 regulation apps are evaluated on different terms, that the correctness and portrayal of complex rules are the main challenges and that developers and editors are underusing users' potential for crowdsourcing.

With its global reach, high embeddedness in daily life, and high conflict between physical safety and other values, the COVID-19 pandemic differs significantly from other crises that are typically studied in crisis informatics, such as natural disasters or more limited health dangers (Gui et al., 2017; Pine et al., 2021; Reuter & Kaufhold, 2018). Interpretations of the virus as an overblown, inevitable, or acute risk have varied by country and over time (Chater, 2020). Citizens' information needs increased and remained high, particularly as the first measures against COVID-19 were announced (Ohme et al., 2020), but uncertain, contradictory, overly complex, changing, and inaccurate information posed challenges (Pine et al., 2021). Information about restrictions was strongly sought out, surpassed only by inquiries about the spread of the virus (Dadaczynski et al., 2021), but Germans were particularly challenged by the differences in regulation across the country (Haunschild, Pauli, & Reuter, 2021) since measures are mainly implemented federally in a decentralised manner leading to great local variation (Hegele & Schnabel, 2021). Germans perceived a responsibility to stay informed about current regulations, but held agencies as co-responsible and expected them to provide adequate information (Haunschild, Pauli, & Reuter, 2021; Reuter et al., 2019). Particularly when perceiving a need to stay informed and lacking an information strategy (Schmitt et al., 2018), information overload can occur. This can lead to withdrawal from information seeking (Chon & Park, 2021) and a reduced intention to self-isolate in the pandemic (Farooq et al., 2020).

Crisis informatics (Reuter & Kaufhold, 2018; Soden & Palen, 2018) has shown the relevance of ICT for communication between agencies and citizens (Reuter & Kaufhold, 2018) and volunteerism (Starbird & Palen, 2011): However, reliability is often a challenge (Laato et al., 2020; Marchal & Au, 2020). Emergency apps are one solution that provides information about emergencies from trusted agents, such as research institutes or state agencies. Some mobile applications are specific to one type of emergency, some include warnings only as a supplement to daily information (e.g. extreme weather warnings in weather apps), other apps are built to warn about multiple hazards (Tan et al., 2017). While such apps are widely regarded as important (as far as they centralise many relevant warning types in one authoritative app), they are rarely adopted (Dallo & Martí, 2021; Kaufhold, Haunschild, & Reuter, 2020). Usage intention is positively influenced by risk perception, trust, and perception of using warning apps as a subjective norm (Fischer et al., 2019). An analysis of warning apps revealed that malfunctions and the temporal and spatial relevance of warnings are main concerns (Kotthaus et al., 2016). Furthermore, dependability, avoidance of advertisement, resource efficiency, appropriate audio interface for alerting, and avoidance of in-app browsing are usability requirements that are particular to warning apps (Tan et al., 2020a). Research suggests that even during the COVID-19 crisis, pandemics were infrequently mentioned as hazards that should be included in a multi-hazard warning app (Dallo & Martí, 2021). Such warning apps, like *NINA*, were largely not perceived as filling the information needs in a study in April 2020 (Haunschild, Pauli, & Reuter, 2021). Studies suggest that warning apps should on the one hand contain all relevant topics, while at the same time notifications that are perceived as irrelevant strongly reduce usability (Tan et al., 2020c). In light of this tight rope walk of too much and too little information, it is unclear how the inclusion of COVID-19 into a warning app and the specifically designed COVID-19 regulation apps are perceived.

App store reviews, which contain bug reports, feature strengths or shortcomings, user requests, praise, complaints, and/or usage scenarios (Guzman et al., 2015) have been successfully used to gain insights into warning app usability issues (Kotthaus et al., 2016; Tan et al., 2020a). Written by users who are specifically motivated to share their experiences, as a crowdsourced task to identify the best app for a specific purpose (Khalid et al., 2015) or to increase pressure on the developers (Pagano & Bruegge, 2013), they are not necessarily representative of the average app's user. At the same time, because a large segment of reviews contains aspects of software requirements, feature requests, and use scenarios, they are used to inform future development of missing features, errors, etc. (Gao et al., 2018; Martin et al., 2017; Pagano & Maalej, 2013). Research shows that amateur reviews can be as good as expert reviews for predicting long-term popularity (Santos et al., 2019).

Some reviews about COVID-19 technology exist but they either portray the very beginning of the pandemic (Bassi et al., 2020; Davalbhakta et al., 2020; Golinelli et al., 2020; Ming et al., 2020), digital technologies generally (Islam & Najmul

Islam, 2020; Vargo et al., 2021), or health apps (Chidambaram et al., 2020; Ming et al., 2020). Tools for the general public mainly concern data sharing and contact tracing (Davalbhakta et al., 2020; Vargo et al., 2021). While news media and news apps become particularly popular in crises, including during COVID-19 (Ohme et al., 2020), they include debate and discourse and may therefore contribute to information overload when searching for current rules. With this lack of studies on warning and information apps for the protracted COVID-19 crisis with its particular information challenges (Pine et al., 2021), it has remained unclear whether users' preferences are similar or different to those expressed for multi-hazard warning apps. We therefore ask: *RQ1) What are the similarities and differences between the reviews of COVID-19 regulation apps and the multi-hazard warning app?* and *RQ2) What are citizens' perceived challenges, gaps, and advantages?*. In addition to the formal state agencies' crisis response, convergent informal activities have been identified which include supporting others (Hughes et al., 2008), sharing local information (Gui et al., 2017), "voluntweeting" on social media (Starbird & Palen, 2011), crowdsourcing (Mulder et al., 2016), and crowdmapping (Shahid & Elbanna, 2015). Through expert networks or software development communities (Van Gorp, 2014), volunteers are also involved in creating new online applications, e.g. in the COVID-19 hackathon #WirVsVirus (Haesler et al., 2020). Challenges for volunteered and technical communities often include shortage of resources and volunteers, but also collaboration with formal organisations (Van Gorp, 2014). The data revealed that particularly the developers of the volunteered apps were active in responding to the reviews. We therefore ask: *RQ3) What challenges are expressed by the developers of the COVID-19 information apps?*.

11.2   Method

To identify apps that show updates of local regulations in the dynamic crisis, we performed a market analysis. In app stores, we searched for the (German) keywords "COVID", "Corona", and "incidence", resulting in 249 apps. We excluded 154 apps not related to the topic (e.g., Snapchat), 11 dedicated only to tracking COVID-19 infection chains (e.g., Corona-Warn-App), 20 for educating and documenting symptoms (e.g. Corona Health), 7 apps only about the vaccine (e.g., STIKO-App), 18 general health apps (e.g., WHO Info), 12 city or agency apps (e.g., Darmstadt) and 13 news apps (e.g., Tagesschau). We then manually screened the description of the 14 remaining apps which all provide regional incidence numbers regarding COVID-19 infections. Only three apps provide the local rules in addition to statistical information. The three apps are the multi-hazard warning app "NINA" (BBK, 2021), which is the most widely used warning app in Germany (Kaufhold, Haunschild, & Reuter, 2020) and run by the Federal Office of Civil Protection and Disaster Assistance, and the privately run COVID-19 regulation apps "DarfIchDas" (InTradeSys GmbH, 2021) (which translates to "AmIAllowed"); and "CoroBuddy" (CoroBuddy GbR, 2021). While many of them report current statistical data, such as incidence number, vaccination rates and intensive care availability, we selected only those apps that, similar to warning notifications, map infection events to current local restrictions. NINA (10,000,000+ downloads) was launched in 2015 and introduced COVID-19-related aspects in April 2020 (BBK, 2020). *DarfIchDas* (500,000+) was launched in September 2020 and *CoroBuddy*'s (10,000+) first review appeared on March 15, 2021 (abbreviated N, C, and D in the source of quotes). The apps share the portrayal of local regulations aiming to inform about what is currently (not) allowed in different regions and regarding specific areas of life (see Fig. 11.1). Whereas *NINA* and *DarfIchDas* list users' favourite regions, *CoroBuddy* only displays one selected region at once. *DarfIchDas* allows searching the list of measures with keywords and added an incidence history of the past 14 days during the course of the study. *CoroBuddy* and *NINA* represent the threat situation through a colour scheme. The warning app *NINA* is the only one of the three apps to show a map of Germany, with the regions colour-coded according to their incidence levels. *NINA* also sends push notifications about government announcements regarding the COVID-19 pandemic and provides general information about COVID-19 (e.g., basic knowledge, vaccinations, etc.).

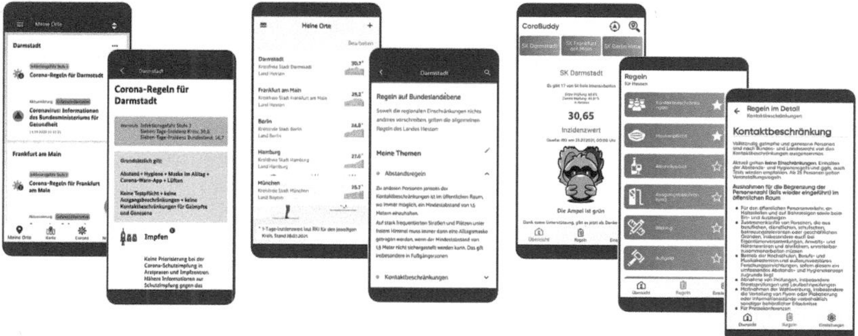

Fig. 11.1 Screenshots of the warning app *NINA* (left) and the COVID-19 regulations apps *DarfIchDas* (middle) and *CoroBuddy* (right), from July 28, 2021 (see the electronic supplementary material for more details)

To compare these apps, we analyse the praise, and complaints, while mentioned feature aspects, requests and usage scenarios are coded as users' perceived advantages, information challenges, and gaps. Two researchers iteratively discussed and generated a suitable coding scheme, which was built abductively with some categories deduced from previous usability assessments on crisis apps (Dallo & Martí, 2021; Kotthaus et al., 2016; Tan et al., 2020a) and an analysis of review responses (Vu et al., 2019), while other codes emerged from the text. Due to the novelty and the speed of the development of updates, the analysis does not show the critiques of the most current versions, but rather users' needs and assessments regarding information in the protracted crisis. Since the three apps are non-commercial in nature, they are also instructive for volunteering and non-profit app development in crises. We coded all reviews from March 15 to May 31, 2021, a time when all three apps were published and which covers the full third wave from its uptake in March, its peak in April, and the decline and end in May, marked by the German Federal Institute of the Ministry of Health's (RKI) downgrading of Germany's risk status from "very high" to "high" on June 01, 2021 (RKI, 2021). The observed time period includes a shift in German policy with the entering into force of a national law on the protection against Infection ("Bundesnotbremse") on April 24, 2021. The law for the first time set mandatory minimum measures for regions above an incidence rate of 100, requiring regulatory changes in most areas in Germany. We include reviews from the Google Play Store (N = 75, D:402, C:235, total: 712), the Apple App Store (N = 21, D:86, C:-; total 107) and the HUAWEI AppGallery (N = 12, D:1, C:-; total: 13, the app store source is abbreviated as G, A and H in the given quotes). This results in 234 reviews and 319 coded segments from *CoroBuddy* (G:235, A:-, H:-), and 438 reviews with 704 coded segments from *DarfIchDas* (G:402, A:86, H:1). Since *NINA* warns about a wide range of emergencies and was analysed previously (Kotthaus et al., 2016), we filtered the 295 reviews to contain only those related to COVID-19 information with a wide set of keywords surrounding the pandemic, resulting in 106 reviews (G:75, A:21, H:12) and 166 coded segments. This results in a total of N = 832 reviews and 1164 coded segments (Fig. 11.2).

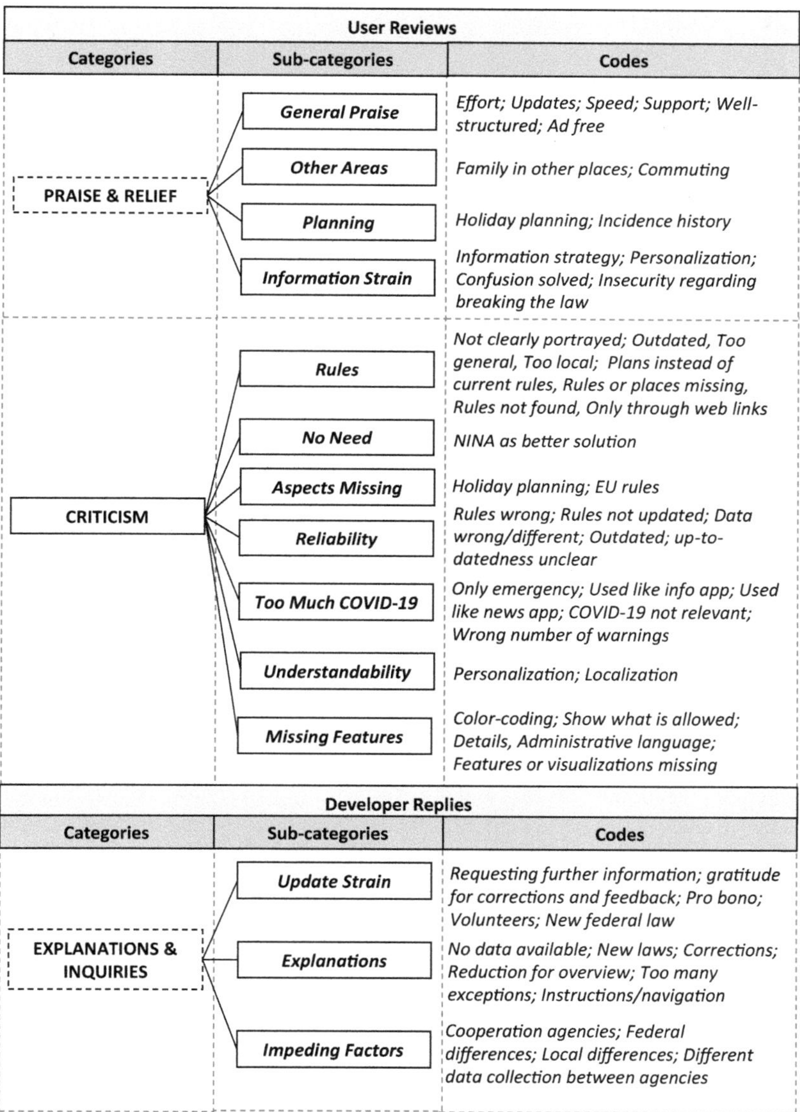

Fig. 11.2 Coding scheme

11.3   Analysis

In the following, we answer the research questions by analysing first the user reviews and then the developers' replies.

11.3.1 Analysis of User Reviews

Differences and Similarities (RQ1). CoroBuddy's reviews are marked by a great number of general praise and gratitude (94 reviews) for the initiative and the volunteers' efforts. Most praise refers to the clarity of the design which enables a quick overview: *"One look is enough and you are up to date"* [CGa8]. Other reviews reveal personal challenges that the app solves, speaking about the *"jungle of rules"*, *"patchwork of rules"* or *"chaos"* (19), and about trouble in keeping an overview (8). Some express that they would have expected state agencies to provide such an overview (5). Complaints relate to the accuracy and lack of updates of rules (51 reviews). The granularity of the information is a challenge (6), with people wishing for municipality or city-level information (instead of county-level information) or not finding rules in the app (5). A traffic light colour scale was used to visually express the local severity of the pandemic in the app. However, the state in the application of its federal COVID-19 law, as well as the federal health agency RKI used different traffic light colour schemes to denote the severity of the spread of COVID-19. This caused confusion among the users who were often unaware of this duality and thus perceived the information be contradicting other official information (11).

DarfIchDas's evaluations are also marked by wide-spread praise and thanks (198), often mentioning the ease and speed of getting an overview, the good support, and the absence of advertisement. Similar to *CoroBuddy*, the accuracy and validity of the portrayed rules and data are often questioned (118). A dominant complaint concerned that the rules displayed were too general (22) and that local rules were only available through website links (16) or could not be found at all (14). At the same time, the many details listed required extensive reading: (*"Only the texts of the regulation are reproduced, but not what specifically applies at my location today"* DGb32). After the app was featured on TV, functionality failures were reported due to server overload. 24 comments across both apps address relief for commuters or travelers in general. While for *DarfIchDas* and *CoroBuddy* the topics were similar, the reviews for *NINA* revealed different aspects.

Among the reviews related to COVID-19, **NINA** received significantly less general praise and thanks (16). The appearance of pandemic-related information was sometimes generally contested because it was not seen as an emergency (12). When

COVID-19 information was generally approved, it was often regarded as clustering the app with older general information at the top, making it difficult to identify new notifications (16). Some users had trouble setting their often loud alarm sounds so that they would exclude COVID-19 notifications. Functionality issues seemed to be more prevalent for *NINA* than for the other apps. Across all apps, around 25% of the reviews mentioned problems with the reliability of the portrayed rules and data, showing that this is the most significant challenge. This is especially true for the COVID-19 regulation apps, for which the number increases to almost 50%. Only a few reviews indicated that the COVID-19 regulation apps were performing "*the state's job*" (DG258) while *NINA* reviews more commonly expressed disappointment or confirmation of low expectations towards state agencies: "*I can't understand how the state app can't manage to update the rules*" (NG44). Few reviews mention the lack of liability and fear of sanctions if the rules are incorrect (5). This suggests that the COVID-19 regulation apps are not regarded as unreliable per se.

Citizens' Perceived Advantages, Challenges, and Gaps (RQ2). The most commonly mentioned use case for the COVID-19 regulation apps was as part of an information strategy to get an overview. Some of the most positive features in this scenario were having "*everything at a glance, a good companion through the chaos of rules*" (CG186), "*without a thousand other unnecessary facts*" (CG88). Some reviews showed that the app helped with a perceived information strain: "*I'm tired of having to find information all the time and that often requires a long search*" (CG18). Specific scenarios that were mentioned included people who are mobile or commuters ("*I am in different regions of the republic several times a week*" (DG132)), have family or other interests in different places ("*I can directly have the districts displayed that are relevant to me*" (DG106)). Similarly, *DarfIchDas'* personalisation feature for saving locations as favourites was often mentioned as helpful and dearly missed before it was included in *CoroBuddy*.

Since the incidence number has come to determine mandatory national measures, incidence trends have become important to enable planning for the future. Some people missed an overview about places where certain activities or vacationing are allowed, possibly extended throughout the European Union or including neighbouring countries. Sometimes a map, filter or search function was missed. Dissatisfaction and insecurity are often caused by a perception of wrong or outdated information when the incidence numbers of the apps are not in line with the ones that users find elsewhere. Indeed, primarily due to delays in the process of transferring data from local agencies to the federal one (RKI), these are often not identical, especially in regions with smaller populations [59]. But many are also dissatisfied with finding incorrect rules, stating that if the app is not fully reliable, it is not useful: "*Unfortunately, however, the information on the limitations lags far behind. And*

precisely this would be absolutely necessary for the now very confusing situation"
(DG220).

11.3.2 Analysis of Developers' Replies

Developers' and Maintainers' Challenges (RQ3). Supporting previous findings on app store review answers (Vu et al., 2019), the review responses of *DarfIch-Das* and *CoroBuddy* often consist of appreciation for the feedback or supportive comments, revealing reviews as a source of motivation for volunteer activity. When users complain about errors or perceived inconsistencies, the COVID-19 regulation app providers often explain the updating strain, pointing particularly to a small team of editors or volunteers. *DarfIchDas* maintainers also mention the many changes required by the federal law, partly transferring blame to the regulatory freedom of federal states in implementing the law. According to the review answers, these differences impede the rule-based automation of incidence trends and resulting restrictions. *DarfIchDas*'s replies mention a lack of interest in cooperation on the part of agencies. This supports past findings of the difficulties of cooperation between formal and informal agencies (Van Gorp, 2014). *DarfIchDas*'s review answers often seek further information, especially about the locations that users report as outdated. When a location is named, the answers often contain gratitude and the promise of correction. *CoroBuddy*, being a very new app, often explained which features have been implemented or will be implemented in the near future to solve the issue mentioned by the users.

Only *NINA* responses provide further contact information and solutions for how to change settings, especially with a view to (de-)activation of GPS or sounds for COVID-19 warnings. While *NINA* responses focus on mobile phone specifications to understand the reported bugs and direct reviewers to customer support (e.g. *"If this tip does not help, I would be very grateful for a short info. If you have any questions, please do not hesitate to contact me at [e-mail]."*, the other two apps often seek to identify places that are reported to contain wrong information (*"Unfortunately, you did not tell us which place your case refers to, [otherwise] we could have taken a look at it"* (DA14)).

11.4   Discussion and Conclusion

The analysis of user reviews allows insights into new apps that portray current COVID-19 related regulation information and that have emerged during the protracted COVID-19 crisis, as well as into the integration of such information into an established warning app. In addition, the analysis of developers' responses to the reviews reveals the challenges they are facing. We can derive the following key findings:

- Overviews about the regulations put in place to limit the spread of the COVID-19 pandemic, provided both by private actors and state agencies, have been gratefully received by many users.
- Warnings and prioritised COVID-19 information in warning apps, however, are often regarded as obstructing information about other emergencies.
- Receiving a concise overview over legal requirements with accurate, timely and location-specific information remains a challenge.
- Uncertainty about the quality of the information provided negatively affects users' trust.
- Users' comments about missing or wrong information in the reviews are appropriated by the developers of the COVID-19 regulation apps to improve the accuracy of the apps' content.

From these findings we can derive implications for design that improve transparency, accuracy and reduce the strain on developers. Since the accuracy of the provided information is often contested, developers should include more features that can help users judge the information's reliability. Transparency and trust could be increased by showing the date and time of the latest update, while a feature could allow users to contest or support the correctness of the information and possibly provide reasons and references.

Looking at the requests for further information about reported errors, none of the apps direct users to a formal mode for reporting or correcting errors. Implementing a system for crowdsourcing gaps and updates may be feasible and can build on insights from digital crisis volunteering (Cobb et al., 2014; Park & Johnston, 2017). For instance, inviting feedback instantly after the use of the tool can attract previously inactive users (Masli & Terveen, 2014). Contributions could range from simple tagging of potentially false segments, to correcting them with revisions. Replies to the reviews could include a link for structured input, which has been shown to improve non-expert feedback (Yuan et al., 2016). Other crises have shown that individuals and emergent online communities can be effective at collecting and analysing complex information (Dailey & Starbird, 2015; Gui et al., 2017).

Cooperation with agencies could to some degree relieve the update strain that results from the local implementations and regulatory differences. Local agencies could be in charge of updating their information, making sure that the information could be fully relied upon. *DarfIchDas* responses indicated a lack of interest in cooperation on the part of agencies. This should be further explored through interviews with developers, agencies, and also agencies cooperating with *NINA*, where local agencies are involved in providing information and releasing an alarm. The challenges described by the developers in the responses may indicate a lack of consideration of harmonisation and digitalisation on the part of German agencies, which are only slowly adapting to digitalisation requirements in government (Halsbenning, 2021).

The findings of this study also indicate avenues for future research: The study suggests that usability aspects identified for warning apps, such as dependability and resource efficiency (Tan et al., 2020a) are less relevant for regulation apps. Instead, portraying complex information and reducing administrative text appears to be the bigger challenge. The many reviews that express relief at getting an overview through the apps appear to confirm previous findings, which showed that citizens were more prone to information overload when they felt they needed to keep up with politics for their daily life (Schmitt et al., 2018). A further indicator may be a particularly popular user review which wished for a widget, which would allow to more easily and constantly survey the situation. This would allow staying informed even without opening any app—a feature quite different from the occasional sound and push notifications used in warning apps. Future research should thus explore which features contribute to a sense of being informed without increasing information overload, particularly in dynamic situations.

In light of the different expectations towards regulation apps and warning apps indicated by this study, future work should further explore where the design of tools that inform about regulation can follow guidelines for warning apps (Tan et al., 2020b) and which aspects need to be different. In addition, messenger apps are increasingly used for communication in large anonymous groups and for news delivery (Lou et al., 2021; Newman et al., 2019), including by the German ministry of health which provides WhatsApp and Telegram broadcasting channels on COVID-19 information (BMG, 2021). Due to their widespread use compared with warning apps, research in human-computer interactions could explore messenger apps as multi-purpose tools for communication in dynamic times.

Reviewing Strategies to Motivate Users to Contribute to Resilience

12

Abstract

Smart cities aim at improving efficiency while providing safety and security by merging conventional infrastructures with information and communication technology. One strategy for mitigating hazardous situations and improving the overall resilience of the system is to involve citizens. For instance, smart grids involve prosumers—capable of producing and consuming electricity—who can adjust their electricity profile dynamically (i.e., decrease or increase electricity consumption), or use their local production to supply electricity to the grid. This mitigates the impact of peak-consumption periods on the grid and makes it easier for operators to control the grid. This involvement of prosumers is accompanied by numerous socio-technical challenges, including motivating citizens to contribute by adjusting their electricity consumption to the requirements of the energy grid. Towards this end, this work investigates motivational strategies and tools, including nudging, persuasive technologies, and incentives, that can be leveraged to increase the motivation of citizens. We discuss long-term and side effects and ethical and privacy considerations, before portraying bug bounty programs, gamification and apps as technologies and strategies to communicate the motivational strategies to citizens.

Original Publication Egert, R., Gerber, N., Haunschild, J., Kuehn, P., & Zimmermann, V. (2021). Towards Resilient Critical Infrastructures—Motivating Users to Contribute to Smart Grid Resilience. *i-com—Journal of Interactive Media, 20(2)*, 161–175. https://doi.org/10.1515/icom-2021-0021

J. Haunschild, *Enhancing Citizens' Role in Public Safety*, Technology, Peace and Security | Technologie, Frieden und Sicherheit, https://doi.org/10.1007/978-3-658-46489-9_12

12.1   Introduction: The Smart Grid, Prosumers and Resilience

Smart cities are envisioned to encompass highly digitised and strongly interconnected infrastructures, like the energy grid, water supply, and telecommunication. Some of these systems or organisations, called critical infrastructures (CI), provide vital services for nations' society and economy. Failures or attacks on these CI can impair crucial services and severely threaten public safety and security (Reuter et al., 2020).

This is particularly true for the energy grid, which provides a continuous supply of electricity for millions of consumers, despite the ever-increasing energy-technical demand of modern societies (Hines et al., 2009). However, since failures and attacks cannot be fully prevented, a high degree of resilience is crucial for minimising hazardous consequences for all involved. Resilience can be defined as "the intrinsic ability of a system to adjust its functioning prior to, during, or following changes and disturbances, so that it can sustain required operations under both expected and unexpected conditions" (Hollnagel, 2010, p. XXXVI).

In the context of energy grids, conglomerates have always been responsible for maintaining a stable operation by adjusting production to the needs of the consumers. In recent years, energy grids have transitioned from the conventional fossil-fuelled and top-down controlled systems towards increasingly distributed smart grid (SG) that strongly rely on renewable energy sources. Whilst these developments have obvious benefits, they introduce their own share of challenges, such as high volatility as the production of renewable energy depends on ever-changing environmental factors, like wind and solar radiation. On top of that, the role of consumers changes. Traditionally, their role within conventional energy grids was a passive one, where they freely consumed electricity provided by large producers, whilst being largely unaware of the impact of their actions and the feats that need to be accomplished by operators to provide a continuous supply with electricity. These consumers have evolved into increasingly active *prosumers*, supported by information and communication technology (ICT) and capable of producing and consuming electricity (Egert, Daubert, et al., 2021). This means that the balancing of demand and supply in SG becomes more complex as the set of interdependent participants in a SG strongly increases.

Consequently, notions of safety and security in these complex and interconnected systems cannot solely be based on technical aspects but increasingly depend on the behaviour of citizens. It is thus important to consider citizens as active contributors to safety and security, in terms of stabilising the SG to prevent hazardous incidents as well as in terms of coping with and recovering from emergency situations such as blackouts, e.g., through emergency preparation and coping strategies (Wethal, 2020).

Numerous options for citizens to contribute to the safety and security of SG are already available in current energy grids, e.g., adapting the consumption behaviour to avoid high consumption during peak hours, and studies have investigated the conditions under which citizens are prepared to cease control over energy consumption (Fell et al., 2015) or motivations for reducing consumption (Pratt & Erickson, 2020). Yet, some of these aspects cannot be regulated or mandated without compromising the autonomy and freedom of choice of the individuals, while others require the provision of certain technologies and technological knowledge. Thus, one challenge lies in enabling and motivating users to voluntarily contribute to the safety and security of smart grids, especially if their "resilient" behaviour might result in compromises or restrictions for themselves for the benefit of others. An additional challenge lies in sustaining motivation and the targeted behaviours over time.

Across different disciplines, strategies such as nudges or incentives have been explored to shift citizens' decisions or behaviours (Ranchordás, 2020; Sunstein, 2013). Yet, motivating citizens to "resilient" behaviours within complex and interconnected CI remains a new and unsolved challenge.

Applying insights from various disciplines, we will shed light on the following research question: *How can citizens be motivated to actively contribute to the safety and security of connected CI, such as smart grids?* This paper presents and discusses different strategies for motivating citizens to contribute to safety and security of CI from an interdisciplinary point of view. As a prominent representative of CI, smart grids are used as an example to illustrate each strategy with practical examples.

The remainder of the paper is structured as follows: Section 12.2 on motivational approaches presents strategies to motivate citizens to contribute to SG safety and security along with practical examples from an interdisciplinary viewpoint. We then discuss the presented strategies with regard to their long-term effects, ethical challenges, as well as practical implications in Sect. 12.3 and present visions of citizen participation through bug bounty programs, gamification, and the use of apps in Sect. 12.4. Section 12.5 summarises the findings.

12.2   Motivating Users and Citizens to Contribute to Safety and Security in Smart Cities

In general, motivation can be defined as the overall desire or willingness of someone to do something (Oxford English Dictionary, 2020). This section presents theoretical frameworks and approaches for motivating users to choose certain options for action or to show specific behaviours. The approaches represent an interdisciplinary perspective with insights from various disciplines, such as a pedagogy (e.g., information provision), behavioural economics (e.g., nudging), psychology (e.g., persuasive technologies), and politics (e.g., incentives). While the approaches differ in their theoretical background, their perspective on the human being, and their intentions, they rely on similar mechanisms to transfer these intentions. Each approach is presented with some examples for motivating citizens to contribute to SG safety and security. Furthermore, the differences, overlaps, and possibilities for combining approaches will be discussed in Sect. 12.2.5.

12.2.1 Information and Education

Information provision serves to bridge a potential knowledge gap between citizens (or laypersons) and a knowledgeable third-party, such as an organisation, the state, or experts of various kinds to improve the decision making-process (Calo, 2014). Education and information provision can be regarded as essential for achieving transparency and for empowering citizens with different levels of background knowledge and skills to make informed privacy and security decisions where incomplete or asymmetric information is the norm (Acquisti et al., 2017). However, too much information can be overwhelming for the user.

For example, research found that information provision through notices is seldom working in practice (Calo, 2014). Obar and Oeldorf-Hirsch (2020) showed that privacy policies and terms of service are rarely read and Florêncio et al. (2014) found that password creation suggestions are seldom followed, despite year-long advice. Previous work from the security and privacy area thus suggests that information should be provided in a concise format considering the user's mental models (e.g., using known metaphors) (Raja et al., 2011), and that standardisation of information (e.g., standardised labels) (Kelley et al., 2010) may support understanding. Examples for standardised labels related to smart grids are energy labels on electric devices and suggestions for standardised privacy labels indicating how organisations make use of the citizens' personal data (Kelley et al., 2009).

Information and Education Examples Examples for practical information provision to motivate citizens in contributing to SG safety and security are already discussed in the literature (Goulden et al., 2014; Hargreaves et al., 2013). This research confirms that the simple provision of information (e.g., using in-house displays that provide information about the electricity consumption) is insufficient to motivate citizens. In comparison, if information provision is combined with other techniques, such as incentive-based approaches, the probability of citizens contributing to the safety and security of the grid can be increased (Allcott, 2009; Belton & Lunn, 2020; Herter, 2007). For example, it was shown that the decision-making of citizens can be improved by providing information on smart meters and facilitating the choice of time-of-use tariffs in combination with personalised estimated costs as an incentive (Belton & Lunn, 2020).

12.2.2 Nudging

Nudges (Thaler & Sunstein, 2008) are small tweaks of physical or digital decision interfaces aiming to encourage "wiser" decisions, e.g., secure behaviour, without limiting or significantly influencing peoples' choice set. Nudges generally work by activating automatic cognitive processes such as biases or heuristics (Calo, 2014; Hansen, 2016).

Nudges have been successfully applied in a large range of physical contexts such as encouraging healthy nutrition (Kroese et al., 2015), organ donation (Whyte et al., 2012), or physical activity (Van der Meiden et al., 2019). In addition, nudging has found its way into the digital and cyber space, including password creation (Hartwig & Reuter, 2021; Renaud & Zimmermann, 2019), WiFi selection (Turland et al., 2015), and privacy-friendly app choices (Balebako et al., 2011). An example is that of using a position effect to make people choose a secure WiFi option in a public place such as an airport. When the secure WiFi option appeared first in the selection list, more people tended to choose that option (Zimmermann & Renaud, 2021). As such, nudging might be a promising strategy to encourage certain decisions within the smart city context in which physical and digital spaces become increasingly intertwined.

Nudge Examples Nudges towards pro-environmental decisions have often been summarised under the term "Green Nudges" (Schubert, 2017). Previous research on green nudges showed that users can be nudged towards saving electricity by providing them feedback on their consumption behaviour or comparisons with other users (Allcott, 2011; Ayres et al., 2013). However, these nudges may also create unintended effects. For example, those households that usually consumed less than average, unexpectedly increased their energy consumption (Schultz et al., 2007) perhaps because of a tendency to go with the norm. These side-effects could be mitigated by including an indication of desirable behaviour (Schultz et al., 2007). On the level of providers, exemplary nudges include enrolling clients in green energy sources in their contracts per default (Sunstein & Reisch, 2016) and research confirmed that more people select green energy choices when those are presented as the default (Pichert & Katsikopoulos, 2008).

Schubert (2017) suggests the use of an *Ambient Orb*, as a green nudge, which changes its colour to indicate the current smart grid load. Even though imagined in the context of climate change, a similar intervention could serve to increase users' awareness of the grid's load factor and may encourage users to contribute to the grid's stability.

Zhao and John (2021) examined the use of framing nudges to encourage users to build community resilience through preparation for natural disasters, e.g.by storing food or household retrofitting. They used the examples of a hurricane, earthquake, or flood and applied the choices to invest in physical mitigations, an insurance, or to do nothing with the chance that no adverse consequences arise. In general, the percentage of people choosing to invest was higher when the decision was framed as a gain. Even though the contexts of the study slightly differed from the SG one, the results might be transferable.

12.2.3 Persuasive Technologies

Persuasive technologies are interactive technologies that aim to change the attitude or behaviour of the respective user (Fogg, 2002). Whereas nudges can exert influence on a subconscious level, influence from persuasive technologies always comes from a conscious interaction with the technology. Hence, users make an active decision to change their behaviour in a certain way (e.g., to reduce CO_2 emissions) and use technologies that support them in this desired behavioural change. Consequently, persuasive technologies are designed for more far-reaching behavioural changes than nudges, which are not intended to restrict the user's scope for decision-making.

The concept of persuasive technologies was first described by Fogg (2002). He postulates that three factors must be present at the same time for a change in behaviour to occur: motivation, ability, and a prompt (Fogg, 2009b). Possible motivators include gaining pleasure (or preventing pain), hope (or overcoming fear), and winning social acceptance (or preventing social rejection). The aspect of ability focuses on one's resources, e.g., in terms of time, money, or physical strength. However, according to the FBM, even if people score high on motivation and ability, they still need to be prompted to exercise a certain behaviour.

Persuasive Technology Examples Multiple studies examine the potential of persuasive technologies to encourage energy saving. Beheshtian et al. (2020) explore the possibilities of a persuasive social robot aiming to facilitate sustainable behaviour—such as saving energy—in shared living spaces. Most of their 20 participants agreed that social components such as comparing their energy consumption with other flats and competing against others could motivate them to save energy. They also liked the idea of getting points and rewards as well as information about how to save energy. Takayama et al. (2009) describe the successful gamification approach, which implements several social feedback mechanisms, e.g., by allowing families to compare their CO_2 emissions against those of their neighbours.

Bourgeois et al. (2014) developed and tested four different interventions aiming to motivate households who produce their own "green" energy to shift their laundry activities to maximise the use of their self-generated energy. The results of an 8-month field study with 18 households suggest that while feedback mechanisms fail to motivate a behaviour change, proactive suggestions as well as contextual control (e.g., replacement of the control panel with a tablet that offers the opportunity to either start the washing process immediately or automatically at the best time) seem promising for facilitating behaviour change.

12.2.4 Incentives

The participation of laypersons in processes for maintaining the continuous operation of CI is often—if not always—accompanied by some kind of *burden*, such as a partial blackout for prosumers who offer reserve capacity. Therefore, rational choice strategies suggests that such burdens should be adequately compensated. In this context, *incentives* aim to alleviate the participation of laypersons by directly or indirectly providing a compensation. In comparison to persuasive technologies (see Sect. 12.2.3), which aim to motivate people in making active decisions, incentives use direct or indirect mechanism for reimbursing certain user behaviours.

Psychological theories differentiate between intrinsic or extrinsic motivation, i.e., whether the motivator lies within the execution of a certain action or in the consequences of the completed action (Rheinberg, 2009). Exemplary findings analysing the intrinsic motivators behind graffiti spraying, e.g., identified sensation seeking, a flow experience, creativity, and camaraderie (Rheinberg & Manig, 2003). Extrinsic motivators include monetary, social, or other rewards, such as job-related benefits (Lepper & Greene, 2015; Ryan & Deci, 2000). The differentiation can be relevant when designing interventions or choosing incentives to motivate a certain behaviour. Research showed that intrinsically motivated behaviour can be crowded out when offering extrinsic incentives, e.g., when offering children a reward for an action they formerly enjoyed doing (Deci, 1975).

Incentive Examples Incentivisation is a strategy that has been used within the energy grid for decades. Various pricing strategies exist, which aim to encourage the consumers in the grid to adjust their behaviour, like time-of-use prices, real-time pricing, and inclining block-rate pricing (Abushnaf et al., 2015; Allcott, 2009; Belton & Lunn, 2020; Herter, 2007). The majority of these strategies is based on indirect reimbursement strategies, where consumers can manage to *pay less* if they adhere to the behaviour, which is suggested by the strategies (e.g., avoid strong electricity consumption during peak-consumption hours, where prices are high). Such strategies have been successfully used within home energy management model (HEMS) as an incentive for automatically scheduling the use of appliances to maximise the monetary benefits of the consumers (Mohsenian-Rad et al., 2010).

12.2.5 Combination of Approaches

The presented interventions differ with regard to their theoretical background, their disciplinary viewpoints, and their intentions even though they may sometimes make use of similar measures to achieve their aim (see Table 12.1). However, the approaches also show a certain degree of overlap in the mechanisms they use. For example, incentivisation encourages users to pick a certain option and to overcome burdens e.g., increased efforts or downsides. Persuasive technologies instead make use of incentives to support users in their self-chosen behaviour changes and aims. Incentivisation also partially overlaps with nudging. Even though Thaler and Sunstein (2008) exclude "significant" economic incentives from the nudge definition, the criterion is rather vague so that other incentives such as social rewards may still fulfill the nudge definition.

Table 12.1 Overview over the criteria for differentiating the presented interventions

	Information Provision	Nudges	Persuasive Technologies	Incentives
Cognitive process	reflective	automatic	reflective	reflective
Primary target	decision & behavior	decision	behavior	decision
Degree of coercion	low	medium	low	high
Degree of interaction	low/medium	low	high	low
Resources provider	low	low	high	high
Resources user	high	low	high	low

Research showed that the combination of interventions, such as several persuasive technology mechanisms (Matthews et al., 2015) or information provision and nudging (Kroese et al., 2015; Sunstein, 2015; Zimmermann & Renaud, 2021) can even be beneficial.

Yet, researchers and practitioners should be aware of these overlaps to clearly position, design, and purposefully combine their interventions. Additionally, they should bear in mind the different theoretical concepts behind and the implications of different approaches in order to anticipate and avoid potential negative side effects or unintended outcomes.

12.3 Evaluation and Discussion

The analysis of motivational aspects and technologies to implement them shows that there are many avenues to enhance the integration of prosumers to contribute to the resilience of CIs. While the given examples mainly portray the energy sector, similar challenges exist with regard to other CIs. The water sector is also grappling with involving citizens in reducing and steering water consumption and with communicating with consumers (Heino & Anttiroiko, 2016; Laspidou, 2014; Novak et al., 2018). In choosing and adapting motivational strategies, ethics, privacy aspects, as well as potential side-effects should be considered.

12.3.1 Ethical Considerations

Guidelines for ethical psychological research (American Psychological Association, 2016; European Federation of Psychologists' Association, 2005; The British Psychological Society, 2014) suggest to respect persons and their autonomy, to maximise beneficence, to practice justice, to work according to scientific integrity standards, and to take over social responsibility. These meta guidelines can also well inform the depletion of measures to motivate citizens to or to make citizens behave in a certain way.

For example, citizens could be respected by being informed transparently about the measures and their purpose, and by looking for solutions that do not unnecessarily compromise citizens' autonomy. Exemplary strategies following that approach are educational approaches aiming to increase knowledge and awareness, persuasive technologies that foster the active interaction with the user, or nudges that aim to encourage certain choices without limiting the choice set. In terms of nudges, research suggests that transparent "hybrid nudges", i.e., the combination of nudges with information provision, may be especially favourable due to the enhanced transparency of the intervention (Renaud & Zimmermann, 2019).

The beneficence of deployed measures for the safety, security, and welfare of citizens could be weighed against potential costs and risks for individuals, e.g., the need to give up on privacy or certain privileges in emergency situations. Citizens could be motivated to accept potential constraints by highlighting desirable social norms and the contribution to the common good or by compensating downsides with suitable incentives.

Deployed measures could be rated in terms of justice such as whether certain societal groups are discriminated or excluded. This could be the case if certain benefits or functionalities are only available to people with certain devices or levels of expertise. For example, persuasive technologies should be designed to be usable and accessible for different age groups such as children and seniors.

It is reasonable to evaluate all deployed measures in terms of their scientific integrity and quality to ensure effectiveness and to identify possibly unanticipated or unintended consequences. An example is provided by the intervention aiming to reduce energy consumption through social comparison and finding that people below the average surprisingly increased consumption. This outcome could only be identified and mitigated through suitable evaluation (Schultz et al., 2007).

Finally, decision-makers bear considerable social responsibility. They should be aware of the potential short-term and long-term consequences of their actions considering the before-mentioned guidelines. For example, the use of high financial incentives might lead people to engage in activities against their will out of financial despair. This might not only negatively affect the citizens themselves, but may also compromise the long-term effectiveness of the measure, especially when the incentive is taken away at some point.

Yet, the decision to deploy certain measures is definitely not an easy one. Challenges remain, such as when to compromise autonomy and to limit free choice. Which situations justify constraining strategies such as laws or restrictions? In case of limited resources, who should benefit in which way? How can accessibility and non-discrimination be ensured given different age groups, technical equipment, and varying levels of expertise? Which measure should be deployed when scientific evidence is scarce or contradictory? An approach for targeting these important questions, that is already practiced in some cases (Digitalstadt Darmstadt, 2018), is the depletion of an interdisciplinary ethical committee that evaluates processes and deployed measures from various perspectives.

12.3.2 Privacy Aspects

With the ongoing digitalisation of cities and public spaces, the privacy of citizens is gaining in importance. Here, the General Data Protection Regulation (GDPR) is of relevance, which since 2016 is part of EU law on data protection. By addressing data transfer in particular, the aim of the GDPR is to give individuals control over their data and to provide uniform regulations for privacy in IT in the EU. The GDPR holds users accountable for the protection of their data by requesting informed consent for the use of their data (The European Data Protection Board, 2020). However, with increasing data collection in public spaces, e.g., by cameras and sensors, it quickly becomes impossible for citizens to keep track of the collection of their data. It is furthermore hardly possible to decline the collection of one's data in public spaces without restricting one's behaviour, e.g., by avoiding to visit certain public places.

Yet, cities sometimes might need to collect data of their citizens for safety or security purposes. This has been subject of heated discussions before, e.g., when the interior ministry in Germany tested facial recognition cameras at the Berlin-Südkreuz station in 2018 (Delcker, 2018), or recently in the course of the introduction of the German Corona-Warn-App (Burgess, 2020). German citizens tend to express concerns about governmental surveillance, but in an Orwellian manner they usually lack concern in the context of data collection by private organisations like Google or Facebook (Altmann et al., 2020). While this is often explained by the historical experiences of being spied on and prosecuted by the Gestapo and Stasi (Gerber et al., 2018), it remains unclear how governmental agencies should deal with this fear of their citizens of being under surveillance. Research is needed to address this issue and explore possibilities for communicating decisions about technical implementations regarding data collection and privacy protection to the general population. This is especially important in case the population is expected to collaborate in the data collection, e.g., by installing and using the Corona-Warn-App, by reporting information about safety or security related incidents, or by participating as prosumers in SG (Döbelt et al., 2015).

However, research on crisis apps suggests that users are willing to make privacy concessions to increase security (Tan et al., 2020b). A possible solution for supporting citizens in keeping track of and managing the collection of their data could be the provision of a digital privacy assistant (Colnago et al., 2020). This assistant could be implemented, e.g., as a mobile solution in the form of an app. Combining machine learning and manual input, the assistant would be able to semi-automatically identify the privacy preferences of its user and communicate and enforce them to data collecting devices in the public space.

12.3.3 Long-Term and Side Effects

Long-term and side effects of relying on technology to communicate with and motivate prosumers can be the exclusion of segments of the population. Therefore, different preferences of various socio-demographic groups should be considered. While age differences should not be overstated, certain aspects such as internet self-efficacy are often different for older or younger people (Chung et al., 2010). Research on warning apps indicates that while many preferences are universal, women are partly interested in different warning topics than men (Kaufhold, Haunschild, & Reuter, 2020). Similarly, women and men are likely to have deviating considerations for their energy use, leading to differences in peak energy consumption and thus requiring different motivations for desired behaviours. The implementation of motivational strategies and choice of technologies should consider such differences and ensure that no groups are excluded. Such differences may also be relevant for motivating various demographic groups, who may respond differently to motivation strategies.

Introducing technologies that connect prosumers and CIs introduces further vulnerabilities into the system. At the same time, citizens typically behave insecurely online (Zwilling et al., 2022), potentially leading to another weak link when it comes to securing critical infrastructures. Any technology used to integrate prosumers should thus consider the cybersecurity risk and CI resilience (Hollick & Katzenbeisser, 2019). Moreover, nudging may compromise users' autonomy if they are not aware of the subtle or covert influence (Hansen & Jespersen, 2013; MacKay & Robinson, 2016), or cause reactance on the users' side due to the perceived manipulation. This issue could be addressed by making nudges more transparent (Nys & Engelen, 2017; Thaler & Sunstein, 2008), e.g. by combining it with other approaches such as information provision. Initial research indicates that increasing the transparency of the nudge does not necessarily comprise its efficacy but may even be useful (Kroese et al., 2015; Zimmermann & Renaud, 2021). In addition, involving users should ensure that psychological needs are considered. For example, research indicates that even without additional extrinsic motivation such as praise, the mere recognition of the contribution goes a long way in keeping participants motivated (Ariely et al., 2008).

12.4   Visions for Citizen Participation

This section discusses the contribution of citizens to infrastructure resilience as an aspect of citizen participation. As such, it introduces bug bounty programs, gamification, and apps as strategies to foster citizen involvement and highlights the use of apps to engage citizens in different contexts.

12.4.1 Bug Bounty Programs

Bug Bounty Programs (BBP) can be understood as a form of crowdsourced penetration testing, where the identification of IT system weaknesses are financially rewarded (Ding et al., 2019). While using bug bounty rewards is cheaper than employing many cybersecurity experts for penetration testing, individuals who participate are mainly motivated by their contribution to public safety and the recognition of their community (Kranenbarg et al., 2018).

However, the (cyber-)security of any critical infrastructure should be preserved when involving individuals, since they are crucial for a functioning society (Hollick & Katzenbeisser, 2019). Hence, it is prohibited to let civilians interact directly with critical parts in such systems or to test them for crisis scenarios (Alcaraz & Zeadally, 2015). However, when looking at the field of critical information infrastructure (ICC), which are considered a subset of CI (Dehling et al., 2019), we see a shift towards the participation of individuals to secure these systems. For example, BBPs are used by large IT companies like Microsoft, Facebook, and Google. Such programs allow white-hat hackers to hack promoted systems and report vulnerabilities found in exchange for a (usually monetary) reward. This has been shown with great success, since large IT infrastructures are becoming too complex for teams to find vulnerabilities and defend systems against cyber attacks. Not only companies, but also agencies in the U.S. like the Department of Defense (DoD) (Newman, 2017) use this kind of civil participation to strengthen the security of their IT infrastructure. Since then, it has become a trend in the U.S. to use BBPs as a method of discovering vulnerabilities, even in critical infrastructure (Eversden, 2019). With the emergence of more information and communication technolgy (ICT) systems in critical infrastructures, these systems become inherently insecure due to insecure hardware and software components (Ferris & van Renssen, 2021; Onyeji et al., 2014). A bug bounty approach could potentially increase the security of such, thus, increase the reliability and resilience, due to the lack of aforementioned vulnerabilities. IT security experts (or even less skilled participants) would be able to directly interact with the CI, to report vulnerabilities, and eventually to confirm strong defensive measures in terms of IT security. Another benefit of this approach is the potentially increasing trust of civilians in these infrastructures, because of a direct participation possibility.

However, this approach requires a controlled involvement of individuals. Precautions must be taken to exclude criminal actors from interacting with security-relevant information and systems. In addition, when vulnerabilities are triggered, users interact with systems in unexpected ways, i.e., by crashing the system or escalating privileges. Hence, it is imperative to provide enough backup systems to seamlessly swap out compromised systems or even to separate active and inactive (e.g., backup) facilities and replicate a realistic workflow in the systems under investigation (e.g., by replicating old protocols), before enabling BBPs. Moreover, due to the possible installation of backdoor software, a thorough observation of system logs and user behaviour is necessary to mitigate both problems. IT forensics might be necessary to ensure a proper functioning of the system state after investigation before reinstalling the systems into the real-world. Otherwise, *testing systems* might never be allowed to enter as live-system, but may solely be used for security testing in this simulated environment.

Overall, BBPs have a high potential to increase the security of CI. Currently existing bug bounty formats need to be adapted to the high security needs of CI and additional security measures must be installed to be of use in this domain.

12.4.2 Gamification

Gamification refers to a concept that transfers designs, mechanics, and heuristics of games into a non-gaming context to enrich the user experience by invoking feelings like excitement and joy (Deterding et al., 2011). The goal is to leverage the users' intrinsic motivation for playing games in a pragmatic context that aims to fulfill specific goals (Hassenzahl, 2004). To achieve these goals, game-mechanics aim to increase the motivation of users to adjust their behaviour (Stieglitz et al., 2017). Literature shows that gamification has been successfully applied in various domains, like mobile education (Su & Cheng, 2015), redesigning of business processes (Korn & Schmidt, 2015), and cybersecurity (Fink et al., 2013). Furthermore, gamification has been successfully applied in the context of urban infrastructures. For instance, prominent examples aim at influencing the water consumption of citizens and aid water utilities in improving their strategies for system operation (Luca et al., 2015; Rizzoli et al., 2018). In the domain of energy grids, gamification is used to achieve reductions in peak demand and costs for infrastructure operators (AlSkaif et al., 2018; Gnauk et al., 2012; Johnson et al., 2017; Kashani & Ozturk, 2017).

In practise, commonly applied game mechanics are, among others, scoring systems, levels, and achievements (Zichermann & Cunningham, 2011). For instance, Gnauk et al. (2012) proposed a prototypical demand dispatch system for energy grids, which allows citizens to communicate available flexibilities (i.e., variable consumption periods for local appliances) to control authorities. Authorities can leverage these variable consumption times to optimise the operation of the grid by scheduling the consumption accordingly or negotiate changes. For decisions of the citizens that improve the overall demand dispatch operations they are rewarded with *Earth Saver Points*, which can be used to earn titles in the context of the application (e.g., Eco Hero) or gain small extrinsic rewards.

Despite the success of gamification in various fields, the effective application of the concept is challenging (Hamari et al., 2014). Several concepts and methods that have been discussed within this work need to be combined carefully. For instance, the representation of information for citizens must be carefully designed, to prevent overwhelming effects and discourage them (see Sect. 12.2.1). Furthermore, interaction interfaces need to be designed thoroughly to support citizens in conducting beneficial actions intuitively (see Sect. 12.2.2. Furthermore, aspects like the voluntariness of tasks, the nature of the system (e.g., pragmatic or hedonic), and the general citizen involvement and attitude can strongly influence the applicability of gamification (Hamari & Koivisto, 2013; Hamari et al., 2014).

Overall, gamification represents a promising technique to increase the involvement of citizens with actions to improve safety and security of smart city infrastructures. The possibilities for potential applications are numerous, but each concept needs to be carefully designed in order to be effective.

12.4.3 Crisis Communication and Apps

Due to the importance of time in communicating the state of the energy system and potential crises, typical tools for a day-to-day interface would be apps. For example, warning apps typically deal with the communication between emergency services and citizens, but information and communication technologies are also used for coordination among citizens or among agencies (Reuter & Kaufhold, 2018). Due to the dynamic nature of crises, mobile crisis applications are commonly used for (1) gathering data from the crowd, (2) organising collaboration during disasters, (3) spreading official information, (4) collecting and processing data for situational awareness, (5) allowing users to notify others during disasters (Tan et al., 2017). Apps are also already commonplace for owners of solar panels, to adapt their energy use to their own production. Technologies such as home energy management systems are being developed that help prosumers manage their energy reserves and consumption (Romero Herrera et al., 2017). Such tools that enable prosumers to make situation and device specific choices are particularly relevant as prosumers are more resistant to remotely versus personally controlled changes to device operations for balancing the grid (Michaels & Parag, 2016). Typically, such information would be offered on demand, as *pull* rather than *push* information, requiring prosumers to proactively search for the relevant information, e.g., by opening their app. In case of emergencies, *push* messages increase the likelihood that crisis information is noticed immediately. These systems are already used in warning apps. Typical warnings include information about the incident as well as instructions on how to behave (Fischer-Preßler et al., 2020). However, critical infrastructures do not typically have channels particularly for emergency communication. Instead, their communication with citizens relates mainly to public relations and corporate crisis management.

Borrowing from crisis informatics, modes of communication should be established that (1) communicate the state of the system so that prosumers can make informed decisions and (2) communicate urgently needed action or emergency concessions to increase prosumers' timely cooperation. One aspect to consider is whether to use general-purpose or built-for-purpose tools. General purpose tools are familiar and relevant in daily life (Tan et al., 2017), because they primarily fulfill a function not related to warnings, such as weather apps that also warn about extreme weather or the German Ministry of Health's use of the messenger apps WhatsApp and Telegram for COVID-19 related information and behavioural recommendations (BMG, 2021). Built-for-purpose tools, in contrast, are only used in emergencies, which are rare, but they are more adaptable to their specific functions. In warning apps, such as FEMA or KATWARN, this includes the option to determine a geographic area of interest or content areas, such as traffic, weather, or cyber crime (Kaufhold, Haunschild, & Reuter, 2020; Reuter, Kaufhold, Leopold, & Knipp, 2017). Regarding the use of specific apps for communicating with prosumers to increase resilience, we suggest that four main challenges exists. Firstly, despite the fact that a large proportion of citizens regard warning apps as important, they are seldom used. In 2019, only 16,5% of German citizens were using any warning app, although over 60% agreed that they were relevant and 65% demanded that all warnings should be centralised in one app (Kaufhold, Haunschild, & Reuter, 2020). Therefore, secondly, the relevance of a tool is dependent on the number and relevance of the organisations providing it. Hence, a warning tool that involves prosumers should be relevant to users' daily life and include several infrastructures. This could mean combining a general energy consumption and production app with a warning or motivational feature or building on channels that are already wide-spread, such as commonly used messaging apps. The third challenge relates to indications that citizens prefer tools that enable them to be active contributors, e.g., to help as witnesses and in the search for missing people, and that have elements of two-way communication (Kaufhold, Haunschild, & Reuter, 2020). Finally, apps for specific purposes have specific usability requirements. While many insights from crisis informatics can be applied to resilience communication, warning apps' usability requirements (Tan et al., 2020b) may not be identical with the requirements for motivating the co-production of resilience with prosumers. In order to be appealing, prosumers should be involved in studying the particular usability requirements for such tools, e.g. the personalisation and notification options.

Likewise, many nations are currently focusing on the use of tracing apps to combat the COVID-19 pandemic, thereby relying on citizens to collaborate in data collection and notification, i.e., tracing social contacts and notifying potential contacts of a positively tested person (Ahmed et al., 2020). Which features are relevant for the mass acceptance of such an app seems to depend on how citizens feel about the general concept of such an app. While critics seem to place particular emphasis on privacy and societal benefit, the undecided may be mainly convinced by the app's convenience, and advocates are likely to use the app regardless of its features (Trang et al., 2020). All motivating approaches introduced in Sect. 12.2 are suited for implementation in form of a mobile app. Further, these approaches could (1) increase citizen engagement to participate in apps they have already installed, such as contact tracing apps in the context of COVID-19, (2) motivate more citizens to install such apps in the first place, e.g., by educating citizens about the app's functionality and privacy aspects or by rewarding them for the app installation, and (3) serve as a framework for the development of new apps which aim to maximise users' motivation for collaboration. To increase the relevance of such an app, it may combine several aspects of citizens' contributions to the common good, which may require the coordination between several infrastructure providers or the coordination by administrative agencies. In this context, smart city initiatives may be a good place to start, as they are already implementing public digitisation, of which critical infrastructures could be one of several areas for involving citizens.

12.5 Conclusion

The energy sector represents a leading example for increasingly involving citizens to actively participate and contribute to resilience of the system as a whole. With the changing role of consumers towards prosumers—capable of consuming and producing electricity—the potential for participation is further increased. While this introduces numerous technical challenges, social aspects like awareness and motivation become increasingly important. For instance, studies have already investigated prosumers' conditions for ceding different levels of autonomy over their electricity consumption, combined with different compensation schemes. Furthermore, social and intrinsic motivations have been explored in the context of consumption reduction and sustainability. In this work, we showed that a variety of options are available for motivating prosumers in the energy sector. These motivational strategies can be enhanced by information and communication technologies that offer new ways of communicating with prosumers, of informing about the state of the system, about crises, or prosumers contributions. Infrastructure providers should thus explore recreating their communication strategies in manners that involve their users and prosumers. While the examples presented in this work mostly refer to the energy sector, we emphasise that presented concepts and strategies can be applied to other sectors as well.

User Acceptance and Effect of Preparedness Nudging for Warning Apps 13

Abstract

Warning apps are used by many to receive warnings about imminent disasters. However, their potential for increasing awareness about general hazards and for increasing preparedness is currently underused. With a mixed-method design that includes a representative survey of the German population, a design workshop and an app evaluation experiment, this study investigates users' preferences regarding non-acute preparedness alerts' inclusion in crisis apps and the effectiveness of nudging in this context. The experiment shows that while the social influence nudge had no significant effect compared to the control group without a nudging condition, the confrontational nudge increased the number of taken recommended preparedness measures. The evaluation indicates that the preparedness alerts increased users' knowledge and their motivation to use a warning app. This motivation is, in contrast, decreased when the messages are perceived as a disruption. While many oppose push notifications, favour finding persuasively designed preparedness advice in a separate menu or as an optional notification.

Original Publication Haunschild, J., Pauli, S., & Reuter, C. (2023). Preparedness Nudging for Warning Apps? A Mixed-Method Study Investigating Popularity and Effects of Preparedness Alerts in Warning Apps. *International Journal of Human-Computer Studies, 172.* https://doi.org/10.1016/j.ijhcs.2023.102995

Supplementary Information The online version contains supplementary material available at https://doi.org/10.1007/978-3-658-46489-9_13.

13.1   Introduction

Individual preparedness can be crucial to limit or even prevent the damage caused by emergencies. Nowadays, warning and emergency apps are available in many countries and offer a mobile warning system. They are able to reach many app users quickly, providing reliable and targeted information and multi-media content to people. Most emergency apps are geared towards the response phase, focusing on spreading information fast and wide and distributing concise localised warnings and recommendations (Tan et al., 2017, 2020c). For the preparedness phase, recommendations are typically only found in a menu that needs to be proactively sought out (Hauri et al., 2022). However, human judgement of the need to comply with response and preparedness recommendations is influenced by cognitive biases (Meyer, 2006), previous experiences (Diekman et al., 2007) and a country's risk culture (Appleby-Arnold et al., 2020; Cornia et al., 2016; Reuter et al., 2019). Even when citizens are warned and have the knowledge to take preventive actions, they thus routinely underestimate the risks, rely on agencies, or perceive that nothing much can be done (Cornia et al., 2016; Meyer, 2006; Paton, 2019), leading to milling and a lack of preparedness.

Therefore, notifications with general preparedness advice, in addition to acute warnings, could increase safety and emergency apps' utility (Tan et al., 2020c). However, notifications with general preparedness advice may also lead users to abandon the app, particularly those who use the app only to receive alerts, because they perceive that non-acute notifications misrepresent the emergency situation (Bonaretti & Fischer, 2021). Therefore, it is an open and important question whether notifications with preparedness advice increase the utility of warning apps. Persuasively designed preparedness advice might further motivate users to implement preparedness measures by engaging with psychological patterns, such as risk aversion or social norms (Mirsch et al., 2017; Oinas-Kukkonen & Harjumaa, 2009). Nudges, which are easy to integrate in a simple design, "alter people's behaviour in a predictable way without forbidding any options or significantly changing their economic incentives" (Thaler & Sunstein, 2008). While nudging has been applied to different contexts, including safe and secure behaviour (Hartwig & Reuter, 2021; Renaud & Zimmermann, 2019; Zetterholm et al., 2021), its usefulness in the context of warning apps and preparedness has hardly been studied so far. However, designing nudges that are accepted by (potential) users of warning apps is a challenge.

We therefore explore the research question: How should preparedness alerts be designed and integrated into warning apps to increase hazard preparedness? We explore this question with a mixed-methods design. First, we review related work on emergency preparedness and persuasive design (Sect. 13.2), after which we outline our mixed-method research approach (Sect. 13.3). We then follow the digital nudge design process (Mirsch et al., 2018) that consist of four steps: (1) context, (2) ideation and design, (3) implementation, (4) evaluation. In the first step, we discuss the study context in Sect. 13.4 through a representative survey in Germany (N = 1090), looking at user acceptance of nudging in the context of preparedness information and warning apps. This is followed by the second step ideation and design (Sect. 13.5), in which we investigate the design of persuasive preparedness alerts and nudges for emergency preparedness in warning apps, first, by testing different nudge types in the same representative survey (Step 2a), then, through collaborative design workshops (Step 2b, N = 5). Section 13.6 describes the implementation (Step 3) of three prototypes based on the findings: One serves as a baseline prototype that offers general preparedness advice without any nudges, whereas the other two prototypes include either a socially or confrontationally worded nudge for each hazard. Section 13.7 evaluates the nudges' effect on engagement with the prototype in a one-week experiment (N = 76), with the groups using the prototypes on their own smartphones (Step 4a). In addition, short questionnaires capture participants' impressions and behavioural intentions after engaging with the preparedness messages and nudges. As an alternative measure of nudge effectiveness, we captured the experiment participants' implemented preparedness measures in a post-study survey (N = 76). In addition, the post-study survey was used to evaluate the user acceptance of preparedness nudging and to qualitatively capture ideas for improving the nudges (Step 4b). We then discuss the findings, implications for design and the study's limitations (Sect. 13.8) and follow with a short conclusion (Sect. 13.9).

13.2 Related Work

Crisis informatics examines the "intersecting trajectories of social, technical and information perspectives during the full life cycle of a crisis: preparation, response, and recovery" (Hagar, 2013). As an interdisciplinary field, it is rooted in Human-Computer Interactions and employs insights from Psychology to explain and improve how humans interact with information and communication technologies. Interactions surrounding emergencies are particularly important and interesting aspects of that field, because users' safety is at stake and because such non-routine situations are characterised by uncertainty, urgency and a high mental load (Tan et al., 2020a). Therefore, in the following, we present Crisis Informatics findings on designing digital emergency prevention interventions and findings from Psychology and Human-Computer Interactions on cognitive biases and persuasive design, before pointing out existing research gaps.

13.2.1 Crisis Informatics for Emergency Preparedness

The terms disaster, emergency and crisis are commonly used interchangeably to describe sudden and usually unforeseen events that "call for immediate measures to minimize [their] adverse consequences" (UN Department of Humanitarian Affairs, 1992, p. 34). Although certain hazards, e.g., natural disasters, cannot be prevented, their resulting negative impact can be prevented or mitigated through adequate preparedness. Preparedness and prevention are thus related and cannot always be fully separated. Preparedness means "the knowledge and capacities […] to effectively anticipate, respond to, and recover from, the impacts of […] hazard events or conditions" (UN International Strategy for Disaster Reduction, 2009, p. 21). In the context of preparedness, crisis informatics investigates the use of ICT for public safety and security (Reuter, 2022) where they are used as early warning systems, which communicate alerts about non-imminent hazards or warnings about imminent hazards.

Legacy warning systems such as sirens and radio are now complemented by mobile warning components. These mainly include Cell Broadcast, which are text messages sent to all cellular devices by the cell towers in a warning region. Similarly, Location-Based Short Message Service (LB-SMS), can be used to send text-based alerts to mobile phones registered at a cell tower (Hauri et al., 2022). Since smartphones are widely used in many countries, can be customised based on personal preference and geolocation, can alert and store relevant information, and can present multi-media content, warning apps have great potential to support users before, during, and after emergencies. While some apps (such as news and social media apps) are used in day-to-day life and are adapted for emergency mitigation purposes during crises, others are built particularly for disaster purposes (Tan et al., 2017). Studies in several countries have shown that a major user demand exists for there to be only one national app that covers all important hazards (Dallo & Martí, 2021; Haunschild et al., 2022b), including disruptions of daily routines such as bomb disposals, school closures or large traffic accidents (Kaufhold, Haunschild, & Reuter, 2020). This indicates that users are unwilling to have an app that only provides warnings and another that focuses only on preparedness. This could explain the relative insignificance of emergency apps that offer other emergency features such as checklists without also offering warnings (Kaufhold, Haunschild, & Reuter, 2020). However, most apps discussed in research are dedicated to disaster response and only about a quarter to disaster preparedness (Tan et al., 2017). When it comes to relevant information and features of emergency apps, Dallo and Martí (2021) found that participants included behavioural recommendations as desired relevant information.

In Germany, 60% rate warning apps as useful in general and over 50% even support the idea that a government emergency app could be preinstalled on smartphones (Kaufhold, Haunschild, & Reuter, 2020). In addition, warning app usage has been increasing (Kaufhold, Haunschild, & Reuter, 2020), ranging between 16% and 33% in different European countries (Reuter et al., 2019). These previous findings indicate that warning apps are the main crisis tools used to increase safety and security (Kaufhold, Haunschild, & Reuter, 2020). They are used in many countries (Reuter et al., 2019) but rarely include extensive preparedness information (Hauri et al., 2022; Verrucci et al., 2016). In addition, due to the wish to integrate all relevant safety and security features into one app, preparedness advice should be easy to integrate into existing, widely-used apps, rather than introducing a new preparedness app. However, a preparedness feature should not deter users from downloading or continuing to use a warning app, as such apps have important safety and security functions for acute emergencies. Therefore, it is central to design a feature than can be easily integrated without decreasing the usability of warning apps. For emergency apps to be useful to users, they have to "provide users with faithful representations of an emergency [...] [in order] to respond" (Bonaretti & Fischer, 2021). This entails the three dimensions of transparent interaction (activation through alerting, saliency that allows judging the severity and type of emergency, and usability allowing easy interaction with the app), representational fidelity (being relevant, current, exact, complete, consistent, and trustworthy), and situational awareness (supporting prompt and actionable protective actions). One study has found that the perceived risk, perceived trust in the relevance and accuracy, as well as social expectations influence the intention to use a warning app and to comply with its recommendations (Fischer et al., 2019). This is, however, decreased by data security concerns, whereas the intention to use a warning app correlates with the intention to comply with its recommendations (Fischer et al., 2019).

Other usability requirements that may be important not only for the warning app, but also for the design of additional features are, phone resource usage, and minimal external links (Tan et al., 2020a). A study revealed that primarily app utility, app dependability, and output positively influence the intention to continue using a disaster app (Tan et al., 2020c). App dependability means that the app needs to be error-free and output means that critical information must be easy to grasp (Tan et al., 2020c). In contrast, user input and interface graphics have a negative influence (Tan et al., 2020c), which indicates that disaster apps should not require user input and should have a simple design. App utility means that "the more users perceive that the app delivers its intended function, the more likely the users will continue using the app. [...] [Users must] perceive that the app does not deviate from its main function" (Tan et al., 2020c, p. 7). This indicates that if the intended function of an emergency app is to increase safety and security, added preparedness advice could increase app utility and thus emergency app usage.

13.2.2 Emergency Preparedness and Nudging

A large body of research exists that provides recommendations and guidelines for designing effective warnings for imminent dangers, that include recipients' risk perception, social norms and experiences as important elements to spark a prompt reaction (Bean et al., 2016; Kim et al., 2019; Laughery & Wogalter, 2006; Schroeder et al., 2017). As risk perception is a core factor concerning people's reaction to a warning (Wachinger et al., 2013; Wood et al., 2018), a common strategy is to provide reliable information about a hazards' risks and consequences. However, risk perception and the perceived need to prepare for emergencies are prone to a host of biases (Paton, 2019). To name only a few, the risk compensation bias means that when people perceive that other actors, e.g. agencies, are taking precautions, they perceive their environment as less threatening and may not take necessary precautions (Etkin, 1999). This bias may be particularly prevalent in Germany, where a state-oriented risk culture prevails, in which citizens tend to rely on agencies (Cornia et al., 2016). Due to the unrealistic optimism bias, individuals tend to think that they are less likely to be affected by negative future events than others (Weinstein, 1980). In addition, people are attached to the status quo and tend to prefer immediate benefits, even if they incur costs in the future (Paton, 2019; Thaler & Sunstein, 2008). Due to these biases, information about risks does not always lead to preventive actions (Eiser et al., 2012)). Dual Process Theory is often used to explain systematic deviations from rational judgement, including the systematic underestimation and inertia regarding risks and their prevention (Kahneman, 2011). It stipulates that humans frequently use their automatic system (System 1), which, being fast, reflexive, and unconscious, needs fewer cognitive resources, but also relies on flawed heuristics derived from readily available information (Kahneman, 2011).

To engage with and confront unhelpful biases, persuasive technology design has been explored to support behaviour change and to serve as triggers in a digital environment (Fogg, 2002; Lockton et al., 2010; Oinas-Kukkonen & Harjumaa, 2009). Persuasive system principles are aimed at supporting different elements: users' primary tasks, the computer-human dialogue, the system's credibility, and social elements. Each of these categories consist of further sub-categories, such as reduction, tunneling, tailoring, personalisation, self-monitoring, simulation, and rehearsal for supporting primary tasks (Oinas-Kukkonen & Harjumaa, 2009). These support principles can be implemented through different features. For example, gamification typically includes elements of human-computer dialogue (rewards) and social support (competition and social comparison). However, this example shows that many persuasive design interventions require user input and data collection, significantly influence the functioning of an app and cannot easily be integrated into existing apps without changing their nature.

In contrast, as a simple design intervention, nudging uses human's predisposition to cognitive biases to support them in making better choices (Thaler & Sunstein, 2008). Nudges that engage reflective thinking (System 2 nudges) are perceived as more acceptable because they are regarded as more transparent and as respecting of free choice (Jung & Mellers, 2016; Reisch & Sunstein, 2016). Harrison and Patel (2020) order the amount of influence exerted by nudges in health care from framing information, to prompting an implementation intention, to enabling a choice to guiding a choice through defaults. While the latter should be used at the time that a decision is made, the former are to be employed in cases in which the decision cannot be influenced at the time of the decision (Harrison & Patel, 2020, p. 797). This is the case with many emergency preparedness measures, which often need to be implemented before the emergency. Many measures even need to be taken before an acute emergency warning, because they require time, infrastructure and a purchase to be implemented. Framing information and prompting an implementation intention are, therefore, the main nudge elements that can be leveraged in emergency preparedness nudging.

When applying reflective text-based nudges aimed at framing information, language and wording are important. Text-based nudges, often delivered via SMS, are popular, because they are low-cost interventions that can reach many people and that simultaneously serve as a reminder prompt. So far, research has investigated wording of text-message nudges primarily in finance and health (Avery et al., 2021; Dai et al., 2021; Page et al., 2020). A study of text-based vaccination reminders showed a significant effect of ownership prompts in the texts, whereas added video information did not have a significant effect (Dai et al., 2021). Interestingly, this effect observed in an experiment was unlike the self-assessment made by study participants in an online survey, supporting the value of testing intervention in the wild (Dai et al., 2021). A similar study of text-based interventions confirmed the findings and additionally found that interventions performed better when they were "congruent with the sort of communications patients expected to receive from their healthcare provider (i.e., not surprising, casual, or interactive)" (Milkman et al., 2021, p. 1). While this shows that framing and communication style are important, a study on flood preparedness found that the way that a descriptive nudge was worded and whether concrete percentages were given, did not influence the outcome (Mol et al., 2021). Similarly, a study investigating application for financial college support found that differently worded SMS-based nudges (including social comparison and commitment) did not have an effect (Page et al., 2022). However, since a similar text-based reminder had increased timely application by 3% (Page et al., 2020), the authors attribute this to the lack of a trusted relationship between SMS recipient and senders. This is in line with a study on SMS text-based emergency warnings which found that unfamiliarity with the sender lead participants to dismiss warnings (Kim et al., 2019). These insights suggest that when it comes to text-based nudges, communication must rely on a trusted relationship and should be congruent with styles expected from trusted agencies, such as emergency management agencies.

In a review on nudging in Human-Computer Interaction (HCI), Caraban et al. (2019) show that nudges are used as a facilitator, motivator, or trigger for behavioural change by using the mechanisms of facilitating, confronting, deceiving, socially influencing, creating fear, and reinforcing. Fields that commonly employ nudges are sustainability and health (Coskun et al., 2015; Orji & Moffatt, 2018), in which a large number of apps use digital nudging or persuasive technology design to educate about, remind of and help manage healthy and sustainable behaviours (Johnston et al., 2018; Nkwo et al., 2021; VanAhn et al., 2019). Concerning safety, recent research on COVID-19 apps discusses the use of nudges to increase behaviour that reduces the spread of the virus (Michalek & Schwarze, 2020; Zetterholm et al., 2021). Nudging has recently also been explored with regard to online safety and security by nudging towards better password creation (Hartwig & Reuter, 2021; Renaud & Zimmermann, 2019) and software updates (Frik et al., 2019). Mirbabaie et al. (2020) investigate Twitter posts that communicate emergency-related content with regard to which contain nudging elements and whether these lead users to share the post. Other studies have explored persuasive design and risk, using Virtual Reality to try to change users' attitude towards risks (Chittaro & Zangrando, 2010) or helping emergency management practitioners, e.g. in overcoming cognitive biases through a series of questions when responding to oil spills (Brooks et al., 2020).

Only one recent study investigates nudging in relation to preparedness (Mol et al., 2021). The researchers use an online laboratory experiment to investigate social norm nudges to increase flood preparedness among homeowners. They test two descriptive social norm nudges, in which they either state the exact statistic of other homeowners' investment decision or that 70% of previous participants had made an investment, before asking participants to decide about making an investment. The study finds no significant effects of the social norm nudge on flood preparedness. Similarly, they found no effect when eliciting participants' beliefs about the percentage of other participants that would invest (norm-focusing). However, both in the norm-focusing group and in the control group, the believe that many other participants would invest correlated positively with participants' own investment decision. Exploring other factors that predict positive flood preparedness, the authors found previous flood investments, personal norm (the degree to which one finds oneself morally obliged to perform an activity), immediate gratification bias (a preference for immediate benefits that incur higher later costs), response efficacy and expected effects of flooding (Mol et al., 2021). Developing nudges that target these aspects could thus be promising for increasing preparedness.

The only study looking at warning apps with a persuasive design lens is an analysis of app store user comments concerning widely-used warning apps in Germany, which identifies persuasive design elements which were praised or desired by users (Kotthaus et al., 2016). Studies looking at the effect of usability on the intention to comply with warnings in warning apps to some degree address aspects related to persuasiveness. Fischer-Preßler et al. (2021) suggest that due to the role of social influence for warning app use, social responsibility should be used to promote the adoption of warning apps. Perceived response efficacy, and to a lesser degree perceived severity and vulnerability, are also relevant for usage adoption and should thus be leveraged.

13.2.3 Research Gap

With their use of push notifications for acute warnings and advice about imminent hazards, warning apps have to some degree already been functioning as persuasive technology that aim to trigger the taking of precautions. However, these are so far only used to persuade citizens to react to imminent dangers. In contrast, persuasive design targeting the increase of preparedness has so far only been implemented in selected online games and is absent in warning apps (Verrucci et al., 2016). This is problematic since research shows that cultural norms in countries like Germany lead to preparedness measures being neglected (Cornia et al., 2016) cognitive biases influence risk assessment and lead to unsafe behaviour (Eiser et al., 2012). A study analysing web-based and mobile preparedness ICT found that "none of the resources analyzed has implemented any means to remind users about their need to remain prepared over time nor do any monitor progress towards enhanced preparedness. The lack of these dynamic interactive features reduces the chances of users returning to the websites or applications for more continuous and sustained learning" (Verrucci et al., 2016, p. 1598).

While an analysis of user reviews has shown users' interest in persuasive design elements in warning apps (Kotthaus et al., 2016), there is no research dedicated to persuasiveness in warning apps. Social norms, risk perception and social support have been suggested to increase compliance with recommendations of warning apps

(Fischer et al., 2019), but they have not been tested empirically. In addition, one study suggests that personal norms and a belief that others are implementing preparedness measures can increase flood preparedness (Mol et al., 2021). However, whether these elements could be used as nudges to enhance preparedness notifications has not been explored yet.

Another research gap concerns the user acceptance of preparedness notifications. While studies of user expectations and warning app usability requirements exist, they focus on warnings about acute emergencies (Appleby-Arnold et al., 2019; Dallo & Martí, 2021; Fischer-Preßler et al., 2020; Tan et al., 2020a). So far, whether preparedness notifications could be integrated into these apps to increase their usefulness or whether preparedness notifications may be perceived as distracting from the main function of alerting about acute emergencies has not been explored. If the nudges are perceived negatively, this could lead to reactance (Lukoff et al., 2022; Ward et al., 2021) and a decreased interest in warning apps, which could negatively affect public safety. While preparedness features could increase warning apps' utility and public safety, preparedness information is currently neglected and not proactively delivered within warning apps. We investigate whether nudges can convince users of the relevance of preparedness and which nudges effectively increase the taking of preventive measures.

13.3   Research Approach

We design a study that tests the effect of nudges on implemented preparedness measures. Building on Fogg (2009a)'s guidelines for designing persuasive technology, the Digital Nudge Design method (Mirsch et al., 2018) offers a four-phased process to guide the design. The phases are (1) defining the context, (2) ideation and design, (3) implementation, and (4) evaluation of the digital nudge.

When it comes to the context, we chose official multi-hazard warning apps as a technology channel (Fogg, 2009a) that users are familiar with. This is the case because warning apps are the mobile component of the German warning system (Hauri et al., 2022) and regarded as important channels by citizens in crisis situations (Kaufhold, Haunschild, & Reuter, 2020). Since users strongly prefer a single app for emergency-related functions (Kaufhold, Haunschild, & Reuter, 2020), the preparedness feature and nudges should be easy to integrate into warning apps and adhere to warning app design requirements, which include a simple design and elements, such as naming reliable sources that increase trust (Tan et al., 2020a).

According to Fogg (2009a), behavioural change requires users to be motivated and able to change their behaviour. In addition, they require a trigger that reminds them to act accordingly. Concerning the trigger function of warning apps, a warning needs to include a distinct, strong notification (e.g. sound or vibration pattern) that disrupts users' routine and draws attention to the imminent hazard (Tan et al., 2020a). When the hazard is non-imminent, an alert can be subtler and less attention-grabbing. Applied to smartphone alerts, we chose silent push-notifications, which present a preview of the information on the home screen, as a trigger.

To gain further insights into users' general motivation to prepare, their preferred modus of receiving preparedness advice and their acceptance of nudging in this context, we conducted a representative survey (N = 1090). In selecting the scenarios, we considered which hazards might be relevant to our user group at the time of our experimental study. We decided on the emergency categories cybersecurity, traffic safety, and fire protection, as all participants own digital devices, are road users, and live in an apartment or house. This ensures that all users have a basic motivation to implement the measures. For the preparedness advice, we reviewed behaviours recommended by official agencies and chose simple and relatively cheap measures, e.g., downloading an anti-software virus app, checking bicycle lights, or buying a fire extinguisher (see the electronic supplementary material for all recommended measures) to ensure that all respondents were able to perform the tasks.

The representative study also contributes to the ideation and design of the preparedness feature and nudges, as we use it to gain first insights into users' preferences for different types of nudges. We use nudges because they are persuasive design elements that can be easily integrated into warning apps while adhering to the warning app design requirement of simplicity, which allows users to interact with the app even under mental load in stressful situations. Based on findings from the survey, we conducted design workshops with potential warning app users (N = 5) until saturation had been reached concerning ideas for implementing the preparedness nudges. The workshops encompassed discussing design strategies that facilitate the processing of preparedness alerts, the effectiveness of the nudges, and suitable push notifications.

The results were used to implement a warning app prototype, similar to official warning apps, but with persuasive preparedness advice and preparedness alerts. This prototype was evaluated in an experimental study (N = 76), focusing on the effect of a confrontational nudge and a social influence nudge. The experiment compared user perceptions, behaviour intentions, implemented preparedness measures, and app interactions of a control group without any nudging condition with those of two experimental groups that interacted with nudges. Finally, to explore users' perceptions of the preparedness messages and the nudges, we conducted a post-study survey (N = 76). Figure 13.1 shows the study's mixed-method research.

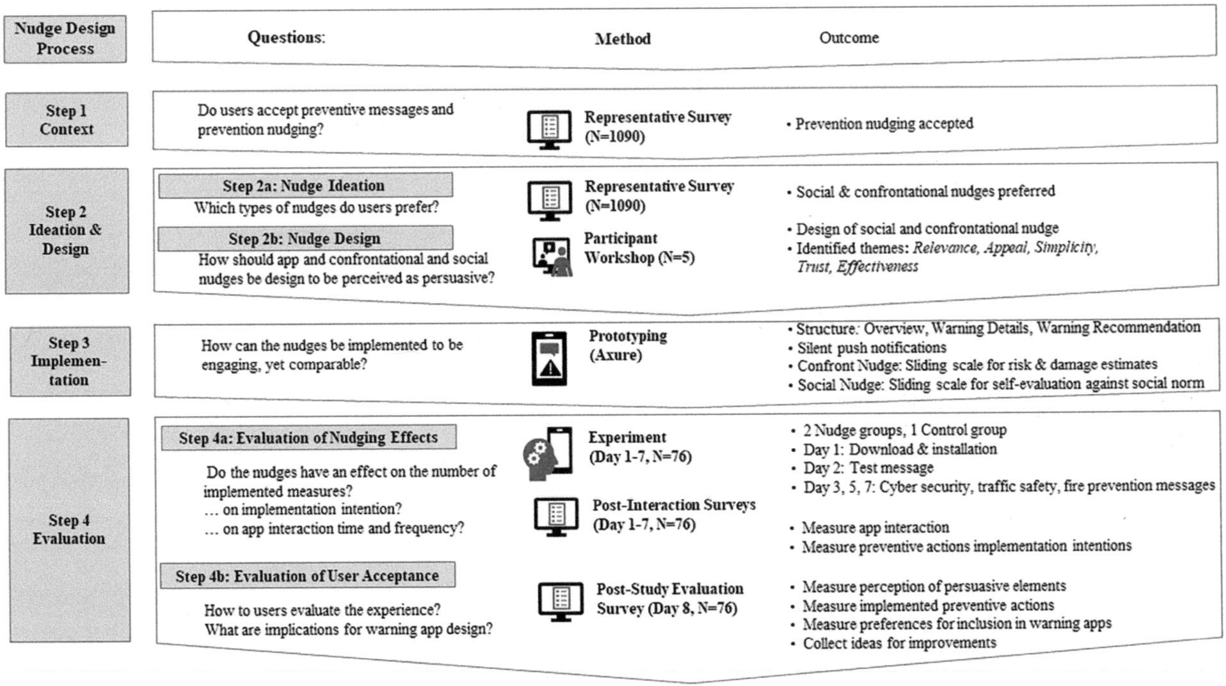

Nudge Design Process	Questions:	Method	Outcome
Step 1 Context	Do users accept preventive messages and prevention nudging?	Representative Survey (N=1090)	• Prevention nudging accepted
Step 2 Ideation & Design	**Step 2a: Nudge Ideation** Which types of nudges do users prefer?	Representative Survey (N=1090)	• Social & confrontational nudges preferred
	Step 2b: Nudge Design How should app and confrontational and social nudges be design to be perceived as persuasive?	Participant Workshop (N=5)	• Design of social and confrontational nudge • Identified themes: *Relevance, Appeal, Simplicity, Trust, Effectiveness*
Step 3 Implemen-tation	How can the nudges be implemented to be engaging, yet comparable?	Prototyping (Axure)	• Structure: Overview, Warning Details, Warning Recommendation • Silent push notifications • Confront Nudge: Sliding scale for risk & damage estimates • Social Nudge: Sliding scale for self-evaluation against social norm
Step 4 Evaluation	**Step 4a: Evaluation of Nudging Effects** Do the nudges have an effect on the number of implemented measures? … on implementation intention? … on app interaction time and frequency?	Experiment (Day 1-7, N=76)	• 2 Nudge groups, 1 Control group • Day 1: Download & installation • Day 2: Test message • Day 3, 5, 7: Cyber security, traffic safety, fire prevention messages
		Post-Interaction Surveys (Day 1-7, N=76)	• Measure app interaction • Measure preventive actions implementation intentions
	Step 4b: Evaluation of User Acceptance How to users evaluate the experience? What are implications for warning app design?	Post-Study Evaluation Survey (Day 8, N=76)	• Measure perception of persuasive elements • Measure implemented preventive actions • Measure preferences for inclusion in warning apps • Collect ideas for improvements

Fig. 13.1 Mixed-method study design following the Digital Nudge Design Process (Mirsch et al., 2018) and key outcomes. Own depiction.

13.4   Context Survey

Defining the context of the preparedness nudges, we chose the explore their integration into warning apps, as this is the ICT that is predominantly used to convey crisis information. To gain first insights into users' preferences and nudges in this context, we conducted a survey in October 2021, which is presented in the following.

13.4.1 Context Survey Design, Sample and Analysis

The commercial and ISO-certificated panel provider Gap Fish implemented the survey, ensuring a sample that is representative of the German population in age, gender, geography, and education. The questionnaire included two quality checks that had to be passed, which resulted in $N = 1090$ responses. The survey contained two questions about: whether one feels that precautionary measures can reduce harm (Q1), whether one feels sufficiently informed about precautionary measures (Q2) and four questions about the relevance (Q3) and way of presenting precautionary measures in warning apps (Q4–Q7). Furthermore, we asked whether participants feel that others (Q8) or they themselves (Q9) could benefit from being nudged towards taking precautionary measures. We also inquired whether they could be prepared to submit information that allows to personalise nudges (Q10) or preventive messages (Q11) (see the electronic supplementary material for the questionnaire).

As some items do not fulfill the criterion of normal distribution, we use the non-parametric Mann Whitney U test to investigate group differences between current users and non-users of warning apps, between men and women, and participants who are older or younger than 55 years. We use the Pearson correlation coefficient r to evaluate the effect size (Cohen, 1988) and correct it using Bonferroni-Holm, to avoid spurious results from multiple comparisons (Holm, 1979).

13.4.2 Survey Results

The representative survey of the German population indicates that people feel that information about preparedness measures should be integrated in a warning app $(M = 3.83, SD = .96)$ rather than conveyed through a different channel $(M = 3.32, SD = 1.02)$. Preventive measures should be event-related and sent along with acute warnings $(M = 3.93, SD = .93)$, rather than unprompted by an imminent hazard $(M = 3.24, SD = 1.18)$. Citizens are largely open towards nudging, with 52% regarding it as generally helpful (12% opposed, $M = 3.57, SD = 1.01$) and 47% stating that they would appreciate being nudged themselves (17% opposed, M = 3.41, SD = 1.11). More opposition (around 30%) is encountered when it comes to providing personal information to improve the preparedness messages, however around 35% of respondents were open to this (M = 3.11, SD = 1.25). Looking at group differences, most significant differences exist between current users and non-users of warning apps, while women and older people also judge some items differently (see 13.2). With $|r| \approx 0.1$ the effects are weak or very weak.

Independent Variables	Significant Differences (Corrected p-Value)	Survey Questions	M (SD)
		Q1: Perceived power of precautions	3.8 (1.0)
Usage: Warning App Users		Q2: Feeling informed	3.1 (1.1)
	$r = 0.11, p < 0.01**$	Q3: Relevance of precautions in apps	3.3 (1.2)
	$r = 0.11, p < 0.01**$	Q4: Precautions as push notifications	3.2 (1.2)
Gender: Women	$r = -0.11, p < 0.01**$	Q5: Precautions along with warnings	4.0 (0.9)
	$r = 0.12, p < 0.01**$	Q6: Precautions only as menu item	3.7 (1.0)
	$r = 0.12, p < 0.01**$	Q7: Other channels for precautions	3.3 (1.0)
		Q8: Nudging others	3.6 (1.0)
	$r = -0.13, p < 0.01**$	Q9: Nudging self	3.4 (1.1)
Age: Older people	$r = -0.15, p < 0.001***$	Q10: Personalization for precautions	3.1 (1.2)
	$r = -0.13, p < 0.01**$	Q11: Personalization for nudging	3.1 (1.2)

Fig. 13.2 Significant correlations, p value corrected with Bonferroni-Holm

When comparing the survey participants who reported to be currently using a warning app with those who were not, current app users agree more that precaution advice should be delivered along with acute warnings (M = 4.10; SD = 0.89, non-users: M = 3.87; SD = 0.93) or for them to appear only as a menu item (M = 3.93; SD = 0.88, non-users: M = 3.64; SD = 0.99). In addition, they are significantly more willing to provide information in order to improve the preparedness warnings (M = 3.36; SD = 1.28, non-users: M = 3.03; SD = 1.23) and the preparedness nudges (M = 3.31; SD = 1.28, non-users: M = 2.99; SD = 1.24). Supporting previous findings (Kaufhold, Haunschild, & Reuter, 2020), we find that socio-demographic factors impact few items and only marginally. Women are less willing to submit information for personalising nudges (M = 2.94; SD = 1.21, men: M = 3.22; SD = 1.29). Participants older than 55 years are similarly less willing to offer information for personalisation of precaution advice (M = 2.93; SD = 1.25, younger people: M = 3.29; SD = 1.21) or of nudges (M = 2.89; SD = 1.25, younger people: M = 3.24; SD = 1.24). They are also less of the opinion that precautionary information should only be available as a menu item that needs to be proactively sought out (M = 3.61; SD = 0.98, younger people: M = 3.83; SD = 0.96). With the correlation coefficient r close to 0.1, the differences are small (Cohen, 1988).

13.5   Ideation Survey and Workshops

To get first insights into which types of nudges are deemed effective and appropriate, we explored first ideas for nudges in the representative survey. The results of the survey lead to a pre-selection of promising nudges. Following this, we conducted participant workshops to explore the concrete design of the nudges and their integration into a preparedness feature of a warning app. The following section describes the survey and the workshops.

13.5.1 Step 2a: Representative Survey for Preparedness Nudge Ideation

The first ideas for nudges that could be effective as preparedness nudges, we considered psychological processes that are relevant for risk evaluation and combined them with insight on persuasive system design. This lead to a first set of differently framed nudges that were assessed by participants in the representative survey, which is presented in the following.

Ideation Survey Procedure, Sample and Analysis   To get first insights into which nudges are perceived as motivating and appropriate, we asked respondents in the same representative survey (see Sect. 13.4.1) to evaluate different nudges according to these two dimensions and asked participants about the reasoning for their choices. As we strive to identify accepted nudges that can easily be integrated into warning apps, we focus on text-based nudges combined with a simple interactive nudge (see Table 13.1).

We include different nudges from different categories based on Caraban et al.'s (2019) review of nudging in HCI, who identify the categories facilitate, confront, deceive, social influence, fear, and reinforce, which each include further subcategories. We design three diverse nudges that target fear, by "evok[ing] feelings of fear, loss and uncertainty to make the user pursue an activity" (Caraban et al., 2019, p. 8), in particular we use the mechanism reducing the distance, which is used when the beneficial outcomes are distant in time. We use two emotional nudges to do this: One shows a picture of a burning apartment, the other shows a quotation from someone who has lost their home due to a fire. A third fear-nudge is more informational and shows the number of people that die daily due to a fire. Within the confronting nudges, the mechanism reminding of the consequences fits well with the aim of increasing preparedness, as it targets the regret aversion bias. These nudges are similar to the fear nudges that use loss aversion. We survey one

Table 13.1 Participant evaluation of appropriateness and motivation of different nudges (sorted by preference (the order was randomized in the survey)

Nudge Characteristics	Nudge Wording	Appropriate (M) *(SD)*	Motivating (M) *(SD)*
Personal social norm, emotional	N1: When you buy a fire extinguisher, you protect your family and neighbors in case of fire. Responsible tenants and owners have a fire extinguisher.	3.74 (1.12)	3.66 (1.15)
Confront, information (response efficacy)	N2: A high-quality fire extinguisher costs less than €50. In 2019, a house fire caused an average building damage of 6639€ and a household damage of 2159€. (Source: GDV)	3.51 (1.20)	3.53 (1.24)
Fear, information (responsibility)	N3: In 2020, 20% of all fires that caused significant damage in buildings were caused by human error. (Source: IFS)	3.38 (1.15)	3.25 (1.18)
Fear, information (risk)	N4: In 2018, an average of one person died per day in Germany due to smoke, fire, or flames. (https://www.feuerwehrverband.de/presse/statistik)	3.31 (1.17)	3.31 (1.22)
Fear, graphic	N5: [Picture of a three-story house with balconies in flames]	3.10 (1.32)	3.20 (1.33)
Fear, quotation	N6: "You just don't realize that this has happened to you. That the house is gone, from one day to the next."	3.06 (1.27)	3.07 (1.28)

such nudge which confronts users with a possible regret resulting from not having implemented a simple measure to avert losses. Finally, we design a social influence nudge that uses "people's desire to conform and comply with what is believed to be expected from them" (Caraban et al., 2019, p. 7) (see Table 13.1). We do not further explore nudges that deceive because they are not transparent, less accepted (Reisch & Sunstein, 2016) and thus likely to diminish trust. In addition, facilitation nudges target the decision at the moment when it is made (Caraban et al., 2019, p. 4). This is difficult to implement when it comes to preparedness, because the natural decision-making moment is typically an imminent hazard warning—when it is too late to implement many measures. Similarly, reinforcement nudges do not lend themselves to text-based interventions. However, as warning apps are already making use of push-notifications as prompts, we will use these in the study, too, albeit in all groups, including the control group, to ensure that all participants are exposed to the preparedness alerts.

A study of secure online behaviour shows that warnings are more effective when they include a coping message that delivers advice about how to avoid a cyberattack, rather than only a threat appeal message that describes the negative consequences (van Bavel et al., 2019). We therefore couple each nudge with the recommendation to buy a fire extinguisher. The nudges were presented in randomised order. Answers were given on a 5-point Likert scale from fully agree to fully disagree.

Design Survey Results   The survey shows that the preferred nudges were the social-normative nudge (N1, 34% rated this as the best nudge) and the confrontational nudge (N2, 24% rated this as the best nudge). Both nudges were worded in a constructive manner, highlighting the positive outcomes of a precaution. They are followed in their ratings by two text-based fear nudges (N3–4) and the picture and quotation-based fear nudges (N5–N6).

Qualitative statements showed that the favoured nudges were perceived as neutral and non-paternalistic. Support for N1 was based on the perception that the statement was unemotional, did not incur negative emotions, but instead shifted the focus to social relations, which are perceived by many as "what is truly important in life". Participants liked that the message sounded caring and responsible, and it conveyed that they could take a simple measure to save lives. N2 was perceived as informative, factual, and convincing by showing a stark benefit in relation to the costs. Some also argued that money was the best way to convince others.

13.5.2 Step 2b: Workshops for Preparedness Nudge Design

While the survey showed some general preferences for the preparedness nudge framing, it was unclear show the nudges should be implemented as a preparedness feature in an app. Therefore, we conducted design workshops which are presented in the following.

Design Workshops Procedure, Sample and Analysis   We conducted individual remote online design workshops (N = 5, convenience sample of students with a background in Human-Computer Interaction and Media Communication, each lasting around 75 minutes) to design social and confrontational nudges that might increase users' motivation to comply with prevention recommendations. To introduce nudging and to provide a reference point for the different types of nudges, we used the Nudge Deck (Caraban et al., 2020), a design support tool consisting of cards which introduces the nudge categories that were also used to devise the nudges used for the survey (Caraban et al., 2019). In semi-structured interviews and with the collaborative online platform MIRO (miro.com), we iteratively designed a warning message for non-acute hazards and nudges based on participants' qualitative feedback (see the electronic supplementary material for the workshop guideline). After transcribing the audio of each workshop session, relevant text sections were iteratively coded using thematic analysis and building the coding scheme inductively (Braun & Clarke, 2006).

Due to the results of the representative survey, the social-normative and informative-confrontational nudges were chosen for further exploration regarding their concrete implementation. In addition to creating persuasive nudges, we also wanted the technology to be persuasive, to avoid a scenario in which users might stop using the app because of its lack of important persuasive design elements, for example with regard to primary task support, dialogue support or system credibility support (Oinas-Kukkonen & Harjumaa, 2009). Asking users about persuasive design elements for both the app and the nudges, helped us create an app that, on the level of the design, motivates both the treatment group and the control group to engage with it.

Design Workshops Results   The design of social and confrontational nudges, as well as persuasive warnings for warning apps, were therefore elaborated in design workshops with new participants who had not participated in the survey. From the workshops, 60 codes emerged, which could be grouped into the five themes (1) relevance, (2) simplicity, (3) appeal, (4) effectiveness, and (5) trust. Each theme includes, to varying degrees, aspects about general design proposals and specific details of the nudges. Table 13.2 shows users' design ideas for integrating nudges into warning apps. Participants suggested that nudges can increase reflection about the personal relevance of a hazard. On the one hand, photos or questions could activate the availability heuristic, bringing instances and reports of accidents to mind. On the other hand, personalised data could make clear what role the hazard plays in one's own geographical and social environment. Another approach was for users to enable neighbourhood collaboration and show positive examples from a relevant social group. The need for simplicity could be met by creating a separate sub-tab for the nudge. Nudging aspects that participants said would be appealing were positive framing, interesting information, and elements that make interaction more fun. Since nudges aim to increase the effectiveness of the information in terms of implementing recommended actions, this was a central aspect. Suggested strategies were gamification and rewards for implemented measures, reciprocity, and social comparison, making commitments for preparedness measures, and pointing out the consequences of the hazard. Finally, trust was an important theme. In addition to the app originating from a trusted agency, participants suggested that statistics could increase trust in the importance of the information and the measures. Furthermore, participants were concerned about data collection by a state-run warning app and about nudges creating excessive pressure.

Because we want to design a preparedness feature that can be implemented in warning apps, we need to consider warning apps' design requirements as meta-requirements. While some of the requirements are specific to acute warnings, such as timeliness, others are relevant for the emergency ICT as a whole, because they influence its perception or its usability. With regard to usability, we strive for a simple design that allows for saliency, meaning the ability to easily identify critical from less relevant information and to navigate the tool effectively. In addition, because we must ensure that warnings are being taken seriously, trustworthiness is a key meta-requirement (Bonaretti & Fischer, 2021). Therefore, we evaluate participant's comments in light of these requirements (see Table 13.2).

Table 13.2 Participant suggestions for persuasively designed nudges and warning apps and reasoning for implementation (✓) and non-implementation (✗)

Participant Suggestions	Design Implementation
Relevance	
Personal experiences: "I think everyone, at least as a kid, has been in this situation with their bike [lying on the road], everyone has certainly had a bike crash and that you show this again [through a photo]". "When it says, 'When was the last time you had a bike accident?'. Then I might think about it for a moment" (P4).	✓ Use examples of issues and incidents that users are familiar with -> selection of fire protection, traffic safety, cybersecurity
Personalized data: "[while seeing a map, you] think 'Yes, I live there in this environment'" (P2).	✓ Display of a map based on user location (similar to many warning apps, incl. NINA and KATWARN)
	✗ No other personalization due to limited agreement with personalization in pre-study
Social norms: "When the Müller family says, 'We need to call the fire department, and they should bring our family a new fire extinguisher'. But that [the app] does this for the whole neighborhood. So that the feeling of togetherness arises. That everything is organized together and then carried out together" (P5).	✗ In conflict with the requirement of *trust—concern about pressure* and *trust—no data collection*

(Continued)

Table 13.2 (Continued)

Participant Suggestions	Design Implementation
Relevance	
Relevant comparison group: "[The comparison should be] maybe not only with friends but also with the neighborhood and maybe all the people in the city who have this app" (P2).	✗ In conflict with the requirement of *trust—concern about pressure* and *trust—no data collection*
Appeal	
Positive framing: "I would rather say something like 'You are on a good track'" (P4), "Those people [who have taken preparedness measures] should be portrayed as neighborhood heroes" (P5). "I don't know whether showing a sad video of a paraplegic after a bike accident would cause defensiveness. I think that quickly becomes too much" (P1).	✓ Encouraging wording, geared towards gains rather than losses
	✓ Neutral rather than emotional
Additional appealing information: "a current statistic that includes bicycle and traffic accidents. […] Just interesting information that stays in your head a little bit" (P1), "include some funny icons that make it a little bit more appealing and not so dry. […] So that not everything looks like an official agency website, but that it is also fun" (P5).	✓ Interesting information
	✓ Appealing icons, modern design
Simplicity	
Use of tabs to give an overview: "I like the idea to divide the message [into] different sub-tabs for general information and countermeasures, links and so on" (P5).	✓ Different tabs are implemented to describe the risk and the preventive measures

(Continued)

Table 13.2 (Continued)

Participant Suggestions	Design Implementation
Trust	
Statistics: "So basically, you can convince me with good statistics of a lot of things" (P1).	✓ Inclusion of statistics
No data collection: "When you set up a point system like that, […] it sounds a bit like the system in China" (P2).	✓ Limited data collection: app uses only location (anonymized)
	✓ Clarify privacy or non-transmission of data in case of data input
Concern about pressure: "[Official agencies] can't force people to do something" (P5).	✓ Low pressure reflective nudges
Effectiveness	
Gamification rewards: "If I could reach some sort of level with points, that would give me a good feeling. Maybe then, I'd like to do something like that […] I would like to see a bar system or a score next to your status or level, but you also get an overview of the level of others" (P4).	✗ In conflict with the requirement of *trust—concern about pressure* and *trust—no data collection*
Social influence: "Reciprocity is always the highest" (P3), "'The average user has checked their bike within the last nine months'. You can do it with an indirect comparison" (P1).	✗ In conflict with the requirement of *trust—concern about pressure* and *trust—no data collection*
Commitment: "That you can set an appointment to be reminded to inspect your bike" (P1).	✗ Not feasible for the study design where triggers are controlled for
Pointing out consequences: "[…], statistics would actually be quite cool, or remind people of past accidents. So that you get the feeling you could change something if you inform yourself" (P2).	✓ Statistics of consequences included in reflective nudges

13.6 Implementation in App Prototypes

We consolidated the findings of the survey and design workshop into a web-based warning app prototype for Android called "PreWARN", which was built using Axure RP 10. By following warning app design guidelines (Tan et al., 2020a) and the design workshop outcomes, we designed a persuasive preparedness alert that serves as the baseline alert for the control group. A neutral blue colour scheme reflects the preventive, non-acute character of the warning. To give an overview upon opening the alert, we designed a window in the middle of the screen displaying all relevant information, including source, date, severity, and location (see Fig. 13.3, left). Below the window, two tabs with "Details" and "Behaviour Recommendations" indicate additional available information (see Fig. 13.3, right).

The recommendation section includes a checkbox list, links to external websites for further information about one of the recommended behaviours, and an embedded video of a state agency's YouTube channel with general preparedness information. We implemented two types of nudges as persuasive elements, a social influence and confrontational nudge (see the electronic supplementary material for all nudges and preparedness advice), which emerged from the survey as the preferred and appropriate ones. While participants in the design study also mentioned other persuasive elements, such as gamification, we decided on nudges because they are more in line with the usability criteria of warning apps, such as simplicity, minimalism, and minimal user input (Tan et al., 2020a). Therefore, they are less likely to conflict with warning app users' expectations and they have a better chance of being adapted in the official warning apps. This is important since users strongly wish for only one warning app (Haunschild et al., 2020). In addition to working with the space constraints in an app, the nudges should increase the motivation for all preparedness recommendations in each hazard category. The alerts in the nudging condition consisted of the improved persuasive design with an additional nudge. For both nudges, we asked users to provide estimates with a slider. To increase trust and elicit realistic estimates, we pointed out that the input of the slider was only collected anonymously.

For the implementation of the nudges, we designed a slider (see Fig. 13.4). Through interaction with which the slider, participants reflect about a hazard and are confronted with a gap between their estimate and the real consequences or preparedness. By then providing preparedness measures that are easy to implement, we seek to enhance self-efficacy and preparedness-efficacy, i.e., a sense that the participants and the measures can increase their safety and the safety of the people who are close to them.

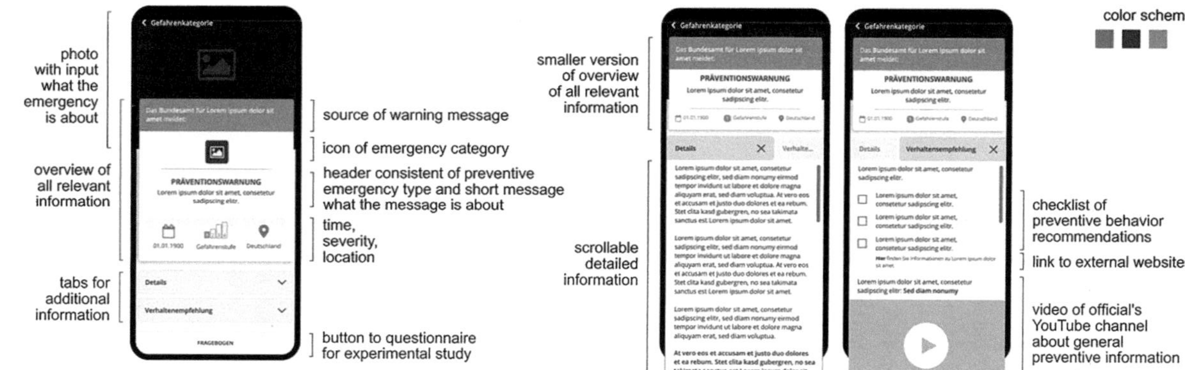

Fig. 13.3 Schematic design of the warning (left: opening screen, right: tabs "Details" and "Prevention Recommendations")

For the social influence nudge, we lead users to compare themselves with a social norm (see the electronic supplementary material for the nudges). Participants assessed their own safety regarding the specific hazard from low to high on the slider, with low ratings resulting in a red and high ratings in a green bar. The classification was calculated conservatively to convey that more could still be done to be fully prepared. The result was followed by a statement about a social norm and social consequences, along the lines of: "People who behave in a particularly safe way not only protect themselves but also others". This aims to trigger a personal social norm, meaning a feeling of responsibility for the people who are close to the users. We designed the confrontational nudge in a similar way but with a wording that reminds of the negative consequences of poor prevention by asking participants to quantitatively estimate a hazard's damage or the role that personal activities play for the negative outcome with the slider. The users' answer was shown as a red bar when they underestimated the danger (see Fig. 13.4, right).

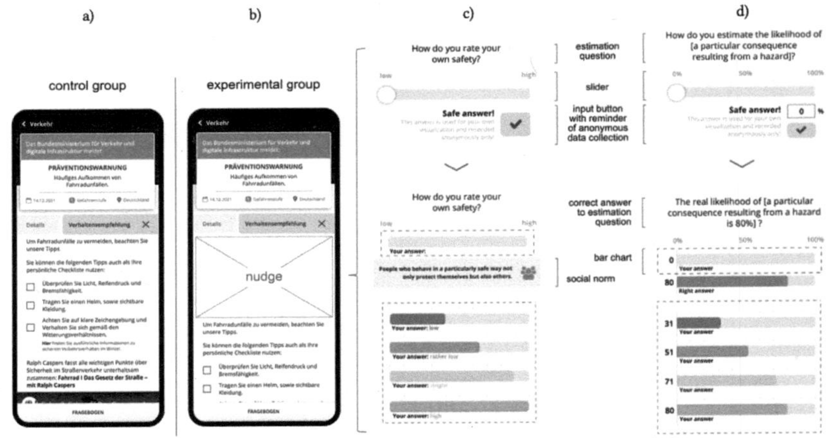

Fig. 13.4 Differences of the (a) control group without a nudge and (b) experimental group, including a schematic visualisation of the (c) social influence nudges and (d) confrontational nudges

13.7   Experiment and Post-Study Survey

In the final step of the nudge design process, we evaluated the effectiveness of the nudges regarding the reported intention to implement preparedness measures and the reported number of implemented preparedness measures after the study. In addition, we analysed the interactions with the app, participants' reactions after receiving the alters and their attitudes after the one week of receiving recommendations to assess user acceptance of the nudges and the preparedness feature.

13.7.1 Step 4a: Evaluating Preparedness Nudge Effectiveness: A One-Week Experiment Interacting with a Preparedness App

To compare whether adding a social or a confrontational nudge to the persuasively designed preparedness alerts increases users' preparedness, we conducted an experiment with another new sample of participants. In the experiment, a control group and two nudge groups interacted with three preparedness alerts in the prototype on their own smartphones in the course of several days.

Experiment Procedure, Sample and Analysis   Participants were randomly assigned to a control group (CG) without a nudging condition, or to one of two experimental groups, which get a warning with a confrontational nudge (EG Confront) or a warning with a social influence nudge (EG Social). The Chi-square χ^2 tests confirmed that the randomised assignment of participants was successful, which ensures that potential group differences stem from the experimental setting instead of socio-demographic factors or personal experiences. The experiment was conducted between 10 to 17 December 2021. On day 1, participants received instructions for the installation of the prototype, followed by a welcome and silent test push notification on day 2. On day 3, they received the first preparedness notification concerning the topic of cyber security. Two days later, preparedness advice on traffic safety was sent, which was followed by a notification on fire prevention on day 7 (see the electronic supplementary material for the warnings' and push notifications' content and for the timeline). On day 8, participants received instructions for deinstalling the prototype and a post-study questionnaire.

Because it is difficult to measure whether the nudges had an effect on the outcome, we used several approaches. As a measure of general interest in the issue, we tracked participants' interactions with the app, operationalised as number of interactions with the warnings and time spent interacting with the warning. These measures were tracked using Matomo (matomo.org). To capture participants' impressions of the interactions directly after the warning, we administered short post-interaction surveys after participants had interacted with each preparedness alert. These included a measure of whether the warning had been perceived as helpful, whether participants planned to implement the measures and two questions controlling for any technical issues with the prototype. Another two questions controlled for whether the hazards chosen were relevant to the users and whether they had any own previous experience with them (see the electronic supplementary material for the post-interaction questionnaire).

Due to the large number of required participants, we used convenience sampling for the study, with participants mainly originating from the student body of the universities involved. In addition, we used mailing lists and Facebook and Twitter to circulate the study invitation. Participants who did not give answers about the dependent variables had to be excluded from the analyses. The analyses are thus based on data from N=76 participants, 35 females (46.1%) and 40 males (52.6%). The participants' age ranged from 18 to 71 years (M = 29.71, SD = 12.10). 31 participants have a high school diploma or lower (41%), while 44 hold a university degree (58%). Around a third of the participants has experience using a warning app, with 22 participants currently using one (28.9%) and seven participants having used one in the past (9.2%). Another third (35.5%, 27 people) indicated that they did not use a warning app but planned to do so.

We analysed differences between the nudging conditions (between-subject factor). Even though we could not assume normality of the model's residuals, we used a one-way ANOVA as it is relatively robust to minor violations of the normality assumption (Schminder et al., 2010). We therefore generated a QQ plot, which confirmed that deviations were only minor. When the prerequisites of normal distribution of residuals, homogeneity of the error variances or covariances were violated, we computed a robust mixed ANOVA, which is robust against violations by using location and dispersion measures (Mair & Wilcox, 2020) but can be interpreted analogously. As studies suggest that men and women react differently to loss and gain framing (Chittaro & Buttussi, 2016), we analyse the influence of gender, as well as of other socio-demographic factors.

Experiment Results   No Nudging Effect on Preparedness Behaviour Intention but on Reported Behaviours: Participants who were shown the confrontational nudge underestimated the likelihood of the occurrence and consequences across all emergency categories according to their slider input. Similarly, participants with the social influence nudge condition estimated their safety to be quite high across all emergency categories but somewhat lower for that of cybersecurity. Participants rated all emergency categories as rather relevant (cybersecurity M = 4.09, SD = .98; traffic safety M = 4.39, SD = .59; fire protection M = 3.69, SD = 1.01).

To examine the effect of different nudging conditions on the intention to take preventive measures across different emergency situations (see Table 13.3), we calculated a robust mixed ANOVA (N = 59, n = 31 CG, n = 13 EG confront, n = 15 EG social). There was neither a statistically significant interaction between the nudging conditions and the emergency categories (F(4, 15.72) = 2.37, p = .096) nor a statistically significant main effect of the between-subject factor (F(2, 15.44) = 1.26, p = .310). This means that the different nudging conditions have no effect on the compliance intention across the emergency categories.

Table 13.3 Mean values and standard deviation for compliance intention for each hazard type (7-point Likert scale)

Experimental condition	N	Cybersecurity M *(SD)*	Traffic safety M *(SD)*	Fire protection M *(SD)*
CG no nudge	31	4.32 (1.30)	5.03 (1.33)	3.60 (1.11)
EG confront	13	5.02 (1.32)	5.42 (0.94)	3.99 (1.01)
EG social	15	4.16 (1.39)	5.06 (0.87)	4.57 (1.16)

A one-way ANOVA was conducted to determine the effects of different nudging conditions on the number of taken preparedness measures, as reported on the final day of the experiment, five (cyber), three (traffic), and one day (fire) after the alerts (N = 72, n = 39 CG, n = 15 EG confront, n = 18 EG social). That analysis showed a statistically significant effect with a medium effect size of the nudge condition on reported preparedness measures (F(2, 69) = 3.26, p < .05, η^2 = .08). Tukey post-hoc analysis revealed a significant difference (p < .05) in post-reported preparedness behaviour between the control group (M = 1.51, SD = 1.47) and the experimental group with confrontational nudges (M = 2.73, SD = 1.22) but no difference compared to the social influence nudges (M = 1.56, SD = 1.58). Tukey analysis was calculated with a 95% confidence interval of the difference between means from 0.03 to 2.41 points on a 0 to 12 scale. We can conclude that participants with a confrontational nudge reported significantly more implemented recommendations than participants without a nudge (see Fig. 13.5).

In a multiple regression, we tested exploratively whether gender, age, experiences with the warning app, experiences with the emergency category, and relevance of the emergency category predict post-study reported implemented measures, controlling for our experimental conditions no nudge and confrontational nudge. The analysis showed no statistical significance (F(9, 44) = 1.25, p = .289), which indicates that the effect is not influenced by any of these variables.

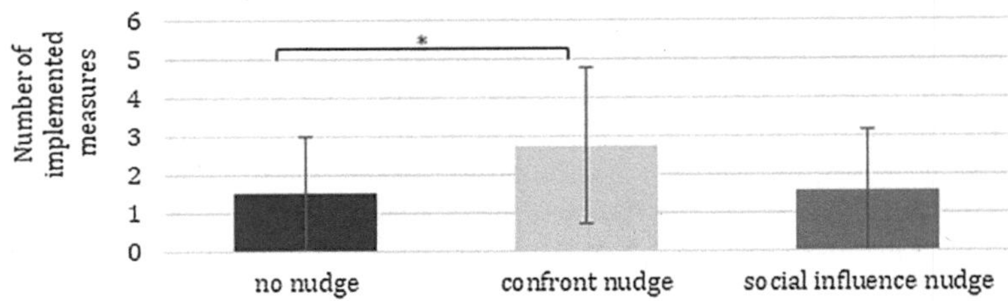

Fig. 13.5 Number of reported implemented preparedness measures after one week of warning app use (M, SD, * significant group differences)

No Nudging Effect on Time Spent and Interactions with the Prototype App: We calculated a robust mixed ANOVA to examine the effect of different nudging conditions on the time spent with the preparedness alerts for the different hazards (N = 76, n = 43 CG, n = 15 EG Confront, n = 18 EG Social, see Table 13.4). This revealed no statistically significant effects, neither for the within-between interaction (F(4, 14.18) = .56, p = .689) nor for the main effect of the between factor (F(2, 14.05) = .57, p = .578) or the within factor (F(2, 10.91) = 3.36, p = .072).

Table 13.4 Mean values and standard deviation of time spent (in minutes) with each warning message

Experimental condition	N	Cybersecurity M *(SD)*	Traffic safety M *(SD)*	Fire protection M *(SD)*
CG no nudge	43	2.23 (1.78)	2.82 (3.58)	2.58 (4.21)
EG confront	15	4.84 (6.13)	5.21 (6.69)	3.24 (3.65)
EG social	18	2.78 (2.28)	3.63 (7.30)	2.65 (4.37)

In the same manner, we examined the effect of different nudging conditions on interactions with each preparedness alert (see Table 13.5). Any clicks on links or tabs are counted as an interaction. The robust mixed ANOVA showed no statistically significant interaction effect (F(4, 14.30) = .81, p = .536) or main effect of the between factor (F(2, 14.22) = .49, p = .691), which indicates that nudging did not influence the interaction with the app.

Table 13.5 Mean values and standard deviation of the number of interactions within the warning messages

Experimental condition	N	Cybersecurity M *(SD)*	Traffic safety M *(SD)*	Fire protection M *(SD)*
CG no nudge	43	5.40 (2.83)	3.53 (2.42)	3.23 (2.34)
EG confront	15	5.93 (3.12)	5.27 (3.24)	4.13 (2.29)
EG social	18	4.61 (2.22)	3.94 (2.10)	4.17 (3.14)

13.7.2 Step 4b: Post-Study Survey Results Investigating User Perceptions

In the fourth phase of the nudge design process, the nudges are evaluation regarding the desired outcome. We evaluate a) the nudges' effectiveness measures by the intention to implement preparedness measures and number of self-reported implemented measures and b) users' acceptance of the nudges and the way they were implemented.

Post Study Survey Procedure Sample and Analysis Intentions often do not predict behaviour (although this is more likely for a single action such as a concrete preparedness measure than a for goal, which requires multiple actions) (Sheeran, 2002). Therefore, we also administered a survey one day after the last preparedness alert, which included questions about which of the recommended behaviours had been implemented concerning each hazard (see the electronic supplementary material for the post-study questionnaire and all percentages). In addition to the planned implementation of preparedness measures, this provided us with a measure of realised activities. The post-study questionnaire also served to reveal participants' perceptions of the preparedness feature and the nudges. Therefore, it also enquired about the feature's usability and utility, participants' evaluations of their own security behaviour and a reflection about the effects of receiving preparedness alerts. Besides exploring the overall utility of the preparedness warnings, we also asked concrete questions concerning external links and push notifications, as these features, while enhancing the warnings' effectiveness, contradict warning app usability recommendations (Bonaretti & Fischer, 2021; Tan et al., 2020a). The questionnaires were administered online with SoSci Survey. Data collecting services were self-hosted, and data was stored only on university servers.

The post-study survey was conducted with the experiment participants. For the analysis of user acceptance and interaction experience, we included only those participants who had interacted with all three alters and answered the post-study survey

(N = 76). Since these answers are not dependent on the experimental condition, we included all participant comments made relating to open feedback for improving the nudges and the preparedness feature.

Answers were typically on a 5-point Likert scale from "fully disagree" to "fully agree" and for behaviour intentions on a 7-point Likert scale. First, we analysed whether preventive warnings increased warning apps' utility. We use Kendall's tau-b (τb) as a non-parametric measure of correlations, because the Shapiro-Wilk test and the histograms show that the data are not normally distributed. Secondly, we qualitatively investigated participants' suggestions for improving the preparedness feature and the nudges. The open questions were analysed qualitatively with thematic coding along deductively derived categories (Braun & Clarke, 2006), using MAXQDA. The categories were deduced from the Effective Use Theory for Emergency Warning Systems (Burton-Jones & Grange, 2013; Fischer-Preßler et al., 2020), allowing comparison of the requirements for preparedness alert with those for emergency warnings.

Post-Study Survey Results   Interesting New Information Increases Warning App Usage Motivation: The results show that respondents largely appreciated the preparedness advice, but did not welcome the use of silent push-notifications. This is reflected by the findings that forty-four percent of the study participants would be somewhat more motivated to use a warning app if it included preventive information, while a similar number of people (36%) disagreed (M = 3.00, SD = 1.12). This motivation significantly correlates with having learned something new through the warnings (τb = .29; p < .01), rethinking one's behaviour (τb = .49; p < .01), and reporting to have implemented more measures than without the app (τb = .27, p < .01). However, having felt disturbed (15%, M = 2.17, SD = 1.14) is associated with being less motivated to use a warning app (τb = −.34, p < .01). Interestingly, having felt pushed towards a behaviour (11%, M = 2.04, SD = 1.03) has no significant effect, even though the two correlate strongly (τb = .42, p < .01). A third of respondents also felt motivated by the experiment to use an app dedicated to preparedness advice, whereas 52% disagreed (M = 2.65, SD = 1.21). This motivation does not correlate with any of the variables, except for the motivation to use a warning app (τb = .20, p < .05). Because warning app users might be more motivated to increase their safety, we tested whether current users of warning apps differed with regard to their preferences. Surprisingly, the only significant effect that we identified is that they agree less that preparedness information should be sent in a similar fashion as regular warnings in warning apps (τb = −.24, p < .05). However, since 77% of the people who are currently using a warning app in our sample were male, this may also be due to gender differences, as women are more open to the use of push notifications ($\tau b = .24$, $p < .05$) and to the integration of preparedness information into warning apps ($\tau b = .21$, $p < .05$).

Sixty percent of participants stated that they had learned something new and the majority (61%, M = 3.48, SD = .97) agreed that they should do more for their safety and security and had started to rethink their behaviour. Over a quarter (27%) claimed that due to the app, they had done more for their safety or security. Analysing the group differences by using eta for a nominal and metric variable, we find that hardly any variation in this question can be attributed to the experimental groups, meaning that the nudging warnings were perceived similarly as the plain warnings. The nudging conditions had no significant effect on these evaluations. Despite these positive evaluations of the preparedness information, the results confirm that integrating preparedness prompts into warning apps is a challenge: While 87% of the respondents would welcome the preparedness information in a menu item in a warning app, a majority (57%) disapproved of the use of silent push notification for the warnings. However, 35% were open to these preparedness prompts in warning apps. When specifying that these push notifications would have fewer permissions than acute warning alerts (e.g., a less forceful tone), agreement rose to 64% (29% opposed, M = 3.47, SD = 1.26) (Table 13.6).

Table 13.6 Reflection of security behavior and app experience after the study (5-point Likert scale), CG = control group, EG1= experimental group with confrontataional nudge, EG2 = experimental group with social nudge, Rethought = Rethought behavior, Increased = increased behavior, Learned= learned something new, Disturbed = felt disturbed, WA = Motivation to use a warning app, Prev. app = Motivation to use a crisis prevention app

Group	N	Reflection of Security Behavior and App Experience, M *(SD)*					
		Rethought	Increased	Learned	Disturbed	WA use	Prev. app
CG control	39	3.28 (1.15)	2.33 (1.25)	3.31 (1.13)	2.33 (1.20)	2.85 (1.16)	2.54 (1.27)
EG1 confront	14	3.43 (1.23)	2.86 (1.17)	3.71 (1.20)	2.14 (1.03)	3.20 (1.15)	3.01 (1.16)
EG2 social	17	3.71 (0.70)	2.82 (0.88)	3.35 (1.06)	1.82 (0.88)	3.12 (0.99)	2.53 (1.07)

Since ethical nudges require including information to target reflection, and since increasing trust requires adding sources to this information, we included external links with further information, which should usually be minimised in warning apps (Tan et al., 2020a). A great majority (80%) found the extent of external links acceptable (M = 3.00, SD = .49), which supports the initial argument that preparedness alerts have different usability requirements than acute warnings and thus merit separate design interventions.

Participants' Suggestions for Improvement: To look at where the requirements are similar or differ and at how preparedness alerts and nudging could be designed to increase users' personal motivation to implement the recommendations, we asked participants to describe how the messages might be improved to be more motivating. Participants mentioned some aspects similar to those relevant for warning systems (see Table 13.7). Most comments concerned ideas for improving the ability to take

Table 13.7 Most frequently mentioned topics (% of all respondents) for increasing the motivation of preparedness notifications and nudges

Situational Awareness	Promptness (immediate actions)	Simplicity of proposed measures/increasing difficulty (11%)
	Actionability (coping countermeasures)	More tips and measures for preventing hazards (12%)
		(Links to) recommended programs, links to relevant settings (for cyber fraud) (6%)
		Providing checklists (10%)
Transparent Interaction	Activation (alert)	No notification, but included in a separate menu item (22%)
		(Optional) push notifications (5%)
		Reminder (in combination with checklist) (5%)
	Saliency (severity and type)	Brief information (11%)
		Examples and indicators (for cyber fraud) (8%)
	Usability (easy interaction)	(Not) including graphics, images, or videos (8%)
		Should not require a separate app (6%)
Representation Fidelity	Relevance (matters to user)	Preventive warnings irrelevant or not requiring an app (10%)
		Different channel (news, local information, web app, social media) (7%)
		Personalization (personally relevant topics, locations) (6%)
		More informative content (5%)
	Trust (reliability)	Providing statistics (8%)
	Exactitude (correct, precise)	*Not mentioned*
	Consistency (with other channels)	*Not mentioned*
	Currency (up-to-date)	*Not mentioned*

prompt and actionable measures, e.g., by suggesting simple measures and measures increasing in difficulty, or by providing checklists. Another dimension concerns transparent interaction, which includes adequate alerts, which for many participants means no or only optional push notifications.

Many also mentioned that the message should be brief and allow to judge the severity. Another main aspect concerned representational fidelity, which indicates that those users were unconvinced that the hazards represented a cause for warnings. Some said they generally were not interested in warnings and in using a warning app. Others found that different channels were more adequate. Suggestions for increasing the relevance of preventive alerts included personalisation and generally more informative information. Statistics were mentioned both to prove the relevance of a hazard, as well as to increase trust in the suggested measures.

13.8 Discussion

In the following, we discuss the results with regard to user preferences and the effectiveness of nudging towards hazard preparedness and with regard to user acceptance of nudging and preparedness feature in warning apps. We further discuss implications for design, the limitations of the study and future work.

13.8.1 Informing the Design of Persuasive Preparedness Nudges

We assessed two different nudge framings that both used a slider to engage users and made them reflect on their own preparedness. The experiment shows that the confrontational nudges, which confront users with the negative consequences of a hazards and their contribution to it (and thus, in reverse, also their own role in decreasing the chances of a negative outcome), had a small significant effect on increasing preparedness. This is in line with another experimental study that investigated nudging to increase flood preparedness in homeowners (Mol et al., 2021). This study found that nudges would be effective, if they influence factors that are related to positive investments. These include the absence of bias that favours immediate benefits that incur higher later costs, response efficacy and expected effects of a hazard. Due to the limited number of experimental groups that we could explore with our sample size, we were unable to further differentiate these effects and nudges in this study, which should be done in future studies.

The social nudge that made users reflect on their own preparedness and its consequences for family, friends and neighbours, and then triggered a personal norm, i. e., a moral obligation to those close to the user, did not have a significant effect. While other social nudges were also ineffective in the previous flood preparedness study (Mol et al., 2021), that study found personal norms to be relevant for preparedness. We therefore sought to trigger such a personal norm in the way we designed our social nudge, by appealing to the general responsibility that one has for one's social environment in the message that followed the slider interaction. However, it is possible that because it was not interactive, that personal norm message had less weight for the users than the slider interaction, in which users compared their preparedness to that of others. While previous studies suggest that under specific circumstances, men and women are differently affected by loss- and gain-framed nudges (Chittaro & Buttussi, 2016), we did not identify any gender differences with regard to the effectiveness of the nudges.

In addition to the effectiveness of the nudge, another relevant aspect is user acceptance. With regard to the social nudge, we find a gap between what participants suggested would be a motivating nudge and a lack of effect of the social nudge. Such a perception-effectiveness gap has also been identified in previous studies (Dai et al., 2021) and might thus support the need for pilot-testing nudges instead of relying on user prediction of their effectiveness. The confrontational nudge, which included a positive wording by emphasising the low costs of an easy measures, as compared to the large costs of a house fire, was also perceived as highly appropriate and motivating in the survey. Particularly the emotional nudges, which presented a personal quote of a victim and a picture of a burning building, were perceived as less motivating and acceptable. This is in line with previous research that shows reflective nudges to be more acceptable than less transparent ones (Reisch & Sunstein, 2016).

The representative survey and the design workshops generated first insights for how to design persuasive preparedness messages. These are in some regards similar, in others different from the persuasive elements that are proposed by Kotthaus et al. (2016) in the only other study of persuasive elements for warning apps. That study used online reviews by warning app users to deduce relevant persuasive elements for warning apps, based on the ones suggested by Oinas-Kukkonen and Harjumaa (2009). They identified the persuasive strategies reduction, tailoring, personalisation, liking and trustworthiness. The evaluation also suggested that rather than reducing content, users wanted more interesting information, likely because they already had basic knowledge about the general hazards. Instead of reduction, our design workshops rather pointed towards simplicity for persuasive preparedness warnings. In the implementation, simplicity was achieved by implementing different tabs that offer details and behavioural recommendations, and by using nudges as a simple design element instead of, e.g., gamification or more complex social facilitation elements. While the previous study suggested that tailoring and personalisation may be good persuasive strategies, the survey, in line with previous research (Tan et al., 2020c), shows that not everybody is willing to share information. The representative survey suggests that participants generally, and older people and women in particular, are not very open to sharing information in order to receive personalised preparedness advice or personalised nudges. However, the results also suggest that it is feasible to differentiate between people who are currently using a warning app and those who are not (Fischer et al., 2019). While the effect is small, we find that warning app users are significantly more open to sharing such information for personalisation.

In the design workshops, participants expressed privacy concern about entering information while interacting with the nudges. Integrating personalisation or user input should thus be done with caution and without or only with anonymous data collection and making this evident to users. Appeal (i.e. liking) is also identified by Kotthaus et al. (2016) and is achieved through a positive framing of the suggested safety measures in the social nudge. Interesting information, especially in the form of statistics, were mentioned as increasing appeal. Trust was also found to influence the persuasiveness (Kotthaus et al., 2016), while mistrust regards the collection of data about users' behaviour, further stressing the importance of data privacy.

In addition to creating a message that is persuasive, our design study participants suggested that social influence, commitment, and confrontation with negative consequences can increase the effectiveness of the preparedness alerts. In line with this, we used checkboxes, social comparison, and a confrontation with the causes and consequences of the hazards. Gamification elements were also suggested but engaging gamification elements conflicts with other requirements, such as little user input, privacy concerns, and simplicity (Chittaro, 2019). While many participants opposed the use of push-notifications as prompts, others suggested occasional optional reminders to remind of the checklists and their completion. Overall, the evaluation indicates that preparedness information should be implemented as a menu and any further emphasis of preventive measures should be an optional feature. Future work should consider alternatives to push notifications as triggers and reminders of preparedness measures. Statistics were deemed helpful, possibly because they increase trust and perceived relevance, which may explain the confrontational nudge's success. While the information's source was not questioned, the mentioning of statistics may point towards a mistrust in the relevance of the hazards, which needs to be further proven to the users. Statements from our participants implied that the assessment of relevance depends on their perception of risk. Many participants considered their own preparedness behaviour to be sufficient and therefore did not see themselves as at risk. However, the survey participants perceived other people as more endangered than themselves, which corroborates the optimistic bias and systematic underestimation of own risks (Weinstein, 1980). This underlines our initial argument that overcoming barriers to taking preparedness measures can be an important aspect of warning apps (Fischer et al., 2019).

The fact that users routinely underestimate the risk and damage of hazards leads to users' evaluation that preparedness advice reduces salience and thus makes it more difficult to identify acute warnings. This poses the challenge that, to be interacted with, the messages need to immediately establish the relevance of the general hazard to users, while also being immediately differentiated from acute warnings. Whereas push notifications suggest unwarranted urgency, a neutral colour scheme and the inclusion of the preparedness alerts in a separate optional menu can be used to differentiate preventive content. On the other hand, participants particularly suggested that general hazards could be made relevant through statistics. This further supports the relevance of hybrid nudges, which combine the persuasive element with information and thus target reflective reasoning while making use of automated cognitive processes (Renaud & Zimmermann, 2018). Due to their transparency, they are deemed an ethical type of nudge (Renaud & Zimmermann, 2018), and due to their high informational content, they fit well with warning apps. Similarly, providing sources for the information that the nudge uses (e.g., the risk, the effect on others) increases trust and is important both for the nudged preventive content and regular warnings (Meyer, 2006).

While we opted for a simple design that could be seamlessly integrated into existing warning apps, other types of persuasive elements are possible and were mentioned by users in this study, including gamification elements and scenarios. Indeed, one study has shown that the use of emotional scenarios in Virtual Reality triggers an attitudinal change in risk evaluation (Chittaro & Zangrando, 2010). However, the most effective strategy also increased users' anxiety (Chittaro & Zangrando, 2010). This is a strong reminder that when it comes to safety and security, precaution is not everything. Instead, other factors like anxiety and trust in reliable warning apps need to be considered when planning the inclusion of preparedness measures and persuasive elements.

13.8.2 Acceptance of Preparedness Features in Warning Apps

The second aspect that this study addresses is the question whether persuasive preparedness information should be integrated into warning apps. The evaluation indicated that while preparedness information makes warning apps more attractive, particularly for those who state they learned something new, those who felt disturbed by the warnings were deterred from using a warning app. This can be explained by Effective Use Theory and its element representational fidelity, which is "the extent to which notifications provide users with faithful representations of an emergency" (Bonaretti & Fischer, 2021). One element of that dimension is relevance, which appears to be violated by notifications about non-immediate threats (Bonaretti & Fischer, 2021). The analysis of users' comments about how to improve the preventive warnings and nudging elements revealed that compared with the criteria for warning systems, some aspects, particularly exactitude, consistency, and currency, appear less relevant. By contrast, major issues consist in proving the relevance of preparedness measures and finding an appropriate activation method to trigger the reflection of preparedness. While the use of push notifications was criticised, as preparedness information is often perceived as not warranting alerts, a majority would be open to occasional notifications. Agencies seeking to increase awareness and emergency preparation could consider occasional or seasonal notifications, or a themed month for preparedness measures. These should be easy to opt out of. Many users also appeared open to creating self-commitment through checklists and occasional reminders.

The fact that relevance of preventive alerts appears to be contested may to some extend be explained by the German state-centric risk culture (Cornia et al., 2016). Since people who use a warning app are interested in improving their safety and intend to follow the app's recommendations (Fischer et al., 2019), it could be argued that nudging could be a good avenue for overcoming aspects of risk culture that increase the risk of emergencies. The evaluation indicates that providing interesting new information is the factor that most strongly correlates with participants rethinking their behaviour and also being open towards receiving prompts, such as push-notifications and using a separate preparedness app. In general, however, participants rather wanted the preparedness information to be included as a menu item, that needs to be proactively sought out. To some degree, this is already offered by some warning apps (Hauri et al., 2022). Most did not prefer a separate app dedicated to preparedness, which is in line with previous findings that integration in one app is a major user demand (Dallo & Martí, 2021; Kaufhold, Haunschild, & Reuter, 2020).

Nudges and warnings can also lead to reactance (Dillard et al., 2021; Lukoff et al., 2022). We find that this might be the case in a subset of participants. While we cannot determine any causal relations, those who felt more disturbed by the preparedness warnings supported the use of push-notifications less, thought less that they should do more for their safety and had not started to rethink their preparedness behaviour. However, we did not find a connection between having felt disturbed by the notifications and a reduction of implemented preparedness measures. In addition, the nudges, as chosen in this study, did not lead to a feeling of having been pushed towards a behaviour, and even that feeling did not decrease the self-reported motivation to use a warning app. These findings encourage further exploration of preparedness messages in warning apps, because even when the messages are perceived as unhelpful, they do not lead to a reduction of preparedness activities.

13.8.3 Limitations and Future Work

While this study's insights offer important first insights into exploring persuasive preparedness measures and their integration into warning apps, it is subject to several limitations. Because we could only compare participants who had engaged with all scenarios and answered all questionnaires for the scenarios, the individual groups were relatively small. To avoid even smaller groups, we could not further differentiate the nudges into more fine-grained categories. As a result, we cannot determine which aspects in the confrontational nudges are the ones that had the biggest impact on users. For future research, we suggest that the confrontational nudges we used are further differentiated and tested. At the same time, such a study should measure the effect of the nudges on other aspects that have been found to increase preparedness, such as response efficacy and the immediate gratification bias (Mol et al., 2021). As the study indicates that the inclusion of non-imminent warnings can have positive effects on preparedness and largely does not demotivate the use of warning apps, further and larger studies are feasible and could be done without jeopardising the perception of warning apps.

With regard to the social nudge, we identified a divergence between participants' prediction of the effectiveness and the true effectiveness of the nudge. While such a divergence has also been found in other studies (Dai et al., 2021), it could be due to the particular design and wording of the nudge as implemented in this study. By asking participants to rate their own safety, our participants on average reported a relatively high state of safety for the various emergency categories. This may have reinforced the optimistic bias, potentially mitigating the effect that the social reminder would otherwise have had (Rahn et al., 2020). This would be in line with research that shows that people who (falsely) feel that they perform better than average when it comes to ecological behaviours, also feel less threatened by climate change (Leviston & Uren, 2020). Since participants in the survey and design workshops emphasised that social influence would have a greater effect on them, other designs of social influence nudges should be explored. While we included an appeal to a personal norm with regard to taking responsibility vis-à-vis one's family, friends and neighbours, the text-based nudge might have been too weak in comparison to the slider interaction. Future social nudges might thus be designed to more strongly target reflection of why preparedness helps others (Knowles et al., 2014). The representative study suggests that positive wording might be preferred to loss-focused frames. Since we focused on testing nudges in the wild and on designing an appropriate tool, future research should also focus more on varying the wording in the text-based nudges, e.g. by testing more nudges in a survey with regard to their acceptability.

Another limitation might result from the measuremeriable, the implemented preventive actions, which was based on participants' expressed plan to implement actions directly after interacting with the nudges, their own statement about whether or not they had implemented more measures than they would otherwise have done, and a check of whether they had implemented a list of measures in the post-study survey. While this adds a measure of (self-reported) implemented measures to the planned measures, participants may over-state the number of implemented measures to meet perceived expectations. This effect may be stronger in participants who have been nudged, because they have been presented with further reasons and norms. Even though we asked respondents to report about specific measures, likely increasing the reliability of the answers, the study was unable to verify participants real activities. In addition, the questionnaire was administered only one day after the survey, which gave participants little time to implement some of the measures. However, the measures were designed to be able to be implemented quickly and it can also be assumed that the likelihood of participants implementing measures in reaction to the preparedness advice decreases as time passes.

While we involved participants in the design of the preparedness warnings and in the choice and design of the nudges, we chose the safety and security scenarios top-down. Instead, users might already have identified preparedness measures that they have been putting off. Another option would be to further involve emergency managers in the choice of recommendations that are most relevant from their perspective. In addition, while it made the nudges more acceptable to users, participants may not have chosen the most effective nudges. In a previous study, participants overestimated the effect of an informational video and underestimated the effect of a text-based commitment nudge (Dai et al., 2021), possibly believing themselves to be less prone to the effect of less transparent System 1 nudges than to reflective System 2 nudges.

In addition, future work could explore further persuasive design mechanisms. To do justice to the wish for integration into one app and to thus facilitate the integration of persuasive preparedness alerts into mainstream warning apps, we only considered persuasive elements that fit to the user requirements for warning apps (Tan et al., 2020a). However, other persuasive elements to increase preparedness, such as gamification, are feasible and may be adequately included in other apps, such as motivational apps or news apps, which could benefit from insights of this study concerning the design and nudging effects. In addition, optional reminders were mentioned that could serve as additional triggers of the planned measures. Along with checklists, which were also mentioned, they can serve to increase users' commitment (Münscher et al., 2016). In addition, rather than activating individual preparedness as a value in a top-down manner, future research could explore the effect of users' reflecting on the reasons for improving individual preparedness (Knowles et al., 2014; Maio et al., 2001). For example, asking users who perceive that they are overusing their smartphones to write down a more valuable way of spending their time has been found to decrease screen time (Xu et al., 2022).

Despite these limitations, the study succeeds in exploring warning apps as a new technology in which nudging might be feasible. It shows that preparedness alerts increase warning app users' knowledge about general hazards, that a confrontational nudge can moderately increase users' general preparedness, and that users are open towards receiving preparedness alerts as an optional feature, thereby increasing warning apps' utility.

13.9   Conclusion

In this study, we have investigated whether and how warning apps can be used to motivate users to preventively prepare for emergencies. Based on insights from a representative survey and design workshops, persuasive preparedness warnings and warning app prototypes with a confrontational and social influence nudge were developed. The prototypes were employed for a one-week experiment, during which participants were sent three preparedness alerts. The control group received the persuasively designed message, while two groups received either a confrontational nudge or a social influence nudge in addition to the persuasively designed message.

The study shows that while both confrontational and social nudges are deemed motivating and appropriate, only the confrontational nudge significantly increases the compliance with preparedness advice. The nudges did not affect the perception of the warnings, the interaction with the app, the intention to increase preparedness behaviour, or the motivation to use a warning app, thus encouraging further exploration of nudging in this context. However, the qualitative analysis of users' perceptions indicates that while users want preparedness information to be integrated in warning apps, non-acute notifications appear to reduce the representational fidelity of warning apps. The motivation to use a warning app positively correlated with the perception of having learned something new and negatively correlated with having felt disturbed by the preparedness alerts. This indicates that interesting non-acute information can increase warning apps' utility, while over-use of alerts that are perceived as unimportant could lead users to abandon warning apps.

Concerning how persuasive prevention messages should be designed, positively framed rational-confrontational and social responsibility nudges were preferred to other nudges. Ensuing workshops, which explored how the ICT for preventive advice and the nudes should be designed in order to be perceived as persuasive, indicated that relevance, simplicity, appeal, effectiveness, and trust are relevant themes. The evaluation indicates that similar to warning messages generally, trust can be enhanced by pointing out the sources of the information. In addition, statistics can be used to underline relevance while increasing trust. The study indicates that particularly checklists and reminders should be explored as persuasive elements in future studies. The high acceptance of nudging both in the representative survey and the experiment, and fact that many participants indicated that they learned something new and have begun reevaluating their preparedness encourages further exploration of preparedness nudging in warning apps.

Breaking Down Barriers to Warning Technology Adoption: Usability and Usefulness of a Messenger App Warning Bot

14

Abstract

In crisis situations, citizens' situational awareness is paramount for effective response. While warning apps offer location-based alerts, their usage is relatively low. We propose a personalised messaging app channel as an alternative, presenting a warning bot that may lower adoption barriers. We employ the design science research process to define user requirements and iteratively evaluate and improve the bot's usability and usefulness. The results showcase high usability, with over 40% expressing an interest in utilising such a warning channel, stressing as reasons the added value of proactive warnings for personalised locations while not requiring a separate app. The derived requirements and design solutions, such as graphically enhanced user interface elements as guardrails for effective and error-free communication, demonstrate that a suitable warning chatbot does not necessarily require complex language processing capabilities. Additionally, our findings facilitate further research on accessibility via conversational design in the realm of crisis warnings.

Original Publication Haunschild, J., Henkel, M., & Reuter, C. (2025). Breaking Down Barriers to Warning Technology Adoption: Usability and Usefulness of a Messenger App Warning Bot. i-com 2025; 24(1):1–25, https://doi.org/10.1515/icom-2024-0067

Supplementary Information The online version contains supplementary material available at https://doi.org/10.1007/978-3-658-46489-9_14.

14.1   Introduction: Problem Identification and Motivation

Informing the population of impending hazards is crucial for crisis response. Events, such as the European floods in 2021 that killed over 200 people, demonstrate the devastating effects of insufficient crisis awareness. Due to the wide-spread use of mobile phones, cell broadcast and warning apps are increasingly used to alert the population. Warning apps combine multi-hazard information in a single platform (Tan et al., 2017), which can be personalised with regard to personally relevant hazards and locations (Hauri et al., 2022). By contrast, cell broadcast (CB) distributes warnings to all mobile phones connected to a cell tower, offering resilience through low bandwidth and independence from the Internet. CB achieves almost perfect reach in countries with the necessary infrastructure (Hauri et al., 2022), but lacks personalisation. Whereas this is offered in warning apps, these have a low to medium reach (Hauri et al., 2022), as users need to actively download them. This incurs costs in choosing an app, battery drainage and by taking up data space (Li et al., 2021), reducing the intention to use and to continue to use a warning app (Fischer-Preßler et al., 2021). Additionally, some countries' risk cultures lead to a lower perceived individual responsibility to prepare (Appleby-Arnold & Brockdorff, 2018; Cornia et al., 2016; Reuter et al., 2019), which may further reduce the motivation to download a warning app. To address these concerns, additional warning channels should be explored. Instant messaging (IM) apps appear particularly viable due to their widespread use.

This idea is supported by current trends in news delivery, which can inform the spread of crisis information. In some countries, IM apps, which are widely and increasingly used around the globe (Statista & eMarketer, 2021), have become the main avenue of news consumption (Newman et al., 2023). Major news outlets have developed distribution bots for social media platforms and messaging apps (Jones & Jones, 2019). Following this trend, agencies have begun circulating crisis related information using messenger apps, such as the German Ministry of Health, informing about the COVID-19 pandemic via Telegram (BMG, 2021).

We stipulate that a warning channel within an IM app could lower adoption barriers and reduce the maladaptive rewards associated with downloading and using warning apps that appear to deter users with a low protection motivation (Fischer-Preßler et al., 2021; Li et al., 2021). Indeed, a representative study has found that one third (34%), and especially current non-users of warning apps, would prefer such a channel to a warning app (Haunschild et al., 2022b). However, warning apps have specific design requirements resulting from the use in critical situations

(Tan et al., 2020a), and users want to personalise warnings and locations (Kaufhold et al., 2020).

An analysis of chatbot design guidelines identifies three studies that offer moderate re-usability, however, they are in the domains of health, education, and customer service rather than the public service or crisis management domains (Elshan et al., 2022). One study compared the effect of using an app or a chatbot for reducing problematic drinking (Dulin et al., 2022). As the bot increased the intention to change the behaviour but not the action itself, app and bot should be used complementarily. As human-bot interactions are highly context dependent (Elshan et al., 2022) and no studies exist related to public crisis warnings via IM apps made for that purpose, we design a chatbot that mimics warning apps by helping users set up warning subscriptions for relevant locations, warning types and severity levels. Due to the particular design requirements for warning channels (Tan et al., 2019), existing guidelines for designing UI and conversational interfaces (Shneiderman, et al., 2016; Sugisaki & Bleiker, 2020) must be adapted to the warning domain.

We follow the design science research (DSR) approach (Hevner et al., 2004; Peffers et al., 2007) and its six steps: (1) identifying a problem, (2) defining objectives for a solution, (3) designing and developing an artifact, (4) demonstrating and (5) evaluating it, before (6) communicating the findings (Wobbrock & Kientz, 2016). Figure 14.1 shows the iterative development informed by user evaluations and the sections that describe each step. The DSR process closely aligns with human-centered design principles (ISO 9241-210), as both emphasise understanding user needs, iteratively designing solutions, and evaluating them. Specifically, our work adheres to these principles by analysing usability requirements and assessing the roles of usability and usefulness for IM apps as warning channels.

After identifying the problem and motivating the topic in Sect. 14.1, we analyse usability requirements for warning apps and bots to define our solution's objectives in Sect. 14.2. Sect. 14.3 describes a conversational design workshop conducted to design the user journeys, personas and conversational interface. Next, we present the iteratively developed prototype, a *Telegram* warning bot, which uses an official warning app's API to send alerts and allows users to personalise warning subscriptions (Sect. 14.4). Sect. 14.5 presents two evaluation studies: First, we conducted interviews focusing on task fulfilment and usability. Secondly, participants performed tasks and evaluated usability, reward and, usage intention through an online survey. The results and limitations of our work are discussed in Sect. 14.6. Sect. 14.7 summarises and concludes the paper.

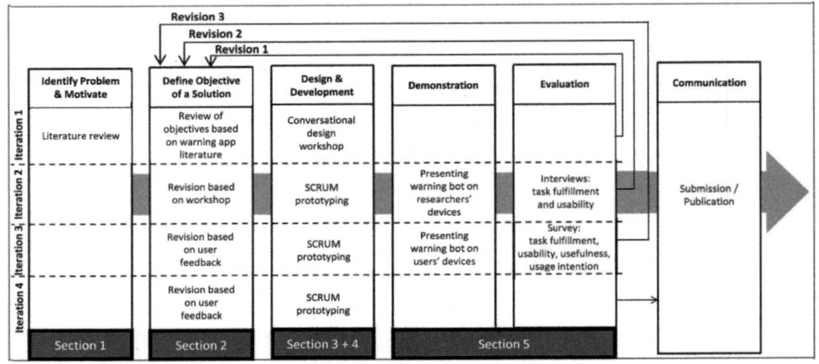

Fig. 14.1 Research approach of our paper using the design science research (DSR) process (Hevner et al., 2004; Peffers et al., 2007), own figure adapted from Meier et al. (2019)

This study makes several contributions: through the DSR process, we contribute findings based on exaptation (Gregor & Hevner, 2013), meaning the application of existing findings to a new domain based on the design and evaluation of an artifact. We use existing findings on the design of warning apps and chatbots and apply them to the domain of mobile public warning systems. The study results in the development of an artifact, a *Telegram* warning bot, which is available for further studies and development. Furthermore, by conducting user studies with a warning channel that aims to lower adoption barriers, our work offers insights into the effects of complementing smartphone apps with IM "botplications" and into the perceived advantages and disadvantages, utility and usability of such tools (Klopfenstein et al., 2017).

14.2   Objectives of the Solution

The following section summarises the findings of the literature that describe the user requirements for warning apps, followed by a review of requirements for chatbots, from which we derive the objectives for the warning bot.

14.2.1 Mobile Warnings and Warning App Requirements

Warning apps provide a mobile channel that allows to personalise warnings based on locations, by selecting relevant areas or via GPS-tracking (Tan et al., 2017). In addition, relevant hazard types can be selected, e.g. severe weather warnings, bomb disposal, school closure or policing information (Hauri et al., 2022). As such, they are distinct from cell broadcast, which is used only for severe civil protection warnings and sends alerts only based on the current radio cell of the mobile device. We analysed the literature on warning apps to identify warning and design requirements that might also be relevant for other warning channels. The literature has analysed user preferences regarding warning apps (Dallo & Martí, 2021; Haunschild et al., 2022b; Kaufhold et al., 2020), and reasons for warning app (non-)adoption and (dis-)continuance of use (Fischer-Preßler et al., 2021; Tan et al., 2020c). Other studies have contributed warning app design requirements (Bonaretti & Fischer, 2021) and usability guidelines (Tan et al., 2020a). Compared to social media, people value warning apps for their fast and reliable information (Haunschild et al., 2020). For those who have used a warning app in a crisis, this information channel is perceived as more helpful than other online sources (Haunschild et al., 2022b). Users prefer an app that centralises various types of warnings and allows personalisation based on personally relevant warnings (Kaufhold et al., 2020; Haunschild et al., 2022b). In line with these findings, a study shows that the key factor that affects the duration of use of a warning app is its perceived usefulness. Users are more likely to continue using the app if they feel that it effectively serves its intended purpose—reliably providing critical information in an easy-to-use manner—and is free of errors (Tan et al., 2020c).

A study of app reviews suggests that users of warning apps discontinue using warning apps due to reliability issues, with users not receiving relevant warnings or receiving false alarms at other times (Tan et al., 2020b). Usability factors also play a role: users discontinue using warning apps that have complex user interface graphics or require too much user input (Tan et al., 2020c). In addition, warning apps have specific design requirements that differ from the requirements of other apps, which relate to simplicity and ease of use, as well as demanding little energy, bandwidth, and storage space, to account for the specific use case in crises (Tan et al., 2019).

Looking at users' motivations regarding the use of warning apps, Fischer-Preßler et al. (2021) find that maladaptive rewards—advantages derived from not using a warning app, such as saving battery and data space—have a negative influence on app adoption and continued use. Similarly, inhibition theory posits that potential losses are perceived more strongly than potential gains (Kahneman & Tversky,

1984), stressing the importance of effort expectancy and other perceived costs for digital service adoption (Distel & Ogonek, 2016). Finally, citizens in state-oriented risk cultures have a low motivation to engage in crisis management and preparedness (Appleby-Arnold & Brockdorff, 2018; Reuter et al., 2019), stressing the importance of low adoption costs.

Bonaretti and Fischer (2021) point towards the importance of timeliness and representational fidelity of warning apps, meaning that warnings need to be easily and quickly noticed and understood by users, e.g. through good visual representations of the crisis and a noticeable auditory and visual alarm signal. Furthermore, the app must be dependable with regard to its warnings, sending an adequate alert level for each relevant and only for relevant emergencies.

Finally, Tan et al. (2020a) present seven higher-level constructs with usability requirements, some specific to warning apps. They suggest (1) app dependability, consisting of instant start, data preservation, error-free operation and clean exit; (2) app design, including branding, orientation, minimal advertising, phone resource usage; (3) app utility, consisting of collaboration, content relevance and search; (4) UI graphics, with the requirements aesthetic graphics, realism and subtle animation; (5) UI input, including control obviousness, de-emphasis on settings, effort minimisation, fingertip size controls; (6) UI output, requiring concise language, standardised UI elements, user-centric terminology and audio output; and (7) UI structure with the elements logical path, top-to-bottom structure and few external links. In addition, Reuter et al. (2017) show that users are reluctant to share private information with state agencies, highlighting the need for privacy by design.

We identify a set of objectives for mobile warnings:

- *Usefulness*:
    - *Adaptability*: offer many relevant instances of interaction through personalisation (Haunschild et al., 2022b; Kaufhold et al., 2020; Tan et al., 2020c)
    - *Representational fidelity*: reliably represent the warning status, is free of errors (Bonaretti & Fischer, 2021)
    - *Timeliness*: noticeable warnings, fast delivery (Bonaretti & Fischer, 2021)
    - *Criticality*: reliably present critical warnings (Tan et al., 2020b)
- *Resource efficiency*: minimise used data space, energy resources, learning and maintenance effort (Fischer-Preßler et al., 2021; Tan et al., 2019)

- *Simplicity*: little user input, simple UI graphics, fast information extraction (Tan et al., 2020c)
- *Usability*: optimised for critical situations and privacy (Reuter et al., 2017; Tan et al., 2020a)

14.2.2 Bot Requirements

Chatbots are used in a large variety of fields, particularly in customer service support and e-commerce, but state agencies are also increasingly using them to offer information about regulations and public services (Følstad & Bjerkreim-Hanssen, 2023). A study analysing the interactions with a public service bot has shown that users largely use very short, utilitarian rather than social inputs and that this brevity is technically justified, as it results in a vastly higher rate of adequate responses by the bot (Følstad & Bjerkreim-Hanssen, 2023). Another study looking at older adults found that Internet competency had a great effect on whether social-oriented or task-oriented interaction styles were more beneficial in e-commerce, with task-orientation performing better for people with lower Internet competency scores (Chattaraman et al., 2019).

Within the public service category, a warning bot pertains to the category of content curation chatbots, which "have a chatbot-driven Locus of Control where the user initiative is limited to accepting or rejecting content offers, or requesting specific content types, serving to filter the presented content selection" (Følstad et al., 2019). These are increasingly developing from one-off interactions to longer-term relationships with regular content provision. Because browsing and searching are less well-suited to dialogue interfaces, these types of chatbots "actively guide users to recommended content" by presenting options, often by using visual elements (Følstad et al., 2019).

Klopfenstein et al. (2017) identify that bots are increasingly run within IM apps to serve as functional replacements of mobile applications, e.g. as news bots instead of news apps. Such "botplications" offer a set of advantages for users and programmers: They are instantly available inside IM apps, require little data traffic, their push notifications appear as messages, the messages can be easily shared within the app and beyond, most messaging apps are independent of the operating system and the hosting platform offers user authentication (Klopfenstein et al., 2017).

However, bots have some significant differences compared to apps, especially regarding the User Interface (UI). Conversational user interfaces (CUIs) are sequential, characterised by turn-taking (Moore et al., 2017; Sugisaki & Bleiker, 2020). In addition to text, they support pictures, stickers, videos, audio, generic data files and packaged data, such as geolocations or contact information (Klopfenstein et al., 2017). In messaging bots, services and information are provided as streams of messages and notifications. Dialogues can take the form of text, voice, video input, or images (Janssen et al., 2022). They can be deployed on various platforms, including websites, messaging applications or social media platforms. In a bot, each message "should be conceptually seen as a micro application, while the conversation is a timeline of past application screens" (Klopfenstein et al., 2017, p. 559). A challenge also lies in guiding the conversation: "because of the free-form nature of the medium, it is easy for bot users to get lost and not to be certain of what commands or what exact syntax is required to perform the desired action" (Klopfenstein et al., 2017, p. 560). Therefore, the bot should suggest next actions, use buttons or suggested syntax as guardrails (Klopfenstein et al., 2017). To ensure guidance, simplicity, effectiveness, and to avoid errors, research also suggests avoiding natural language processing (Klopfenstein et al. 2017, p. 559). This rules out unexpected user input (Chaves & Gerosa, 2021) and avoids misunderstandings by the bot (Li et al., 2020), which would lead to incorrect answers or a lack of responses, and in turn decrease trust in the bot, and thus negatively affect the user experience (Følstad & Brandtzaeg, 2020). Users also often prefer buttons to entering text as this provides them with clear input options enhancing control (Zamora, 2017, p. 257).

Conducting a literature review of task-oriented conversational agents, Göbel et al. (2024) identify five design requirements (goal orientation, usability, communication efficiency, human-likeness, user-centric integrity) and eleven design principles (goal extraction, conversation recovery, accessibility, adaptability, feedback and responsiveness, cultural sensitivity, natural communication, adaptive personality, emotional engagement, transparency, ethics and compliance).

The literature points to the following challenges and requirements for bots and CUIs:

- *Onboarding*: introducing function and interactions (Chaves & Gerosa, 2021; Jain et al., 2018; Janssen et al., 2022)
- *User support and learning*: providing operating aids and allowing users to learn about functions on the fly (Chaves & Gerosa, 2021)
- *Context maintenance*: saving previous inputs and being able to refer to them a at a later point in time (Chaves & Gerosa, 2021; Følstad & Brandtzaeg, 2020; Jain et al., 2018; Janssen et al., 2022; Stieglitz et al., 2022)
- *Guided conversation*: proactively suggesting follow-up actions, offering alternative choices, offering UI enhancements such as buttons or syntax commands (Klopfenstein et al., 2017)

- *Social intelligence*: dealing with inappropriate and unspecified input (Chaves & Gerosa, 2021)
- *Manage conversation breakdown*: offering options of potential intentions after unfamiliar input (Ashktorab et al., 2019)
- *Transparency*: being clear about the process, especially for complex actions (Chaves & Gerosa, 2021)
- *Autonomy*: ensuring control over the interactions (Nguyen et al., 2022)
- *Efficiency*: satisfying results in short time (Moore et al., 2018), quick responses to minimise user input (Jain et al., 2018; Piccolo et al., 2018)
- *Effort minimisation*: reducing cognitive effort by structuring process intuitively (Abbas et al., 2023; Makasi et al., 2022a).
- *Privacy*: following data protection laws and collecting minimal user data (Janssen et al., 2022)

14.3 Warning Bot Design

To decide how to translate the collected requirements into a design, we conducted a conversational design workshop with the involved researchers and programmers ($N = 5$), lasting two hours. The decisions were based on the researchers' understanding of the knowledge base related to user preferences and user needs (see Sect. 14.2.1) as well as bot requirements described in the research literature (see Sect. 14.2.2). We applied the framework of conversational design (Moore & Arar, 2019), using workshop material for conversational interface design from a UX consulting agency, to translate the knowledge base into design decisions. The workshop focused on understanding user needs, motivations and behaviours; exploring the usage context and the expected user group(s); crafting language and tone by determining the appropriate language style, tone of voice, and the bot's personality to align with the context of warnings; defining conversation flows by creating user journeys and ideating the graphical elements of the user interface. The implemented design is discussed in more detail in Sect. 14.4. The literature review and the design workshop culminated in a collection of issues, derived meta-requirements and design implications, and settled on design solutions, which are presented in Fig. 14.2. The key steps and exemplary outcomes are described in the following.

First, we discussed the key problem to be solved for users to achieve crisis awareness based on crisis informatics literature and user-centered studies on public warnings. For example, since people prefer to combine all relevant crisis functions in one tool, we chose to offer both warnings and preparedness information to increase utility (MR1).

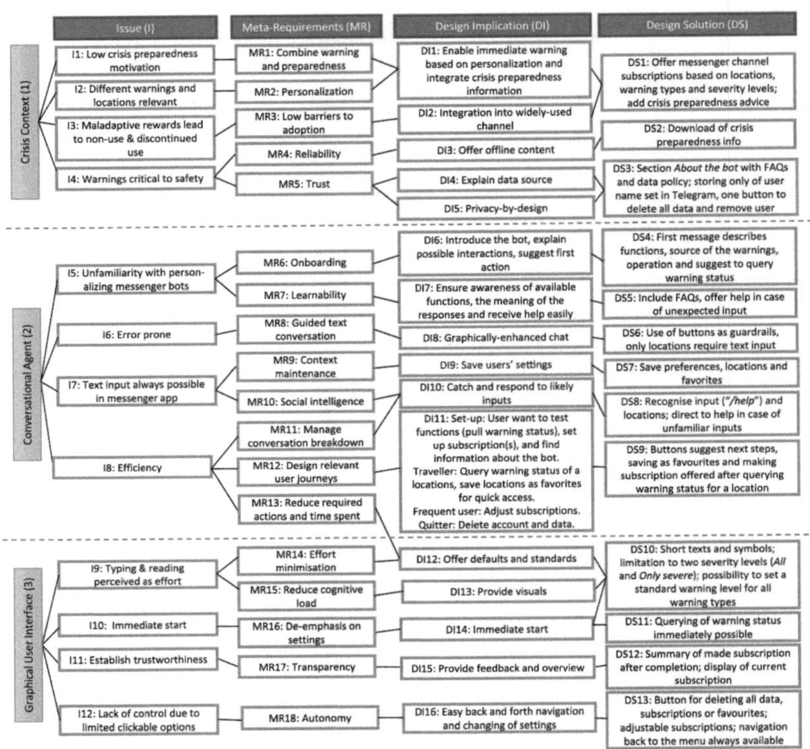

Fig. 14.2 Issues, meta-requirements, design implications and design solutions for the warning messenger channel prototype. Categories inspired by Meier et al. (2019)

Then, we described our targeted user group(s) and the usage context. We aim for situational awareness for the general population, i.e. people who have varying levels of technology affinity and who are moderately interested in receiving crisis warnings but not motivated enough to download a dedicated warning app (I1). By allowing users to subscribe to alerts via a messenger channel (DS1), we lower the barrier to adoption (MR3) and avoid maladaptive rewards, as users do not need to give up storage space for the installation of a standalone warning app (I3). In addition, predominantly, people have little experience with warning apps and personalised messenger bots, and we expect them to be unfamiliar with the set-up process (I5). Related to the conversational agent, this leads to the requirement of onboarding (MR6) and learnability (MR7) as well as a first message that introduces the bot and suggests the first next step (DS4) and the inclusion of an FAQ section (DS5). Related to the context, we expect the set-up of the bot to take place in a calm environment, whereas queries, warnings and looking up of advice may occur in stressful situations, when users may be worried about a crisis (I4), leading, for example, to the requirement of reliability (MR4) and the offering of offline crisis management advice (DS2).

Then, we discussed the persona of the bot, focusing on its personality, its identity, role, and relationship with the user, and the expected communication style. Due to the serious nature of the topic, we orient the bot towards providing a *service conversation*, which is "constrained by rules and roles (e.g. customer service agent/customer)" (Moore et al., 2017). Extrapolating from user requirements for warning apps, we assume that users want an error free (I6) and efficient process (I8) and focus on a bot-guided conversation (MR8) that can be conducted through buttons (DI8) and little text input (DS6), which is supported by this bot persona.

Finally, we devised the conversational flow based on the expected user journey (MR12), assuming that users would follow specific paths in their interactions with the system. We assume that users want to set up personalised (MR2) subscriptions for proactive warnings at specific locations (*push mode*); that they will request information about the warning status at a specific location (*pull mode*) to familiarise themselves with the bot and its output (I2); that they will want to look up crisis management and preparedness advice; that they will want to manage, adjust, or delete their subscriptions and settings; and that they will inquire about the bot, its functions and data (DI11). As the bot immediately starts (I10), we made sure to allow users to query the warning status as soon as the chat window opens (DS11).

By simulating the user journeys, we defined the conversation flow, seeking to increase intuitiveness, control, and transparency (MR17) to establish trustworthiness (I11). We also designed the graphical user interface, deciding on texts that would best describe the buttons' functionalities (see Sect. 14.4). As typing and reading is perceived as effort (I9), we decided to use a visual-centric conversation style (DI13) to reduce error rate, in which "graphical elements, such as buttons, images, emojis, and other visual elements, are mixed into the interaction alongside natural language inputs" (Moore & Arar, 2019, p. 12). Therefore, buttons are used as the primary input channel to steer the conversation and make suggestions for how to proceed, reducing error, recovery time, and typing effort (MR14, MR15) (Klopfenstein et al., 2017; Zamora, 2017). The use of buttons, for example, for deletion of data or navigation back to the menu, also allowed us to steer the conversation and take autonomy (MR18), dealing with the issue of limited clickable options in messengers compared to warning apps (I12). Natural language is only used to input locations. Users can use the command "/help" to request assistance related to their previous interactions at any point (DS8). While natural language is not necessary very often, users always have the ability to enter text (I7), requiring a level of social intelligence (MR10). Unfamiliar inputs were, therefore, also dealt with by directing users to the "/help" command.

14.4   Warning Bot Implementation

The following will describe how the challenges, requirements and defined objectives are implemented in the bot.

14.4.1 Crisis Context

To ensure reliability and trust, we integrate the API of the official and, in Germany, most widely used warning app *NINA*, which combines information from several relevant safety and security sources, such as the weather service and the modular warning system (MoWaS), the unified warning system of the German civil protection authorities (Bund.dev, 2022). The requirements also implied that an already installed application would reduce the resources and maladaptive reward. In addition, the chosen platform should enable auditory and visual alerts. For visibility and quick access, it should be able to send push alerts. Due to the requirement of personalisation, users also need to be able to specify and save their preferences through a transparent process. Analysing the rise of conversational interfaces, (Klopfenstein et al., 2017) finds that bots within messaging platforms fulfil many of these functions, making them attractive platforms for delivering news (Newman et al., 2023). Additionally, Telegram is widely used as a news outlet in some countries, such as Singapore, with users particularly describing efficiency, convenience and effortless access as key benefits (Lou et al., 2021). IM apps are widely used. Globally, *WhatsApp* has 2 billion monthly users, *WeChat* has 1.3 billion, Facebook 1 billion, *Telegram* 0.9 billion, *SnapChat* 0.8 billion, and *QQ* 0.54 billion (Statista, 2024a). While *Telegram* has fewer users, offers an open-source API (Telegram, 2022) as well as many features that are important for the usability of bots (Janssen et al., 2022; Klopfenstein et al., 2017), such as the possibility to create buttons, define quick replies, customise text messages, and publish media for groups or channels for a large number of members (see the online supplementary material for a comparison of IM apps). Despite the lower usage, we chose *Telegram* as the IM to implement our prototype in, as it offers most relevant affordances and thus allows us to build a solution that is more comparable to a warning app. Within the IM, a bot manages the subscriptions and message delivery based on users' settings.

To implement the system, we selected Python 3 and the pyTelegramBotAPI library to facilitate communication with Telegram. We also use the *NINA* API to retrieve relevant information on warnings and disasters. The bot's logic comprises these main tasks: managing the database, generating meaningful responses, providing a user interface, and forwarding slightly edited information from the *NINA* API. Users can proactively request the warning status of locations or set subscriptions to be automatically alerted by the bot (see Fig. 14.3). The bot guides users through multi-step processes, such as creating subscriptions, by providing appropriate prompts. The bot outputs text messages in the *Telegram* chat. The bot's start is managed through the `bot_runner`, which creates three threads. In the first, the subscription mechanism checks for new warnings every two minutes. In the second thread, the `receiver` waits for user input in the chat and then calls the appropriate methods in the `controller`. In the third thread, the `warning_handler` runs. It processes all active warnings upon the initial start of the bot and then scans for new warnings every two minutes by default. The *controller* then accesses various other modules, such as `place_converter`, `nina_service`, `data_service`, `text_templates` and `sender`. The `sender` then sends the chat message to the user. In the `place_converter`, suggestions for requested cities are generated. The `nina_service` serves as the interface to the NINA API, and the `data_service` represents the interface with our database. The appropriate text outputs are created by `text_templates`. For the information architecture, the system employs a database consisting of three JSON files. The first file contains personal information, including the chat ID from Telegram, completed subscriptions, locations saved as favourites, any muted subscriptions and default warning level. The second JSON file stores the postal codes for which active warnings apply, while the third saves which alerts have already been sent to a specific chat ID to prevent multiple warnings.

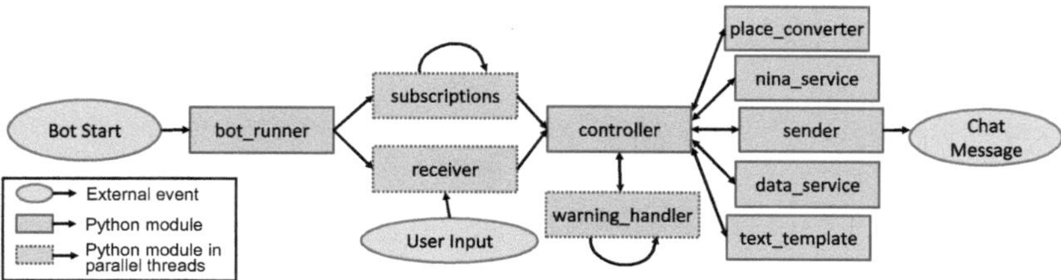

Fig. 14.3 Model of the bot's functional logic

14.4.2 Conversational Agent

Figure 14.4 describes the expected user inputs and resulting states based on the results of the workshop. It shows that warnings can be queried by typing the name or postal code of a location or submitting the GPS location. The query can be turned into a subscription for the queried location, or it can be saved as a favourite for future quick access to the location's warning status. In addition, subscriptions can also be made for specific or all warning types, and all or only severe warnings. The subscriptions can be managed or muted. In addition, data can be managed and deleted and information about the bot can be found. Finally, emergency tips can be downloaded.

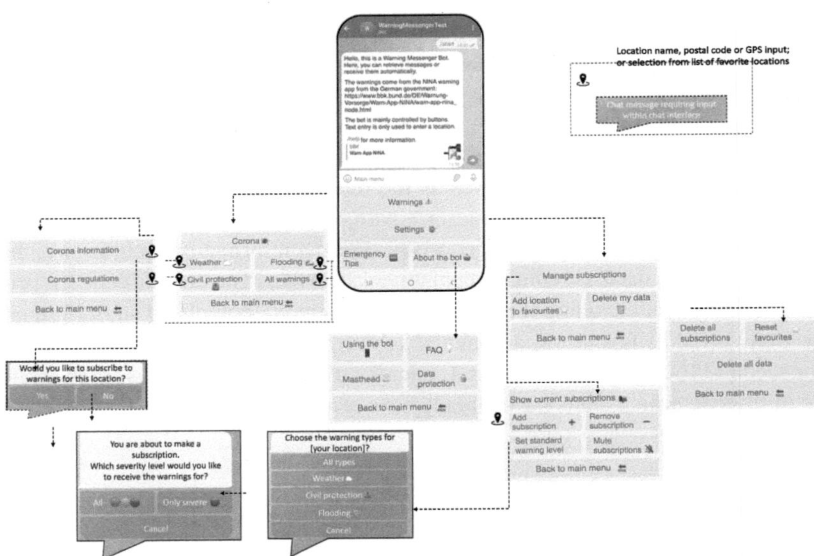

Fig. 14.4 Model of guardrails used to suggest further actions, with button inputs expected for navigation, and text, GPS or postal code for locations

14.4.3 Graphical User Interface

While offering wide usage and many useful features, some design options within *Telegram* are limited. The messages have a fixed design, which can be personalised in the app's settings. The notification sounds can be set individually within the app and specified for different chats, including by uploading ringtones. Contrary to the requirements of warning apps, the latest messages appear at the bottom of the application and not at the top. Each message is automatically timestamped, and the day is also shown. Regarding the text, some options are available to format the text and highlight keywords via the API. Colouring the text is not possible, but messages can be highlighted by using visualisations such as coloured images or emojis. The official icons of the *NINA* API (Bund.dev, 2022) are used to represent the event code. The warning message contains the title and a description of the warning, including the affected areas. If available, the alert also includes the date and time, severity, and version. At the end of the message, there is a link for more information that leads to the source of the message on the website of the German Federal Office for Civil Protection. When the warnings are transmitted by another agency, such as the weather service, no link is available. After the final round of evaluation and implementation of resulting changes, the bot offered the main features depicted in Fig. 14.5.

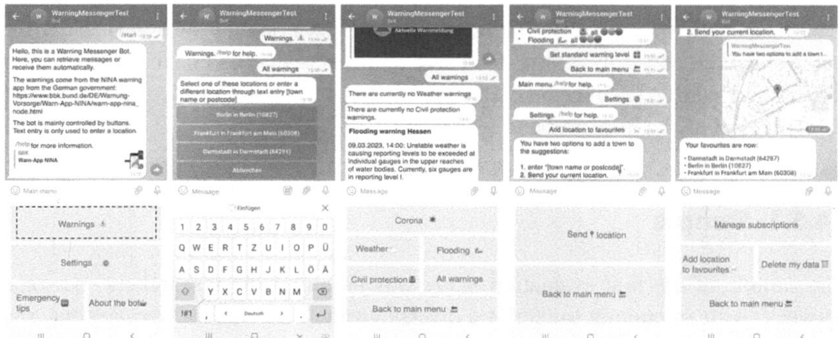

Fig. 14.5 Screenshots of the design and key features of the warning bot, including pulling the warning status for locations and receiving push messages with warnings for saved locations: a) Starting the bot; b + c) requesting and receiving the warning status for a city or GPS location, d) saving the city or GPS location into favourites for frequent future warning pulls, e) setting up locations for immediate warning updates as push messages

14.5 Demonstration and User Evaluation

Following the principles of human centered design, we conducted two rounds of user evaluation, each leading to the implementation of changes to improve our solution. Our iterations focus on the set-up process, as this provides targeted insights into central usability aspects, such as learnability, which were goals that we aimed to achieve with our prototype. In addition, technology adoption is complex and, ideally, requires several instances of data collection, which allows incorporating aspects such as the impact of action on judgement or habit (Kim, 2006). Such a longitudinal study was not conducted due to ethical concerns: we could not ensure the reliable delivery of warnings through the prototype and thus wanted to avoid a false sense of safety in participants. While this means that we cannot judge the usefulness of the warnings in a real setting, the focus on the set-up is justified because it is the instance that requires the most intense user interaction with the bot. By contrast, at a later stage, users are expected to mainly receive alerts passively based on their subscriptions and to make only minor adjustments to their subscriptions. Participants had the right to leave the evaluation at any point.

For the evaluation, we first conducted interviews ($N = 14$) while observing users who fulfilled a given task related to setting up the bot. Interviewers noted users' comments on the tasks and asked follow-up questions where feasible. Finally, users evaluated the systems' usability through a short SUS questionnaire. The second study consists of an online survey ($N = 26$), in which users reported their experiences of performing similar tasks, using the bot on their own devices. Some questions were added to receive feedback on aspects that had been changed after the first evaluation, such as the inclusion of emojis. The survey also focused more on aspects of utility in addition to usability.

14.5.1 Sample

The sample of the first evaluation study was based on the researchers' personal network, with an attempt to recruit a diverse group. Participants' age, gender, and their affinity for technology interaction (assessed via the ATI, 9 items measured on a 6-point Likert scale (Franke et al., 2019), German version (Wessel et al., 2019)) are presented in Table 14.1. It shows that sample 1 represents a large range of age groups and levels of affinity for technology. In the second sample, possibly due to it being an online survey, the sample was younger—with no one older than 39 years—and the median ATI was higher and less varied. While in sample 1, the average ATI score is somewhat smaller than the mean of $M = 3.61$ found in the German population, that of sample 2 is somewhat higher (Franke et al., 2019).

Table 14.1 Research sample details, where N represents the sample size and ATI stands for Affinity for Technology Interaction

Eval.	Recruitment	N	Age	N/Age Group	Gender	ATI M (SD)
1	Personal network	$N = 14$	< 20 20 – 29 30 – 39 40 – 49 50 – 59	3 6 1 2 2	5 women, 9 men	3.44 (1.36)
2	Online ads	$N = 26$	< 20 20 – 29 30 – 39	3 17 5	15 women, 10 men, 1 diverse	3.94 (0.63)

14.5.2 Measurement of Utility and Perceived Usefulness

To test the success rate of users performing specific functions with the bot, in both evaluations, users were asked to perform a set of tasks (see Table 14.2). The selected tasks represented the main functions of the bot and expected challenges in the set-up process. They ensured that users had experienced the available functions before evaluating usability and usefulness. In addition, the tasks can reveal challenges related to specific interactions.

Table 14.2 Evaluations and tasks used for evaluating the warning bot

Task	Evaluation	Description
1	1 & 2	Add the warning channel in Telegram.
2	1 & 2	Check the warning status and add a subscription.
3	1	Add your location to favourites and set up a subscription.
	2	Check for a warning in the location "Testcity".
4	1	View your personal subscriptions.
	2	Add your current location to your favourites and set up a subscription.
5	1 & 2	Find and read the privacy policy, then delete your data from the bot.
6	2	Find and read the bot's crisis preparedness information.

Several approaches exist for measuring whether the artifact is meeting users' needs and offers a positive user experience. The most commonly used aspect of user experience is usability, defined as the extent to which a product can be used by specified users to achieve specified goals with effectiveness, efficiency, and satisfaction in a specified context of use (DIN EN ISO 9241-11, 2018, p. 4). The most commonly used standardised measurement of usability in research and practice is the 10-item System Usability Scale (SUS) (Bangor et al., 2008; Brooke, 1995; Lewis, 2018). By exploring usability, we can judge whether bot-driven personalisation within messenger channels, despite the existing UI limitations, can be designed in a way that allows for a positive set-up experience.

However, research indicates that usability explains only 39% of users' recommendations of a system (Sauro & Lewis, 2016), implying that additional aspects are important. The Technology Acceptance Model (TAM) suggests perceived usefulness as the second main factor (Davis, 1989). It is necessary to account for as it describes whether an artifact fulfils a need related to users' contexts and environments (MacDonald & Atwood, 2014; Toomim et al., 2011). In this study, perceived usefulness is measured through elements of the User Engagement Scale (UES) (O'Brien & Toms, 2010), which consists of the dimensions *reward, perceived usability, aesthetic appeal*, and *focused attention* (O'Brien et al., 2018). The scale offers a good addition to the SUS by focusing on aspects beyond usability (O'Brien et al., 2018). For example, *reward* includes the perception that the experience was worthwhile, or the willingness to recommend an application, which is related to perceived usefulness (see Q25 in the Appendix B). Since user experience has some relevance in safety-critical systems but may be less important than usability (Mentler & Herczeg, 2016), we put usefulness and usability at the centre of our inquiry, including more items on perceived usability (4 items) and reward factors (5 items) and fewer covering aesthetic appeal (3 items) and focused attention (2 items). In addition, we included open questions concerning the reasoning for (not) intending to use the system, as well as perceived advantages and disadvantages compared with other mobile warning channels. Through open questions, we also assessed participants' feedback about tasks and the artifact generally, as well as their reasoning for their adoption intention and perceived advantages and disadvantages compared to other warning channels. Two quality control questions were added to ensure that the participants were really completing the tasks. The full questions and tasks can be found in the online supplementary material.

14.5.3 Data Analysis and Results

We followed the standard for measuring ATI by offering a 6-point scale, all other scales are 5-point Likert scales. In addition, we used thematic analysis to analyse the qualitative answers (Braun & Clarke, 2006) and deduce any challenges related to handling the bot, its structure, visual interface and conversational interface. To analyse the SUS, following common practice, the negatively framed items are inverted, and the data is normalised to represent a scale from 0 to 100 for the whole score consisting of 10 questions, as this allows for a better comparison of the results with other systems (Bangor et al., 2008). The SUS results are analysed with non-parametric tests, as the QQ-plot indicates a violation of the assumption of normal distribution, which is corroborated by the positive skewness of the histogram. Despite the Shapiro-Wilk test resulting in a significance of $p < .01$, we decided to use non-parametric tests due to the relatively small sample size, which limits the robustness of analyses against non-normally distributed data.

The Likert-scale data is thus analysed using Kendall's tau-b coefficient, which is robust to skewed data and suitable for analysing ordinal data on 5-point Likert scales, as it accounts for tied ranks, which are common in such data, and provides a measure of association that aligns with the ordinal nature of Likert responses (Rovai et al., 2013). The qualitative answers were coded independently by two researchers through thematic analysis (Braun & Clarke, 2006), with the coding scheme developed inductively from the text.

Interview Evaluation

In the interview, users interacted with the bot on the researchers' devices. They were encouraged to act as they would using their own devices and to try to use the bot on their own. However, in case they could not perform the task on their own, they could seek out the help of the interviewer who was monitoring the process and taking notes about any unexpected behaviour or user comments. The task-based interviews showed that three out of $N = 14$ participants needed support in order to accomplish the first task, one person needed assistance with task 3. Even the people with a low ATI score felt very or rather well-informed about the functions ($M = 4.43$, $SD = 0.51$) and operation of the bot ($M = 4.43$, $SD = 0.65$). Users predominantly stated that the tasks were very easy or easy to accomplish. In their qualitative answers, they showed a good understanding of the core functions. The overall usability rating, using the common 10-point scale for each item, is 90, with all items rated above 8, indicating an excellent usability (Bangor et al., 2008), which, however, can only be used as a rough indication as participants had the assistance of the interviewers. Asked about the likelihood of using a similar tool in the future, the mean rating was $M = 3.5$ ($SD = 1.02$). When asked specifically whether they would use this tool, just over one third ($n = 5$) stated that they would. These prospective users were all in the two youngest age categories (< 29 years) but had varying experiences with warning apps, ranging from not being aware of their existence to being current users. The logistic regression analysis—to be taken with caution due to the small sample size—revealed that neither participants' mean ATI ($p = .83$) nor SUS ($p = .19$) significantly predicted the binary outcome of planned future use, indicating the relevance of other factors. Looking at users' reasoning, arguments against using the tool were mainly disinterest in warnings generally ($n = 4$), lack of need due to use of a warning app ($n = 2$), not using *Telegram* ($n = 2$) or preferring the design of an app rather than the messenger ($n = 1$). Positive reasoning related to the advantage of receiving automatic warnings was not requiring a separate app and the intuitive design. With regard to design improvements, users requested shorter texts by using symbols and emojis, and a short confirmation upon successfully setting up a warning subscription. For the next iteration, we made the suggested changes.

Online Survey Evaluation

The survey confirmed that all tasks are rather or very easy to accomplish ($M = 4.32$). The first task of finding and starting the bot was perceived as the most difficult ($M = 4.08$). Only one person failed to retrieve the warning status for a given and current location. The mean usability rating, again using the 10-point scale, is 76.88 ($SD = 17.58$), indicating good to excellent usability (Bangor et al., 2008). Fig. 14.6 depicts the evaluation of each item on a 5-point scale for easier comparison with the other evaluations. Aspects that received a particularly positive rating concerned being able to use the system right away and without needing expertise or training. Due to the conversational design, which results in previous inputs becoming less visible, we introduced an additional item asking about the transparency of the system's status, resulting in a relatively good rating ($M = 3.88, SD = 1.10$).

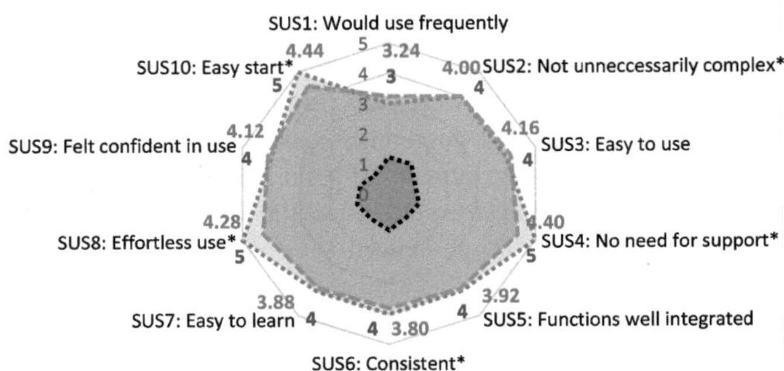

Fig. 14.6 Mean (orange, middle line), median (blue, outer line) and standard deviation (black, inner line) of each SUS item's 5-point scale rating, *inverted items

Regarding privacy, users perceived that the bot's handling of private information seemed trustworthy ($M = 3.69$, $SD = 1.30$), that they did not have qualms using the bot with regard to data security ($M = 3.73$, $SD = 1.31$) or privacy ($M = 3.73$, $SD = 1.06$). The download of the official prevention brochure as crisis prevention advice was perceived as helpful by all but two participants ($M = 3.88$, $SD = 1.41$). Most users appreciated the offline availability, although they would have preferred a more concise summary (the used one has seven pages). The new design iteration included emojis to denote the warning topics and severity. These were largely seen as helpful ($M = 3.88$, $SD = 1.08$). Asking about improvements, 10 users made comments which related to small design improvements (such as in the way the names of locations were presented) and to a technical issue that resulted in longer loading times (which was fixed in the course of the study).

Regarding the adoption intention, 20% fully agreed that they would like to use the bot frequently and almost half rather or fully agreed. If the bot were available on users' favourite messenger app, 20% more planned on using the bot to make subscriptions. Fewer people would use the bot to query the warning status of specific locations. Finally, another important indicator of the overall usefulness of a system is the likelihood of recommending a system to friends or family. The results are depicted in Table 14.3 and show that almost half of the participants would recommend the bot to friends and family.

Table 14.3 Usage and recommendation intentions, rated on a scale from 1 (fully disagree) to 5 (fully agree)

Intention to	M	SD	Fully agree	Agree or fully agree
… use frequently	3.25	1.22	20%	44%
… use on favourite messenger				
… for subscriptions	3.83	1.37	40%	68%
… for inquiries	3.25	1.15	12%	44%
… recommend system	3.50	1.10	20%	48%

In order to explain these preferences, we correlated the intentions to use the system with the individual technology affinity, the usability rating and the dimensions of the user engagement rating through a Kendall tau-b test. The results of the correlation analysis are depicted in Table 14.4. It shows that the intention to use the bot (item 1 of the SUS) correlates with the ATI, SUS and the reward dimension of the user engagement scale. It also correlates with the intention to use the bot in the future if available on users' favourite messenger app.

Table 14.4 Kendall tau-b correlation coefficients between usage intentions and evaluations (ATI = Affinity for Technology Interaction, SUS = System Usability Scale, UES = User Engagement Scale, RW = Reward dimension of the UES). $*p < .05$; $**p < .01$

Intention to	…use for subscriptions	…use for inquiries	…recommend	ATI	SUS	UES	RW
… use frequently	.366*	.359*	.233	.305*	.356*	.131	.404*
… use on favourite messenger							
… for subscription	1	.285	.412*	.259	.297	.351*	470**
… for inquiries		1	.139	.368*	.116	.236	.251
… recommend			1	.400*	.510**	.664**	.724**

People who would be willing to use the bot for its core function, warning subscriptions, rate it high in the reward dimension and would recommend the bot. People who would use the bot to inquire about the warning status, a secondary function introduced mainly to familiarise users with the bot, does not correlate with the perceived usefulness or the intention to recommend the bot. However, it correlated with technology affinity, further stressing that this is a feature for those who like to engage with technology. Affinity for technology, usability, and all dimensions of the user engagement scale (UES Perceived Usability $\tau_b = 0.410$, $p = 0.005$, UES Aesthetic Appeal $\tau_b = 0.531$, $p < 0.001$, UES Focused Attention $\tau_b = 0.615$, $p < 0.001$) all affect the intention to recommend the system.

Looking at users' reasoning, those planning to use the bot in the future mentioned the ease of use and not requiring the download of an additional app. People who do not plan on using the bot mentioned not needing a mobile phone to be made aware of severe crises and rather using different sources, such as browser searches, indicating a perceived low need to be informed in a timely manner, or a general disinterest in crisis warnings, e.g.: "I think that more serious matters would somehow reach me, and I wouldn't need a phone for that. I have never thought about it". Another group who are less interested are those people who are already using a warning app, who described little added value and some disadvantages regarding the user interface, particularly the display of the preparedness information.

Asked about the advantages and disadvantages of the bot compared to a warning app, users mentioned specifically the flexibility offered by the bot and not requiring another app compared to warning apps. One participant stated: "If I'm honest, I haven't used any of the apps yet because I didn't want to install another app and familiarise myself with it. Lack of storage space was also a thing a while ago.… A Telegram bot is more useful for me than another app" (P24). However, other users mentioned that the bot required a messenger app, particularly Telegram. Another perceived disadvantage is the lack of flexibility with regard to the user interface design in the messenger bot. While the handling of the bot received positive evaluations and was perceived as intuitive, one user mentioned that it might pose a challenge for older people: "[…] if the buttons are gone at the bottom, you have to press on a quite small icon to call them up again" (P16). Another disadvantage of the bot compared to apps or CB is that the latter require less or no set-up effort, but they can operate simply based on the live location, which is not the case for the bot (see Table 14.5 for the full list of themes).

Table 14.5 Users' reported advantages and disadvantages (n) of the warning bot compared to other warning channels, $N = 20$

	Advantages	Disadvantages
App requirement	No additional app needed (6)	Requires *Telegram* app (8)
Interface	Easy to use and to learn (4) Topic engagement through interactive design (1)	More difficult set-up compared with one-click GPS-based warnings in warning apps (2) Less attractive and flexible interface (2) - Better presentation of prevention information (1) - Display of maps (1)
Perceived usefulness	Subscriptions for different locations (4) Warning status queries (6)	Disinterest in warnings (4) Use of other channels, such as Internet search (4)
Resilience	–	Less resilient against Internet outages than CB (2) No warnings when in silent mode (1)
Privacy	Better data protection (1)	–

14.6   Discussion

In the following, we discuss the implications of the study for the design of warning channels and for bot-driven personalisation.

14.6.1 Warning "Botplications" Requirements

Based on the requirements for warning apps, the study has shown that messaging channels offer some, if not all, relevant functions that are central to delivering timely, personalised warnings.

Bonaretti and Fischer (2021) analyse warning apps with regard to different layers. Looking at these layers (see Table 14.6), we find that warning apps and warning messaging channels are identical in that they run on the same physical layer on smartphones. However, if an IM app has already been installed, users do not need to invest in additional storage space, and messages typically require a small amount of Internet data. In addition, although not widely available, messenger apps can use peer-to-peer communication, making them, to some extent, resilient to Internet outages. On the deep layer, concerning the alerting object, the bot can send the same alerts as the app, due to the availability of an API of the official alerting app. The crucial difference is at the software level, where one uses a warning app and the other an IM app, whereby the choice of messaging app should depend on the number of users. The surface layer, too, is highly dependent on the software used. Generally, options for sound and visuals are limited by the IM apps' functions and design.

The software layer also influences the provided warning and crisis functions. Regarding the functions, an important difference is that currently, location tracking is not possible within the chosen IM app. Therefore, warnings cannot be made based on the current location of users. However, users can allow short-term access to the location and the bot can use this information to suggest a warning subscription for the respective location. Due to this process, the set-up takes longer than in some warning apps, where the location-based warning is enabled as soon as users grant access to their location information. In addition, the bot cannot send warnings based on the current location, only based on predefined locations of interest. This leads to the bot requiring more set-up than a well-implemented warning app, which in its initial start can ask for the permission to use GPS, enabling warnings with only one click. Another difference is that it is currently not possible to show an interactive map within the IM app. Therefore, it is more difficult to show users that the app is active and to give them some insights into the warning status. We, therefore, implemented

Table 14.6 Layered nature of a warning app, layers based on Bonaretti & Fischer (2021)

Structure	Warning App	Bot
Deep layer (warning object)		
Data	Send official warnings through MoWaS	Send official warnings via API
Software application	Different apps and features offered, in Germany: *NINA, KATWARN, hessenWARN*	Popular messengers: *WhatsApp, Telegram, Signal*
Surface layer (notification)		
Sound	Individually adjustable; can override silent mode, often background activity deactivated due to infrequent use	Can be set for the "channels", but no specific override permissions
Visual	Can be freely designed	Limited design options, depending on messenger
Physical layer (hardware and network)		
Device	Requires data capacity and battery, e.g. for GPS tracking	Requires data capacity for messenger app, though often already installed; no additional battery drainage
Internet connection	Often background activity deactivated due to infrequent use	Requires Internet, possibility of peer-to-peer communication

a feature that allows users to query the warning status of locations, which offers an immediate way for users to interact with the bot and verify its functionality. This feature is not common in warning apps but may provide additional benefit for people who enjoy engaging with technology—especially as "site-seeing", i.e. digitally visiting locations affected by crises, is a common online activity (Hughes et al., 2008). Another important function is offering additional advice. However, the conversational interface appears inadequate for displaying longer texts. Therefore, we opted to offer advice through a file that can be downloaded. While users prefer to have all information in the warning app rather than through external links (Tan et al., 2020a), this allows saving the file on the device, making it available offline.

With regard to the conversational design, the implementation process has shown that many challenges arise from the onboarding process. This is in line with research on voice-based interfaces for visually impaired people, which finds that discoverability of the available functions is a challenge (Pradhan et al., 2018). For onboarding, we have chosen to let users query locations' warning status, as this provides an immediate insight into the capabilities of the bot and the type of output to be expected. In this way, users can experience the bot's warning function before setting up subscriptions. This interaction is used to invite users to set up warning subscriptions for the queried location, which the study confirms to be the key function of the bot. For the set-up process, the chosen settings need to be transparent. In addition, buttons can serve as guardrails to give guidance about the available options. The evaluation has shown that efficiency is a key element with the warning bot. Efficiency was improved by displaying colours and emojis to give immediate visual cues, reducing time spent reading. In addition, texts were shortened substantially. Providing sufficient information to navigate the bot while keeping text to a minimum is another challenge that is intensified in the messaging platform. Users expressly stated that the bot was easy to use. This indicates that the recommendation of buttons to guide the conversation (Klopfenstein et al., 2017) can be confirmed in this context.

While many studies indicate that human likeness increases positive evaluations of bots, this was not mentioned by users in our study. In a study of a municipal service bot, the relevance of human likeness was regarded very differently between users. This may be due to the character of the bot. While the municipal service navigation bot used natural language to support users, the bot developed here is closely related to an application. This supports the differentiation of botplication from other types of conversational bots, with botplications aiming for simplicity and effectiveness, whereas human likeness is irrelevant or secondary (Klopfenstein et al., 2017). The analysis, discussion, and evaluation of chatbots' maturity is mainly based on the use of natural language processing. A typology by (Makasi et al., 2022b) differentiated between first level bots that facilitate service triage (matching user request with rules and linked data sources); second level bots for service information gathering and analysis that access data from multiple sources to personalise the service; and third level bots for service negotiation, which entail predictive analytics and neural networks to evaluate customer circumstances based on various sources such as user profiles and service profiles and providing multiple personalised service options for the customer in their natural language. However, the task-oriented botplication described in this study does not fit into the typology. For example, in the typology, accessing user profiles or continuing service interaction occurs along with NLP components such as sentiment analysis. The bot used here shows that personalised service delivery does not require large language processing capabilities. Indeed, the design recommendations for CUIs within chatbots suggest a limited use of text and an extensive use of graphical UI elements to steer the conversation in an efficient manner (Klopfenstein et al., 2017).

14.6.2 The Role of Usability and Perceived Usefulness for Intended Future Use

Looking at the utility of the warning bot, we find that 68% of study participants stated that they would use the system if it was provided through their favourite IM app. In addition, almost half stated that they would recommend the bot.

The reward dimension of the user engagement scale describes aspects such as whether the time spent with the system was worthwhile and emerged as the most significant, impacting also the intentions to use the bot generally and specifically for warning subscriptions. The intention to recommend the warning bot only correlates with the intention to use it for warning subscriptions in a preferred IM app, indicating that this aspect is perceived as the most useful feature for other people. Thus, this underlines the added value of bots when coupled with proactive notifications, as this is not achieved through other sources, such as Internet searches.

The study suggests that usability, in the context of warning channels, is less relevant for predicting the intention to adopt a system. Instead, the intention to use the bot is more closely related to the utility of a system—aspects that are described in the reward dimension of the user engagement scale (O'Brien et al., 2018). This supports the value of the user engagement scale and its different dimensions, which has also proven useful in the health and news domains (Haque & Rubya, 2023; Holdener et al., 2020; O'Brien & Cairns, 2015). Therefore, future representative user studies that evaluate chatbot designs or compare them to alternative warning channels should include such aspects in addition to usability aspects. This result is in line with previous findings on e-government non-adoption. One study found that users preferred to remain with their status quo, interacting with humans: "The less need citizens feel for public services or the more convenient they find conventional ways of contacting administrations, the less they are inclined to use public e-services" (Distel, 2020). Adapted to the use of warning apps, we also find that many respondents, even when indicating that the tool is useful and has high usability, state that they would continue using public media for warnings. Similar to another qualitative study on e-government, we find that a perceived lack of need or infrequent usage is one of the main barriers (Distel, 2018). In the open statements, it also became evident that, at least for some of the participants, it had not become clear that the subscriptions would automatically alert them in case of a warning. As a result, two participants stated that they might not be querying information just as a crisis was impending, likely due to the unfamiliarity with bot-based subscriptions. Still, more attention should be paid to ensuring that users understand the scope of the offered function, e.g. through a short introductory video (Furqan et al., 2017; Yeh et al., 2022).

In this study, we sought to investigate the usability of a warning messaging bot as an alternative warning channel to warning apps. Therefore, we attempted to emulate features that are offered by the official German warning apps. However, warning apps could conceivably offer additional safety and security functions or include more accessibility features, such as haptic instead of auditory signals or video emergency calls for people with hearing impairments. The structure of the conversational interface introduced in this study can be used to design a fully voice-based process for people with impaired vision, allowing them to query information through natural speech or to set up automatic warning subscriptions. A study shows that conversational assistants provide a range of useful functions for blind people (Pradhan et al., 2018). With added natural language processing functions, the developed conversational design could be used to make personalised warnings accessible to further user groups, who tend to prefer to be able to use mainstream channels rather than specifically designed ones (Martiniello et al., 2022). However, moving towards accessibility would require extensive user studies and changes in the way that information and alert signals are presented (Sherman-Morris et al., 2020).

While the additional steps needed to set up a subscription may be perceived as a disadvantage, one user reported that the process increased engagement with crisis management, possibly by confronting users with the available selection of warning types. This finding may echo an increased engagement with municipal services that was achieved through the use of a chatbot (Abbas et al., 2023).

14.6.3 Limitations and Future Work

The paper has several limitations regarding the interactions that were analysed and the design of the warning bot. Firstly, due to ethical concerns related to ensuring the reliability of the bot "in the wild", the evaluation did not encompass real-world use, restricting the measurement to participants' experiences in the set-up and their intentions for future use. Future work should create an experimental setting that allows the usability of the alerts to be tested. One question addressed the advantages of a messenger bot compared to other warning channels. This question was answered by both current users and non-users of warning apps. Future work should explore the affordances of a messenger warning bot with the affordances of warning apps and cell broadcast, similar to a study which has analysed affordances of news consumption via messenger bots (Lou et al., 2021). In addition, the evaluation is biased towards younger users, who appear to be more willing to use the bot in the future and who are likely to be experienced messenger app users (Statista, 2024b).

With regard to using messenger apps as the containers of bots, we find that some messenger apps offer significant features for open source, low-cost implementations. However, a disadvantage is that certain aspects cannot be adapted. In this prototype, a significant usability limitation is that the text input fields cannot be hidden or disabled. This suggests to users that text inputs may be required. In addition, activation of the text input field can lead to the disappearing of the buttons, requiring some familiarity to recuperate the buttons. In addition, other research has also explored the relevance of humanness of bots with inconclusive results and suggested a dependence on the task and domain (Rapp et al., 2021). Inspired by the usability requirement of simplicity, we did not implement any humanness in the bot in the warning context. Future research could explore the relevance of this attribute through a comparative experiment.

Regarding the conversational design of the bot, due to the importance of achieving correct results in a safety-critical domain, the current prototype does not use natural language processing (NLP). While this does justice to the requirement of minimal input, especially for mobile use, and reduces errors, NLP could enable users to interact more freely with the bot like (Staegemann et al., 2022) have put forth. However, our system's backend, functions and conversational design structure could be used for future accessibility studies, where the messenger or smart speakers may be used to set up the warning bot preferences and where a smart home environment may be used for visual, haptic or auditory signals adapted to users' accessibility needs. This could enable users with visual and hearing impairments to set up and receive noticeable alerts. A previous study has already suggested different signals and intensities to communicate varying warning levels (Haesler et al., 2023). The use of natural language for output would also allow conveying more extensive information regarding crisis preparedness. Using NLP, the bot could answer specific questions related to crisis preparedness upon request. Previous research suggests that such information should include statistical information that stresses the benefits of crisis preparedness (Haunschild et al., 2023).

14.7 Conclusion

The use of *botplications*, i.e. bots that fulfil the tasks currently carried out by apps, holds the promise of lowering adoption barriers by not requiring the download of specific apps for specific functions. Therefore, this study explores the use of a bot run within an IM app as an alternative to warning apps. However, the design of messaging bots, particularly for conversational interfaces, presents distinct challenges and necessitates specific considerations, as these interfaces are characterised by sequential interactions.

Using the design science research approach, this study investigated how functions and usability requirements of warning apps can be transferred to a messaging warning channel. A comprehensive set of requirements for graphically enhanced CUIs in safety-critical contexts was developed. After designing and implementing a warning bot using the API of an official warning app, two rounds of user evaluation provided valuable insights into the bot's usability and user perceptions. The usability analysis shows a positive evaluation, indicating that user-friendly bot-based personalisation, while currently still uncommon in news delivery, can be accomplished. The task-based interviews indicated positive user feedback in general, with only a small percentage requiring some support. Suggestions for design improvements included shorter texts, increased use of symbols and emojis, and the addition of short affirmative messages upon successful actions. These insights guided refinements for the subsequent iteration. Despite the generally high usability rating, the intention to use the system in the future varied. The second round of evaluation, which explored additional elements related to user engagement, showed that particularly elements related to perceived usefulness were relevant to the intention to keep using the system and to recommend it. This indicative finding underlines the relevance of perceived usefulness in addition to usability for technology adoption and should be further verified for warning bots in future studies. The qualitative findings show that a general disinterest in crisis warnings and a low perceived personal responsibility for crisis engagement are the main factors for not wanting to use the tool.

A second strand of reasoning is related to the use of other warning channels. The warning bot prototype and user evaluation indicate that reliable warnings tailored to personal preferences can be provided through IM apps as a viable addition to warning apps or cell broadcast. Though current IM apps offer clear technical constraints like the unavailability of live location tracking or having to adjust auditory alert levels manually. This limits the promise of warning bots and points to the app developers, who could introduce warning-related functions and thus ease the bot set-up process and lower adoption barriers.

In summary, the research demonstrates the significance of tailored conversational design for messaging bots in disaster contexts. The iterative design process, supported by user evaluation, led to a refined interface capable of efficiently delivering trustworthy crisis warnings. The final graphically enhanced CUI shows that a suitable warning bot does not necessarily require complex NLP capabilities. The study hence contributes valuable insights for the development of effective and user-friendly messaging bots in safety-critical applications. Future research should compare the use of a warning bot and a warning app "in the wild" to expand the research focus from set-up to warning perception. The available prototype can furthermore be used for conversational personalisation of warning channels that can increase the accessibility of mobile warnings or warnings in a smart home context.

Bibliography

Abbas, N., Følstad, A., & Bjørkli, C. A. (2023). Chatbots as Part of Digital Government Service Provision—A User Perspective. In A. Følstad, T. Araujo, S. Papadopoulos, E. L.-C. Law, E. Luger, M. Goodwin, & P. B. Brandtzaeg (Eds.), *Chatbot Research and Design* (pp. 66–82, Vol. 13815). Springer International Publishing. https://doi.org/10.1007/978-3-031-25581-6_5

Aboulafia, M. (1991). *Philosophy, Social Theory, and the Thought of George Herbert Mead*. State University of New York Press.

Abushnaf, J., Rassau, A., & Górnisiewicz, W. (2015). Impact of Dynamic Energy Pricing Schemes on a Novel Multi-User Home Energy Management System. *Electric power systems research, 125*, 124–132. https://doi.org/10.1016/j.epsr.2015.04.003

Acquisti, A., Adjerid, I., Balebako, R., Brandimarte, L., Cranor, L. F., Komanduri, S., Leon, P. G., Sadeh, N., Schaub, F., Sleeper, M., Wang, Y., & Wilson, S. (2017). Nudges for Privacy and Security: Understanding and Assisting Users' Choices Online. *ACM Computing Surveys (CSUR), 50*(3), 1–41. https://doi.org/10.1145/3054926

ADAC e.V. (2021). ADAC Umfrage Zu Erste-Hilfe-Kenntnissen. Retrieved July 27, 2023, from https://presse.adac.de/meldungen/adac-ev/tests/adac-umfrage-zu-erste-hilfe-kenntnissen.html

Adam, Marc. . T. P., Gregor, S., Hevner, A., & Morana, S. (2021). Design Science Research Modes in Human-Computer Interaction Projects. *AIS Transactions on Human-Computer Interaction*, 1–11. https://doi.org/10.17705/1thci.00139

Agrawal, A., Abraham, S. J., Burger, B., Christine, C., Fraser, L., Hoeksema, J. M., Hwang, S., Travnik, E., Kumar, S., & Scheirer, W. (2020). The Next Generation of Human-Drone Partnerships: Co-designing an Emergency Response System. *Proceedings of the 2020 CHI Conference on Human Factors in Computing Systems*, 1–13.

Ahmed, N., Michelin, R. A., Xue, W., Ruj, S., Malaney, R., Kanhere, S. S., Seneviratne, A., Hu, W., Janicke, H., & Jha, S. K. (2020). A Survey of COVID-19 Contact Tracing Apps. *IEEE Access, 8*, 134577–134601. https://doi.org/10.1109/ACCESS.2020.3010226

Ajzen, I. (1991). The Theory of Planned Behavior. *Organizational Behavior and Human Decision Processes, 50*(2), 179–211.

Akkaya, C., Fedorowicz, J., & Krcmar, H. (2019). Successful Practices for Using Social Media by Police Departments: A Case Study of the Munich Police. *Proceedings of the 27th European Conference on Information Systems (ECIS)*.

J. Haunschild, *Enhancing Citizens' Role in Public Safety*, Technology, Peace and Security | Technologie, Frieden und Sicherheit, https://doi.org/doi.org/10.1007/978-3-658-46489-9

Alam, F., Ofli, F., & Imran, M. (2020). Descriptive and Visual Summaries of Disaster Events Using Artificial Intelligence Techniques: Case Studies of Hurricanes Harvey, Irma, and Maria. *Behaviour & Information Technology, 39*(3), 288–318. https://doi.org/10.1080/0144929X.2019.1610908

Alcaraz, C., & Zeadally, S. (2015). Critical Infrastructure Protection: Requirements and Challenges for the 21st Century. *International Journal of Critical Infrastructure Protection, 8*, 53–66. https://doi.org/10.1016/j.ijcip.2014.12.002

Allcott, H. (2009). Real Time Pricing and Electricity Markets. *Harvard University, 7.*

Allcott, H. (2011). Social Norms and Energy Conservation. *Journal of Public Economics, 95*(9–10), 1082–1095. https://doi.org/10.1016/j.jpubeco.2011.03.003

AlSkaif, T., Lampropoulos, I., van den Broek, M., & van Sark, W. (2018). Gamification-Based Framework for Engagement of Residential Customers in Energy Applications. *Energy Research & Social Science, 44*, 187–195. https://doi.org/10.1016/j.erss.2018.04.043

Altman, D. G. (1991). *Practical Statistics for Medical Research.* Chapman & Hall.

Altmann, S., Milsom, L., Zillessen, H., Blasone, R., Gerdon, F., Bach, R., Kreuter, F., Nosenzo, D., Toussaert, S., & Abeler, J. (2020). Acceptability of App-Based Contact Tracing for COVID-19: Cross-country Survey Study. *JMIR mHealth and uHealth, 8*(8), e19857. https://doi.org/10.2196/19857

American Psychological Association. (2016). Ethical Principles of Psychologists and Code of Conduct. https://www.apa.org/ethics/code

Appleby-Arnold, S., & Brockdorff, N. (2018). *Culture and Disaster Risk Management—Synthesis of Citizens' Reactions and Opinions during 6 Citizen Summits : Romania, Malta, Italy, Germany, Portugal and the Netherlands.* https://www.um.edu.mt/library/oar/handle/123456789/53192

Appleby-Arnold, S., Brockdorff, N., & Callus, C. (2021). Developing a "Culture of Disaster Preparedness": The Citizens' View. *International Journal of Disaster Risk Reduction, 56*, 102133. https://doi.org/10.1016/j.ijdrr.2021.102133

Appleby-Arnold, S., Brockdorff, N., Fallou, L., & Bossu, R. (2019). Truth, Trust, and Civic Duty: Cultural Factors in Citizens' Perceptions of Mobile Phone Apps and Social Media in Disasters. *Journal of Contingencies and Crisis Management (JCCM)*, 1–13. https://doi.org/10.1111/1468-5973.12282

Appleby-Arnold, S., Brockdorff, N., Jakovljev, I., & Zdravković, S. (2020). Disaster Preparedness and Cultural Factors: A Comparative Study in Romania and Malta. *Disasters.* https://doi.org/10.1111/disa.12433

Ariely, D., Kamenica, E., & Prelec, D. (2008). Man's Search for Meaning: The Case of Legos. *Journal of Economic Behavior & Organization, 67*(3), 671–677. https://doi.org/10.1016/j.jebo.2008.01.004

Ashktorab, Z., Jain, M., Liao, Q. V., & Weisz, J. D. (2019). Resilient Chatbots: Repair Strategy Preferences for Conversational Breakdowns. *Proceedings of the 2019 CHI Conference on Human Factors in Computing Systems*, 1–12. https://doi.org/10.1145/3290605.3300484

Austin, L., Liu, B. F., & Jin, Y. (2012). How Audiences Seek out Crisis Information: Exploring the Social-Mediated Crisis Communication Model. *Journal of Applied Communication Research, 40*(2), 188–207. https://doi.org/10.1080/00909882.2012.654498

Avery, C., Castleman, B. L., Hurwitz, M., Long, B. T., & Page, L. C. (2021). Digital Messaging to Improve College Enrollment and Success. *Economics of Education Review, 84*, 102170. https://doi.org/10.1016/j.econedurev.2021.102170

Ayachi, R., Boukhris, I., Mellouli, S., Ben Amor, N., & Elouedi, Z. (2016). Proactive and Reactive E-government Services Recommendation. *Universal Access in the Information Society, 15*(4), 681–697. https://doi.org/10.1007/s10209-015-0442-z

Ayres, I., Raseman, S., & Shih, A. (2013). Evidence from Two Large Field Experiments That Peer Comparison Feedback Can Reduce Residential Energy Usage. *The Journal of Law, Economics, and Organization, 29*(5), 992–1022. https://doi.org/10.1093/jleo/ews020

Balebako, R., Leon, P. G., Almuhimedi, H., Kelley, P. G., Mugan, J., Acquisti, A., Cranor, L. F., & Sadeh, N. (2011). Nudging Users towards Privacy on Mobile Devices. *Proc. CHI 2011 Workshop on Persuasion, Nudge, Influence and Coercion*, 193–201.

Bangor, A., Kortum, P. T., & Miller, J. T. (2008). An Empirical Evaluation of the System Usability Scale. *International Journal of Human-Computer Interaction, 24*(6), 574–594. https://doi.org/10.1080/10447310802205776

Bassi, A., Arfin, S., John, O., & Jha, V. (2020). An Overview of Mobile Applications (Apps) to Support the Coronavirus Disease 2019 Response in India. *The Indian Journal of Medical Research, 151*(5), 468–473. https://doi.org/10.4103/ijmr.IJMR_1200_20

BBK. (2010). Neue Strategie zum Schutz der Bevölkerung in Deutschland. *Bundesamt für Bevölkerungsschutz und Katastrophenhilfe.* https://www.bbk.bund.de/SharedDocs/Downloads/DE/Mediathek/Publikationen/WF/WF-04-neue-strategie-bevsch.pdf?__blob=publicationFile&v=7

BBK. (2020). Neuer Corona-Informationsbereich in der Warn-App NINA Version 3.1. *Bundesamt für Bevölkerungsschutz und Katastrophenhilfe.* https://www.bbk.bund.de/SharedDocs/Kurzmeldungen/DE/2020/04/nina-corona-infobereich.html

BBK. (2021). Warn-App NINA. *Bundesamt für Bevölkerungsschutz und Katastrophenhilfe.* https://www.bbk.bund.de/DE/Warnung-Vorsorge/Warn-App-NINA/warn-app-nina_node.html

Bean, H., Liu, B. F., Madden, S., Sutton, J., Wood, M. M., & Mileti, D. S. (2016). Disaster Warnings in Your Pocket: How Audiences Interpret Mobile Alerts for an Unfamiliar Hazard. *Journal of Contingencies and Crisis Management, 24*(3), 136–147. https://doi.org/10.1111/1468-5973.12108

Bean, H., Sutton, J., Liu, B. F., Madden, S., Wood, M. M., & Mileti, D. S. (2015). The Study of Mobile Public Warning Messages: A Research Review and Agenda. *Review of Communication, 15*(1), 60–80.

Beermann, V., Rieder, A., & Uebernickel, F. (2022). Green Nudges: How to Induce Pro-Environmental Behavior Using Technology. https://aisel.aisnet.org/icis2022/hci_robot/hci_robot/15

Beheshtian, N., Moradi, S., Ahtinen, A., Väänanen, K., Kähkonen, K., & Laine, M. (2020). GreenLife: A Persuasive Social Robot to Enhance the Sustainable Behavior in Shared Living Spaces. *Proceedings of the 11th Nordic Conference on Human-Computer Interaction: Shaping Experiences, Shaping Society.* https://doi.org/10.1145/3419249.3420143

Belton, C. A., & Lunn, P. D. (2020). Smart Choices? An Experimental Study of Smart Meters and Time-of-Use Tariffs in Ireland. *Energy Policy, 140*, 111243. https://doi.org/10.1016/j.enpol.2020.111243

Bendau, A., Petzold, M. B., Pyrkosch, L., Mascarell Maricic, L., Betzler, F., Rogoll, J., Große, J., Ströhle, A., & Plag, J. (2021). Associations between COVID-19 Related Media Consumption and Symptoms of Anxiety, Depression and COVID-19 Related Fear in the General

Population in Germany. *European Archives of Psychiatry and Clinical Neuroscience, 271*, 283–291.

Bento, A. I., Nguyen, T., Wing, C., Lozano-Rojas, F., Ahn, Y. Y., & Simon, K. (2020). Evidence from Internet Search Data Shows Information-Seeking Responses to News of Local COVID-19 Cases. *Proceedings of the National Academy of Sciences of the United States of America, 117*(21), 11220–11222. https://doi.org/10.1073/pnas.2005335117

Bergram, K., Djokovic, M., Bezençon, V., & Holzer, A. (2022). The Digital Landscape of Nudging: A Systematic Literature Review of Empirical Research on Digital Nudges. *Proceedings of the 2022 CHI Conference on Human Factors in Computing Systems*, 1–16.

Berliner Feuerwehr. (2023). *KATRETTER*. Retrieved October 5, 2023, from https://www.berliner-feuerwehr.de/ihre-sicherheit/praevention/katretter/

Bertot, J. C., Jaeger, P. T., & Grimes, J. M. (2012). Promoting Transparency and Accountability through ICTs, Social Media, and Collaborative E-government. *Transforming government: people, process and policy, 6*(1), 78–91. https://doi.org/https://doi.org/10.1108/17506161211214831

Bijker, W. E., Hughes, T. P., & Pinch, T. (Eds.). (1987). *The Social Construction of Technological Systems: New Directions in the Sociology and History of Technology*. MIT Press.

BMG. (2021). Informationen zum Coronavirus. *Bundesministerium für Gesundheit*.

BMI. (2009). *Nationale Strategie zum Schutz Kritischer Infrastrukturen (KRITIS-Strategie)*. Retrieved March 10, 2023, from https://www.bmi.bund.de/SharedDocs/downloads/DE/publikationen/themen/bevoelkerungsschutz/kritis.pdf?__blob=publicationFile&v=3

BMI. (2022, July 13). *Deutsche Strategie zur Stärkung der Resilienz gegenüber Katastrophen. Umsetzung des Sendai Rahmenwerks für Katastrophenvorsorge (2015–2030)—Der Beitrag Deutschlands 2022–2030*. Bonn. https://www.bmi.bund.de/SharedDocs/downloads/DE/publikationen/themen/bevoelkerungsschutz/BMI22017-resilienz-katastrophen.pdf?__blob=publicationFile&v=2

BMWK. (2021a, September). Schnelle Hilfe beim Wiederaufbau und eine Bessere Warnung der Bevölkerung. *Bundesministerium für Wirtschaft und Klimaschutz*. https://www.bmi.bund.de/SharedDocs/kurzmeldungen/DE/2021/09/%20aufbauhilfegesetz-cell-broadcast.html

BMWK. (2021b, November). Mobilfunk-Warn-Verordnung. Verordnung des Bundesministeriums für Wirtschaft und Klimaschutz. *Bundesministerium für Wirtschaft und Klimaschutz*. https://www.bmwk.de/Redaktion/DE/Artikel/Service/Gesetzesvorhaben/mobilfunk-warn-verordnung.html

Bonaretti, D., & Fischer, D. (2021). Timeliness, Trustworthiness, and Situational Awareness: Three Design Goals for Warning with Emergency Apps. *Forty-Second International Conference on Information Systems*, 1–17.

Bourgeois, J., van der Linden, J., Kortuem, G., Price, B. A., & Rimmer, C. (2014). Conversations with My Washing Machine: An in-the-Wild Study of Demand Shifting with Self-Generated Energy. *Proceedings of the 2014 ACM International Joint Conference on Pervasive and Ubiquitous Computing*, 459–470. https://doi.org/10.1145/2632048.2632106

boyd, D. (2010). Social Network Sites as Networked Publics: Affordances, Dynamics, and Implications. In *A networked self* (pp. 47–66). Routledge.

Braithwaite, I., Callender, T., Bullock, M., & Aldridge, R. W. (2020). Automated and Partly Automated Contact Tracing: A Systematic Review to Inform the Control of COVID-19. *The Lancet Digital Health, 2*(11). https://doi.org/10.1016/S2589-7500(20)30184-9

Braun, V., & Clarke, V. (2006). Using Thematic Analysis in Psychology. *Qualitative Research in Psychology, 3*(2), 77–101.

Brooke, J. (1995). SUS: A Quick and Dirty Usability Scale. *Usability Evaluation in Industry, 189.*

Brooks, B., Curnin, S., Owen, C., & Bearman, C. (2020). Managing Cognitive Biases during Disaster Response: The Development of an Aide Memoire. *Cognition, Technology and Work, 22*(2), 249–261. https://doi.org/10.1007/s10111-019-00564-5

Brynielsson, J., Granåsen, M., Lindquist, S., Narganes Quijano, M., Nilsson, S., & Trnka, J. (2018). Informing Crisis Alerts Using Social Media: Best Practices and Proof of Concept. *Journal of Contingencies and Crisis Management, 26*(1), 28–40. https://doi.org/10.1111/1468-5973.12195

BSI. (2023). *Warn- und Informationsdienst—Bürger-CERT.* Retrieved September 26, 2023, from https://wid.cert-bund.de/portal/wid/buergercert

Buchanan, R. (1992). Wicked Problems in Design Thinking. *Design Issues, 8*(2), 5–21.

Bullock, K. (2018). The Police Use of Social Media: Transformation or Normalisation? *Social Policy and Society, 17*(2), 245–258. https://doi.org/10.1017/s1474746417000112

Bullock, K., Garland, J., & Coupar, F. (2021). Police-Community Engagement and the Affordances and Constraints of Social Media. *Policing and society, 31*(4), 373–385.

Bund.dev. (2022). *Bundesamt für Bevölkerungsschutz—NINA API- OpenAPI Documentation.* https://nina.api.bund.dev

Bundesnetzagentur. (2022, November 23). *Technische Richtlinie DE-Alert.* https://www.bundesnetzagentur.de/SharedDocs/Downloads/DE/Sachgebiete/Telekommunikation/Unternehmen_Institutionen/Anbieterpflichten/OeffentlicheSicherheit/DEAlert/Downlaod_TR-DE-Alert_1.1_Final.pdf?__blob=publicationFile&v=2

Bundesregierung. (2021, October 5). *Warn-App NINA mit lokalen Corona-Regelungen.* Retrieved May 5, 2022, from https://www.bundesregierung.de/breg-de/themen/%20coronavirus/warn-app-nina-1749214

Bündnis Entwicklung Hilft. (2022). *World Risk Report 2022.* https://weltrisikobericht.de/wp-content/uploads/2022/09/WorldRiskReport-2022_Online.pdf

Burgess, M. (2020). Coronavirus Contact Tracing Apps Were Meant to Save Us. They Won't. *WIRED.* https://www.wired.co.uk/article/contact-tracing-apps-coronavirus

Burton-Jones, A., & Grange, C. (2013). From Use to Effective Use: A Representation Theory Perspective. *Information Systems Research, 24*(3), 632–658. https://doi.org/10.1287/isre.1120.0444

Business Insider. (2016). *The Messaging App Report: Messaging Apps Are Now Bigger than Social Networks.* Retrieved August 24, 2023, from https://www.businessinsider.com/the-messaging-app-report-2015-11

Calo, R. (2014). Code, Nudge, or Notice? *Iowa Law Rreview, 99*(2), 773–802. https://ssrn.com/abstract=2217013

Canadian Red Cross. (2012). *Social Media during Emergencies.* http://www.redcross.ca/crc/documents/Social-Media-in-Emergencies-Survey-Oct-2012-English.pdf

Canadian Red Cross. (2021). *Disaster and Emergency Preparedness Survey: Floods, Wildfires, and Earthquake Awareness and Preparedness Objectives.* https://uwaterloo.ca/inclusive-resilience/sites/default/files/uploads/documents/uwaterloo_p4a_climate_change_survey_report_revised_nov_16.pdf

Caraban, A., Karapanos, E., Gonçalves, D., & Campos, P. (2019). 23 Ways to Nudge: A Review of Technology-Mediated Nudging in Human-Computer Interaction. *Conference on Human Factors in Computing Systems—Proceedings*, 1–15. https://doi.org/10.1145/3290605.3300733

Caraban, A., Konstantinou, L., & Karapanos, E. (2020). The Nudge Deck: A Design Support Tool for Technology-Mediated Nudging. *Proceedings of the 2020 ACM Designing Interactive Systems Conference (DIS 2020)*, 395–406. https://doi.org/10.1145/3357236.3395485

Centers for Disease Control and Prevention. (2020). *Crisis and Emergency Risk Communication (CERC) Manual*. Retrieved August 18, 2023, from https://emergency.cdc.gov/cerc/ppt/cerc_2014edition_Copy.pdf

Chater, N. (2020). Facing up to the Uncertainties of COVID-19. *Nature Human Behaviour, 4*, 439. https://doi.org/10.1038/s41562-020-0865-2

Chattaraman, V., Kwon, W.-S., Gilbert, J. E., & Ross, K. (2019). Should ai-based, conversational digital assistants employ social- or task-oriented interaction style? a task-competency and reciprocity perspective for older adults. *Computers in Human Behavior, 90*, 315–330. https://doi.org/10.1016/j.chb.2018.08.048

Chaves, A. P., & Gerosa, M. A. (2021). How Should My Chatbot Interact? A Survey on Social Characteristics in Human–Chatbot Interaction Design. *International Journal of Human–Computer Interaction, 37*(8), 729–758. https://doi.org/10.1080/10447318.2020.1841438

Chen, C. Y., Xu, W., Dai, Y., Xu, W., Liu, C., Wu, Q., Gao, L., Kang, Z., Hao, Y., & Ning, N. (2019). Household Preparedness for Emergency Events: A Cross-Sectional Survey on Residents in Four Regions of China. *BMJ Open, 9*(11), e032462. https://doi.org/10.1136/bmjopen-2019-032462

Chen, L., & Aklikokou, A. K. (2020). Determinants of E-government Adoption: Testing the Mediating Effects of Perceived Usefulness and Perceived Ease of Use. *International Journal of Public Administration, 43*(10), 850–865. https://doi.org/10.1080/01900692.2019.1660989

Chidambaram, S., Erridge, S., Kinross, J., & Purkayastha, S. (2020). Observational Study of UK Mobile Health Apps for COVID-19. *THE LANCET Digital Health, 2*(8), 388–390. https://doi.org/10.1016/S2589-7500(20)30144-8

Chittaro, L. (2019). Exploring the Use of Arcade Game Elements for Attitude Change: Two Studies in the Aviation Safety Domain. *International Journal of Human Computer Studies, 127*, 112–123. https://doi.org/10.1016/j.ijhcs.2018.07.006

Chittaro, L., & Buttussi, F. (2016). Tailoring Web Pages for Persuasion on Prevention Topics: Message Framing, Color Priming, and Gender. *International Conference on Persuasive Technology*, 3–14. https://doi.org/10.1007/978-3-319-31510-2_1

Chittaro, L., Sioni, R., Crescentini, C., & Fabbro, F. (2017). Mortality Salience in Virtual Reality Experiences and Its Effects on Users' Attitudes towards Risk. *International Journal of Human Computer Studies, 101*, 10–22. https://doi.org/10.1016/j.ijhcs.2017.01.002

Chittaro, L., & Zangrando, N. (2010). The Persuasive Power of Virtual Reality: Effects of Simulated Human Distress on Attitudes towards Fire Safety. *International Conference on Persuasive Technology*, 58–69. https://doi.org/10.1007/978-3-642-13226-1_8

Chon, M.-G., & Park, H. (2021). Predicting Public Support for Government Actions in a Public Health Crisis: Testing Fear, Organization-Public Relationship, and Behavioral Intention in the Framework of the Situational Theory of Problem Solving. *Health Communication, 36*(4), 476–486. https://doi.org/10.1080/10410236.2019.1700439

Chung, J. E., Park, N., Wang, H., Fulk, J., & McLaughlin, M. (2010). Age Differences in Perceptions of Online Community Participation among Non-Users: An Extension of the Technology Acceptance Model. *Computers in Human Behavior, 26*(6), 1674–1684. https://doi.org/10.1016/j.chb.2010.06.016

Cinelli, M., Quattrociocchi, W., Galeazzi, A., Valensise, C. M., Brugnoli, E., Schmidt, A. L., Zola, P., Zollo, F., & Scala, A. (2020). The COVID-19 Social Media Infodemic. *Scientific Reports, 10*(16598). https://doi.org/10.1038/s41598-020-73510-5

Civil Denfence NZ. (2020). *Disaster Preparedness Survey 2020.* New Zealand National Emergrency Management Agency. https://www.civildefence.govt.nz/assets/Uploads/public-education/report-disaster-preparedness-survey-en-dec20.pdf

Clark, C., Davila, A., Regis, M., & Kraus, S. (2020). Predictors of COVID-19 Voluntary Compliance Behaviors: An International Investigation. *Global Transitions, 2*, 76–82. https://doi.org/10.1016/j.glt.2020.06.003

Cobb, C., McCarthy, T., Perkins, A., Bharadwaj, A., Comis, J., Do, B., & Starbird, K. (2014). Designing for the Deluge: Understanding & Supporting the Distributed, Collaborative Work of Crisis Volunteers. *Proceedings of the 17th ACM Conference on Computer Supported Cooperative Work & Social Computing*, 888–899. https://doi.org/10.1145/2531602.2531712

Cohen, J. (1988). *Statistical Power Analysis for the Behavioral Sciences* (2nd ed.). Routledge. https://doi.org/10.4324/9780203771587

Coleman, S., & Blumler, J. G. (2009). *The Internet and Democratic Citizenship: Theory, Practice and Policy.* Cambridge University Press.

Colnago, J., Feng, Y., Palanivel, T., Pearman, S., Ung, M., Acquisti, A., Cranor, L. F., & Sadeh, N. (2020). Informing the Design of a Personalized Privacy Assistant for the Internet of Things. *Proceedings of the 2020 CHI Conference on Human Factors in Computing Systems*, 1–13. https://doi.org/10.1145/3313831.3376389

Coombs, W. T. (2007). *Crisis Management and Communications.* https://instituteforpr.org/crisis-management-and-communications/

Cornia, A., Dressel, K., & Pfeil, P. (2016). Risk Cultures and Dominant Approaches towards Disasters in Seven European Countries. *Journal of Risk Research, 19*(3), 288–304. https://doi.org/10.1080/13669877.2014.961520

CoroBuddy GbR. (2021). *Corobuddy—Deine App für die Coronaregeln.* https://www.corobuddy.de/

Coskun, A., Zimmerman, J., & Erbug, C. (2015). Promoting Sustainability through Behavior Change: A Review. *Design Studies, 41*, 183–204. https://doi.org/10.1016/j.destud.2015.08.008

Crawford, K., & Finn, M. (2015). The Limits of Crisis Data: Analytical and Ethical Challenges of Using Social and Mobile Data to Understand Disasters. *GeoJournal, 80*(4), 491–502. https://doi.org/10.1007/s10708-014-9597-z

Cronqvist, M., Farbøl, R., & Sylvest, C. (2022). *Cold War Civil Defence in Western Europe: Sociotechnical Imaginaries of Survival and Preparedness.* Springer Nature.

Crump, J. (2011). What Are the Police Doing on Twitter? Social Media, the Police and the Public. *Policy & Internet, 3*(4), 1–27. https://doi.org/10.2202/1944-2866.1130

Cvetković, V., Roder, G., Öcal, A., Tarolli, P., & Dragićević, S. (2018). The Role of Gender in Preparedness and Response Behaviors towards Flood Risk in Serbia. *International*

Journal of Environmental Research and Public Health, 15(12), 2761. https://doi.org/10.3390/ijerph15122761

Da Silva, T. S., Martin, A., Maurer, F., & Silveira, M. (2011). User-Centered Design and Agile Methods: A Systematic Review. *2011 AGILE Conference*, 77–86.

Dadaczynski, K., Okan, O., Messer, M., Leung, A., Rosário, R., Darlington, E., & Rathmann, K. (2021). Digital Health Literacy and Web-Based Information-Seeking Behaviors of University Students in Germany during the COVID-19 Pandemic: Cross-sectional Survey Study. *Journal of Medical Internet Research, 23*(1), e24097. https://doi.org/10.2196/24097

Dai, H., Saccardo, S., Han, M. A., Roh, L., Raja, N., Vangala, S., Modi, H., Pandya, S., Sloyan, M., & Croymans, D. M. (2021). Behavioral Nudges Increase COVID-19 Vaccinations. *Nature, 597*, 404–409. https://doi.org/10.1038/s41586-021-03843-2

Dailey, D., & Starbird, K. (2015). "It's Raining Dispersants": Collective Sensemaking of Complex Information in Crisis Contexts. *Proceedings of the 18th ACM Conference Companion on Computer Supported Cooperative Work & Social Computing*, 155–158. https://doi.org/10.1145/2685553.2698995

Dailey, D., Starbird, K., Lloyd, M., & Friedland, L. A. (2016). Addressing the Information Needs of Crisis-Affected Communities: The Interplay of Legacy Media and Social Media in a Rural Disaster. In *The Communication Crisis in America, And How to Fix It* (pp. 285–303). https://doi.org/10.1057/978-1-349-94925-0_18

Dallo, I., & Martí, M. (2021). Why Should I Use a Multi-Hazard App? Assessing the Public's Information Needs and App Feature Preferences in a Participatory Process. *International Journal of Disaster Risk Reduction, 57*, 102197. https://doi.org/10.1016/j.ijdrr.2021.102197

Davalbhakta, S., Advani, S., Kumar, S., Agarwal, V., Bhoyar, S., Fedirko, E., Misra, D. P., Goel, A., Gupta, L., & Agarwal, V. (2020). A Systematic Review of Smartphone Applications Available for Corona Virus Disease 2019 (COVID19) and the Assessment of Their Quality Using the Mobile Application Rating Scale (MARS). *Journal of Medical Systems, 44*(9), 1–15. https://doi.org/10.1007/s10916-020-01633-3

Davis, F. D. (1989). Perceived Usefulness, Perceived Ease of Use, and User Acceptance of Information Technology. *MIS Quarterly, 13*(3), 319. https://doi.org/10.2307/249008

Deci, E. L. (1975). *Intrinsic Motivation*. Plenum.

de Graaf, G., & Meijer, A. (2019). Social Media and Value Conflicts: An Explorative Study of the Dutch Police. *Public Administration Review, 79*(1), 82–92. https://doi.org/10.1111/puar.12914

Dehling, T., Lins, S., & Sunyaev, A. (2019). Security of Critical Information Infrastructures. In C. Reuter (Ed.), *Information Technology for Peace and Security* (pp. 319–339). Springer Fachmedien Wiesbaden. https://doi.org/10.1007/978-3-658-25652-4_15

Delcker, J. (2018). Big Brother in Berlin. As Germany Dabbles in State Surveillance, Facial-Recognition Technology Raises Privacy Concerns. *POLITICO*. https://www.politico.eu/article/berlin-big-brother-state-surveillance-facial-recognition-technology/

Denef, S., Bayerl, P. S., & Kaptein, N. A. (2013). Social Media and the Police: Tweeting Practices of British Police Forces during the August 2011 Riots. *Proceedings of the SIGCHI Conference on Human Factors in Computing Systems*, 3471–3480. https://doi.org/10.1145/2470654.2466477

Denef, S., Kaptein, N., Bayerl, P. S., & Ramirez, L. (2012). *Best Practice in Police Social Media Adaptation.* COMPOSITE Project. https://repub.eur.nl/pub/40562/metis_189373.pdf

Dennison, L., Morrison, L., Conway, G., & Yardley, L. (2013). Opportunities and Challenges for Smartphone Applications in Supporting Health Behavior Change: Qualitative Study. *Journal of Medical Internet Research, 15*(4), e2583.

Deterding, S., Dixon, D., Khaled, R., & Nacke, L. (2011). From Game Design Elements to Gamefulness: Defining "Gamification". *Proceedings of the 15th International Academic MindTrek Conference: Envisioning Future Media Environments*, 9–15. https://doi.org/10.1145/2181037.2181040

Diekman, S. T., Kearney, S. P., O'Neil, M. E., & MacK, K. A. (2007). Qualitative Study of Homeowners' Emergency Preparedness: Experiences, Perceptions, and Practices. *Prehospital and Disaster Medicine, 22*(6), 494–501. https://doi.org/10.1017/S1049023X00005318

Dieter, M., Helmond, A., Tkacz, N., van der Vlist, F., & Weltevrede, E. (2021). Pandemic Platform Governance: Mapping the Global Ecosystem of COVID-19 Response Apps. *Internet Policy Review, 10*(3). https://doi.org/10.14763/2021.3.1568

Digitalstadt Darmstadt. (2018). Ethik- & Technologiebeirat—Unterstützung für die Digitalstadt Darmstadt. https://dev.digitalstadt-darmstadt.de/news/ethik-technologiebeirat/

Dillard, J. P., Tian, X., Cruz, S. M., Smith, R. A., & Shen, L. (2021). Persuasive Messages, Social Norms, and Reactance: A Study of Masking Behavior during a COVID-19 Campus Health Campaign. *Health Communication*, 1–11. https://doi.org/10.1080/10410236.2021.2007579

DIN EN ISO 9241-11. (2018). Ergonomics of Human-System Interaction—Part 11: Usability: Definitions and Concepts. *Beuth Verlag.* https://doi.org/https://dx.doi.org/10.31030/2757945

Ding, A. Y., De Jesus, G. L., & Janssen, M. (2019). Ethical Hacking for Boosting IoT Vulnerability Management: A First Look into Bug Bounty Programs and Responsible Disclosure. *Proceedings of the Eighth International Conference on Telecommunications and Remote Sensing*, 49–55. https://doi.org/10.1145/3357767.3357774

Distel, B. (2018). Bringing Light into the Shadows: A Qualitative Interview Study on Citizens' Non-Adoption of e-Government. *Electronic Journal of e-Government, 16*(2), 98–205.

Distel, B. (2020). Assessing Citizens' Non-Adoption of Public e-Services in Germany. *Information Polity, 25*(3), 39–360. https://doi.org/10.3233/IP-190214

Distel, B., & Lindgren, I. (2019). Who Are the Users of Digital Public Services?: A Critical Reflection on Differences in the Treatment of Citizens as 'Users' in e-Government Research. In P. Panagiotopoulos, N. Edelmann, O. Glassey, G. Misuraca, P. Parycek, T. Lampoltshammer, & B. Re (Eds.), *Electronic Participation* (pp. 117–129, Vol. 11686). Springer International Publishing. https://doi.org/10.1007/978-3-030-27397-2_10

Distel, B., & Ogonek, N. (2016). To Adopt or Not to Adopt: A Literature Review on Barriers to Citizens' Adoption of E-Government Services. *ECIS 2016 Proceedings.*

Djalante, R., Holley, C., Thomalla, F., & Carnegie, M. (2013). Pathways for Adaptive and Integrated Disaster Resilience. *Natural Hazards, 69*(3), 2105–2135. https://doi.org/10.1007/s11069-013-0797-5

Döbelt, S., Jung, M., Busch, M., & Tscheligi, M. (2015). Consumers' Privacy Concerns and Implications for a Privacy Preserving Smart Grid Architecture—Results of an Austrian

Study. *Energy Research & Social Science, 9*, 137–145. https://doi.org/10.1016/j.erss.2015. 08.022

Douglasdotter, L. (2023, January). *The Thinking of New Public Governance in Swedish Crisis Preparedness: A Case Study of The Swedish Civil Contingencies Agency (MSB)*.

Dressel, K. (2015). Risk Culture and Crisis Communication. *International Journal of Risk Assessment and Management, 18*(2), 115–124. https://doi.org/10.1504/IJRAM.2015. 069020

Duarte, E. F., & Baranauskas, M. C. C. (2016). Revisiting the Three HCI Waves: A Preliminary Discussion on Philosophy of Science and Research Paradigms. *Proceedings of the 15th Brazilian Symposium on Human Factors in Computing Systems*, 1–4. https://doi.org/10. 1145/3033701.3033740

Dulin, P., Mertz, R., Edwards, A., & King, D. (2022). Contrasting a Mobile App With a Conversational Chatbot for Reducing Alcohol Consumption: Randomized Controlled Pilot Trial. *JMIR Formative Research, 6*(5), e33037. https://doi.org/10.2196/33037

Egert, R., Daubert, J., Marsh, S., & Mühlhäuser, M. (2021). Exploring Energy Grid Resilience: The Impact of Data, Prosumer Awareness, and Action. *Patterns, 2*(6), 100258. https://doi. org/10.1016/j.patter.2021.100258

Egert, R., Gerber, N., Haunschild, J., Kuehn, P., & Zimmermann, V. (2021). Towards Resilient Critical Infrastructures—Motivating Users to Contribute to Smart Grid Resilience. *i-com— Journal of Interactive Media, 20*(2), 161–175. https://doi.org/10.1515/icom-2021-0021

Eiser, R. J., Bostrom, A., Burton, I., Johnston, D. M., McClure, J., Paton, D., van der Pligt, J., & White, M. P. (2012). Risk Interpretation and Action: A Conceptual Framework for Responses to Natural Hazards. *International Journal of Disaster Risk Reduction*, 5–16.

Eismann, K., Posegga, O., & Fischbach, K. (2021). Opening Organizational Learning in Crisis Management: On the Affordances of Social Media. *The Journal of Strategic Information Systems, 30*(4), 101692. https://doi.org/https://doi.org/10.1016/j.jsis.2021.101692

Elshan, E., Engel, C., Ebel, P., & Siemon, D. (2022). Assessing the Reusability of Design Principles in the Realm of Conversational Agents. In A. Drechsler, A. Gerber, & A. Hevner (Eds.), *The Transdisciplinary Reach of Design Science Research* (pp. 128–141). Springer International Publishing. https://doi.org/10.1007/978-3-031-06516-3_10

Etkin, D. (1999). Risk Transference and Related Trends: Driving Forces towards More Mega-Disasters. *Global Environmental Change Part B: Environmental Hazards, 1*(2), 69–75. https://doi.org/10.3763/ehaz.1999.0109

European Federation of Psychologists' Association. (2005). Meta-Code of Ethics. https:// www.bdp-verband.de/fileadmin/user_upload/BDP/website/dokumente/PDF/Profession/ Berufsethik/efpa_metacode_en.pdf

Eversden, A. (2019). The Latest Pentagon Bug Bounty Revealed a Critical Vulnerability [newspaper]. *C4ISRNET*. https://www.c4isrnet.com/dod/2019/10/14/the-latest-pentagon-bug-bounty-revealed-a-critical-vulnerability/

Farooq, A., Laato, S., & Najmul Islam, A. K. (2020). Impact of Online Information on Self-Isolation Intention during the COVID-19 Pandemic: Cross-Sectional Study. *Journal of Medical Internet Research, 22*(5), 1–15. https://doi.org/10.2196/19128

Fathi, R., Thom, D., Koch, S., Ertl, T., & Fiedrich, F. (2020). VOST: A Case Study in Voluntary Digital Participation for Collaborative Emergency Management. *Information Processing & Management, 57*(4), 102174. https://doi.org/10.1016/j.ipm.2019.102174

Fekete, A., & Sandholz, S. (2021). Here Comes the Flood, but Not Failure? Lessons to Learn after the Heavy Rain and Pluvial Floods in Germany 2021. *Water, 13*(21), 1–20. https://doi.org/10.3390/w13213016

Fell, M. J., Shipworth, D., Huebner, G. M., & Elwell, C. A. (2015). Public Acceptability of Domestic Demand-Side Response in Great Britain: The Role of Automation and Direct Load Control. *Energy Research & Social Science, 9*, 72–84. https://doi.org/10.1016/j.erss.2015.08.023

FEMA. (2022). *National Household Survey on Disaster Preparedness. Key Findings.* Retrieved September 13, 2023, from https://fema-community-files.s3.amazonaws.com/2022-National-Household-Survey.pdf

Ferris, N., & van Renssen, S. (2021). Cybersecurity Threats Escalate in the Energy Sector. *Energymonitor.ai.* https://www.energymonitor.ai/digitalisation/cybersecurity-threats-escalate-in-the-energy-sector/

Fink, G., Best, D., Manz, D., Popovsky, V., & Endicott-Popovsky, B. (2013). Gamification for Measuring Cyber Security Situational Awareness. In D. Schmorrow & C. Fidopiastis (Eds.), *Foundations of Augmented Cognition. AC 2013. Lecture Notes in Computer Science* (pp. 656–665, Vol. 8027). Springer. https://doi.org/10.1007/978-3-642-39454-6_70

Fischer, D., Posegga, O., & Fischbach, K. (2016). Communication Barriers in Crisis Management: A Literature Review. *Twenty-Fourth European Conference on Information Systems (ECIS)*, 1–18.

Fischer, D., Putzke-Hattori, J., & Fischbach, K. (2019). Crisis Warning Apps: Investigating the Factors Influencing Usage and Compliance with Recommendations for Action. *Proceedings of the 52nd Hawaii International Conference on System Sciences, 6*, 639–648. https://doi.org/10.24251/hicss.2019.079

Fischer-Preßler, D., Bonaretti, D., & Fischbach, K. (2020). Effective Use of Mobile-Enabled Emergency Warning Systems. *Proceedings of the 28th European Conference on Information Systems (ECIS)*.

Fischer-Preßler, D., Bonaretti, D., & Fischbach, K. (2021). A Protection-Motivation Perspective to Explain Intention to Use and Continue to Use Mobile Warning Systems. *Business and Information Systems Engineering.* https://doi.org/10.1007/s12599-021-00704-0

Florêncio, D., Herley, C., & Van Oorschot, P. C. (2014). Password Portfolios and the Finite-Effort User: Sustainably Managing Large Numbers of Accounts. *23rd USENIX Security Symposium (USENIX Security 14)*, 575–590.

Fogg, B. J. (2002). Persuasive Technology: Using Computers to Change What We Think and Do. *Ubiquity, 2002.* https://doi.org/10.1145/764008.763957

Fogg, B. J. (2009a). A Behavior Model for Persuasive Design. *Proceedings of the 4th International Conference on Persuasive Technology.* https://doi.org/10.1145/1541948.1541999

Fogg, B. J. (2009b). Creating Persuasive Technologies: An Eight-Step Design Process. *ACM International Conference Proceeding Series, 350.* https://doi.org/10.1145/1541948.1542005

Følstad, A., & Bjerkreim-Hanssen, N. (2023). User interactions with a municipality chatbot—lessons learnt from dialogue analysis. *International Journal of Human–Computer Interaction*, 1–14. https://doi.org/10.1080/10447318.2023.2238355

Følstad, A., & Brandtzaeg, P. B. (2020). Users' Experiences with Chatbots: Findings from a Questionnaire Study. *Quality and User Experience, 5*(1), 3. https://doi.org/10.1007/s41233-020-00033-2

Følstad, A., Skjuve, M., & Brandtzaeg, P. B. (2019). Different Chatbots for Different Purposes: Towards a Typology of Chatbots to Understand Interaction Design. *International Conference on Internet Science (INSCI)*, 145–156.

Franke, T., Attig, C., & Wessel, D. (2019). A Personal Resource for Technology Interaction: Development and Validation of the Affinity for Technology Interaction (ATI) Scale. *International Journal of Human–Computer Interaction*, *35*(6), 456–467. https://doi.org/10.1080/10447318.2018.1456150

Fraunhofer FOKUS. (2018). KATWARN—Das Warn- und Informationssystem für die Bevölkerung. https://cdn0.scrvt.com/fokus/a4f8d5daa332c6c3/abd59f309c24/KATWARN-DinA4-6-Seiter-Web-Einzelseiten-Version-April-2019.pdf

Friedman, B., & Kahn Jr, P. H. (2007). Human Values, Ethics, and Design. In *The Human-Computer Interaction Handbook* (pp. 1267–1292). CRC press.

Frik, A., Malkin, N., Harbach, M., Peer, E., & Egelman, S. (2019). A Promise Is a Promise. The Effect of Commitment Devices on Computer Security Intentions. *Conference on Human Factors in Computing Systems—Proceedings*, 36–38. https://doi.org/10.1145/3290605.3300834

Furqan, A., Myers, C., & Zhu, J. (2017). Learnability through Adaptive Discovery Tools in Voice User Interfaces. *Proceedings of the 2017 CHI Conference Extended Abstracts on Human Factors in Computing Systems*, 1617–1623.

Gao, C., Zeng, J., Lo, D., Lin, C. Y., Lyu, M. R., & King, I. (2018). INFAR: Insight Extraction from App Reviews. *Proceedings of the 2018 26th ACM Joint Meeting on European Software Engineering Conference and Symposium on the Foundations of Software Engineering*, 904–907. https://doi.org/10.1145/3236024.3264595

Gao, H., Barbier, G., & Goolsby, R. (2011). Harnessing the Crowdsourcing Power of Social Media for Disaster Relief. *IEEE Intelligent Systems*, *26*(3), 10–14.

Gather, L. (2022). Zivilschutz: Rückkehr der Bunker? *tagesschau.de*. Retrieved May 11, 2023, from https://www.tagesschau.de/inland/gesellschaft/bunker-zivilschutz-101.html

Gerber, N., Zimmermann, V., Henhapl, B., Emeröz, S., & Volkamer, M. (2018). Finally Johnny Can Encrypt: But Does This Make Him Feel More Secure? *Proceedings of the 13th International Conference on Availability, Reliability and Security*. https://doi.org/10.1145/3230833.3230859

Gerhold, L., Peperhove, R., Lindner, A.-K., & Tietze, N. (2021). *Schutzziele, Notfallvorsorge, Katastrophenkommunikation: Wissenschaftliche Empfehlungen für den Bevölkerungsschutz* (Vol. 28).

Gnauk, B., Dannecker, L., & Hahmann, M. (2012). Leveraging Gamification in Demand Dispatch Systems. *Proceedings of the 2012 Joint EDBT/ICDT Workshops*, 103–110.

Göbel, J., Boldt, J., Betke, H., Kuehnel, S., Damarowsky, J., Böhmer, M., & Sackmann, S. (2024). Towards a design theory for task-oriented conversational agents. *ECIS*.

Goersch, H. (2013). Mythen der Notfallvorsorge. *Versicherheitlichung des Bevölkerungsschutzes*, 49–70. https://doi.org/10.1007/978-3-658-02200-6_3

Golinelli, D., Boetto, E., Carullo, G., Nuzzolese, A. G., Landini, M. P., & Fantini, M. P. (2020). Adoption of Digital Technologies in Health Care during the COVID-19 Pandemic: Systematic Review of Early Scientific Literature. *Journal of Medical Internet Research*, *22*(11), e22280. https://doi.org/10.2196/22280

Goulden, M., Bedwell, B., Rennick-Egglestone, S., Rodden, T., & Spence, A. (2014). Smart Grids, Smart Users? The Role of the User in Demand Side Management. *Energy Research & Social Science, 2*, 21–29. https://doi.org/10.1016/j.erss.2014.04.008

Graham, M. W., Avery, E. J., & Park, S. (2015). The Role of Social Media in Local Government Crisis Communications. *Public Relations Review, 41*(3), 386–394. https://doi.org/10.1016/j.pubrev.2015.02.001

Greenlaw, R., Muddiman, A., Friberg, T., Moi, M., Cristaldi, M., Ludwig, T., & Reuter, C. (2014). The EmerGent Project: Emergency Management in Social Media Generation— Dealing with Big Data from Social Media Data Stream. *Workshop on Big Data, Intelligence Management and Analytics Workshop. 7th IEEE/ACM International Conference on Utility and Cloud Computing (UCC)*, 687–689. https://doi.org/10.1109/UCC.2014.111

Gregor, S., & Hevner, A. R. (2013). Positioning and Presenting Design Science Research for Maximum Impact. *MIS quarterly*, 337–355. https://doi.org/10.25300/MISQ/2013/37.2.01

Grinko, M., Kaufhold, M.-A., & Reuter, C. (2019). Adoption, Use and Diffusion of Crisis Apps in Germany: A Representative Survey. In F. Alt, A. Bulling, & T. Döring (Eds.), *Mensch und Computer 2019* (pp. 263–274). ACM. https://doi.org/10.1145/3340764.3340782

Groneberg, C., Heidt, V., Knoch, T., & Helmerichs, J. (2017). Analyse Internationaler Bevölkerungsschutz-Apps: Ergebnisse einer Begleitstudie zu NINA und Smarter. *Magazin Bevölkerungsschutz: Psychosoziales Krisenmanagement*, (1), 565–577.

Guha-Sapir, D., Hargitt, D., & Hoyois, P. (2004). *Thirty Years of Natural Disasters 1974– 2003: The Numbers*. Presses universitaires de Louvain.

Gui, X., Kou, Y., Pine, K. H., & Chen, Y. (2017). Managing Uncertainty: Using Social Media for Risk Assessment during a Public Health Crisis. *Proceedings of the 2017 CHI Conference on Human Factors in Computing Systems, 2017-May*, 4520–4533. https://doi.org/10.1145/3025453.3025891

Guzman, E., El-Haliby, M., & Bruegge, B. (2015). Ensemble Methods for App Review Classification: An Approach for Software Evolution. *Proceedings—2015 30th IEEE/ACM International Conference on Automated Software Engineering, ASE 2015*, 771–776. https://doi.org/10.1109/ASE.2015.88

Haesler, S., Mogk, R., Putz, F., Logan, K. T., Thiessen, N., Kleinschnitger, K., Baumgärtner, L., Stroscher, J.-P., Reuter, C., Knodt, M., & Hollick, M. (2021). Connected Self-Organized Citizens in Crises: An Interdisciplinary Resilience Concept for Neighborhoods. *CSCW '21 Companion: Conference Companion Publication of the 2021 on Computer Supported Cooperative Work and Social Computing*. https://doi.org/10.1145/3462204.3481749

Haesler, S., Schmid, S., & Reuter, C. (2020). Crisis Volunteering Nerds: Three Months after COVID-19 Hackathon #WirVsVirus. *22nd International Conference on Human-Computer Interaction with Mobile Devices and Services*. https://doi.org/10.1145/3406324.3424584

Haesler, S., Schmid, S., Vierneisel, A. S., & Reuter, C. (2021). Stronger Together: How Neighborhood Groups Build up a Virtual Network during the COVID-19 Pandemic. *Proceedings of the ACM: Human Computer Interaction (PACM): Computer-supported Cooperative Work and Social Computing, 5*. https://doi.org/10.1145/3476045

Haesler, S., Wendelborn, M., & Reuter, C. (2023). Getting the Residents' Attention: The Perception of Warning Channels in Smart Home Warning Systems. *Proceedings of the 2023 ACM Designing Interactive Systems Conference*, 1114–1127.

Hagar, C. (2013). Crisis Informatics: Perceptives of Trust—Is Social Media a Mixed Blessing? *School of Information Student Research Journal, 2*(2), 149–153. https://doi.org/10.31979/2575-2499.020202

Halsbenning, S. (2021). Digitalisierung öffentlicher Dienstleistungen: Herausforderungen und Erfolgsfaktoren der OZG-Umsetzung in der Kommunalverwaltung. *HMD Praxis der Wirtschaftsinformatik,* 1–16. https://doi.org/10.1365/s40702-021-00765-5

Hamari, J., & Koivisto, J. (2013). Social Motivations to Use Gamification: An Empirical Study of Gamifying Exercise. *Proceedings of the European Conference on Information Systems (ECIS 2013).*

Hamari, J., Koivisto, J., & Sarsa, H. (2014). Does Gamification Work?—A Literature Review of Empirical Studies on Gamification. *2014 47th Hawaii International Conference on System Sciences,* 3025–3034.

Hansen, P. G. (2016). The Definition of Nudge and Libertarian Paternalism: Does the Hand Fit the Glove? *European Journal of Risk Regulation, 7*(1), 155–174. https://doi.org/10.1017/S1867299X00005468

Hansen, P. G., & Jespersen, A. M. (2013). Nudge and the Manipulation of Choice: A Framework for the Responsible Use of the Nudge Approach to Behaviour Change in Public Policy. *European Journal of Risk Regulation, 4*(1), 3–28. https://doi.org/10.1017/S1867299X00002762

Haque, M. D. R., & Rubya, S. (2023). An Overview of Chatbot-Based Mobile Mental Health Apps: Insights From App Description and User Reviews. *JMIR mHealth and uHealth, 11*(1), e44838. https://doi.org/10.2196/44838

Hargreaves, T., Nye, M., & Burgess, J. (2013). Keeping Energy Visible? Exploring How Householders Interact with Feedback from Smart Energy Monitors in the Longer Term. *Energy Policy, 52,* 126–134. https://doi.org/10.1016/j.enpol.2012.03.027

Harrison, J. D., & Patel, M. S. (2020). Designing Nudges for Success in Health Care. *AMA Journal of Ethics, 22*(9), 796–801.

Harrison, S., Tatar, D., & Sengers, P. (2007). The Three Paradigms of HCI. *Alt. Chi. Session at the SIGCHI Conference on Human Factors in Computing Systems,* 1–18.

Hartwig, K., & Reuter, C. (2021). Nudging Users towards Better Security Decisions in Password Creation Using Whitebox-Based Multidimensional Visualisations. *Behaviour & Information Technology,* 1–24. https://doi.org/10.1080/0144929X.2021.1876167

Hassandoust, F., & Techatassanasoontorn, A. A. (2020). Understanding Users' Information Security Awareness and Intentions: A Full Nomology of Protection Motivation Theory. In *Cyber Influence and Cognitive Threats* (pp. 129–143). Elsevier.

Hassenzahl, M. (2004). The Thing and I: Understanding the Relationship between User and Product. In M. A. Blythe, K. Overbeeke, A. F. Monk, & P. C. Wright (Eds.), *Funology: From Usability to Enjoyment* (pp. 31–42). Springer Netherlands. https://doi.org/10.1007/1-4020-2967-5_4

Hatayama, M., & Nakai, F. (2020). Using Computer Simulation for Effective Tsunami Risk Communication. In K. Yamori (Ed.), *Disaster Risk Communication: A Challenge from a Social Psychological Perspective* (pp. 39–50). Springer. https://doi.org/10.1007/978-981-13-2318-8_3

Haunschild, J., Burger, F., & Reuter, C. (2024). Understanding Crisis Preparedness: Insights from Personal Values, Beliefs, Social Norms, and Personal Norms. *Proceedings of the 21st International Conference on Information Systems for Crisis Response and*

Management (ISCRAM) (Best Paper Award). https://doi.org/https://ojs.iscram.org/index. php/Proceedings/article/view/19

Haunschild, J., Demuth, K., Geiß, H.-J., Richter, C., & Reuter, C. (2021). Nutzer, Sammler, Entscheidungsträger? Arten der Bürgerbeteiligung in Smart Cities. *HMD Praxis der Wirtschaftsinformatik, 58*, 1129–1147. https://doi.org/10.1365/s40702-021-00770-8

Haunschild, J., Henkel, M., & Reuter, C. (2025). Breaking Down Barriers to Warning Technology Adoption: Usability and Usefulness of a Messenger App Warning Bot. *i-com – Journal of Interactive Media, 24*.

Haunschild, J., Jung, L., & Reuter, C. (2023). Dual-Use in Volunteer Operations? Attitudes of Computer Science Students Regarding the Establishment of a Cyber Security Volunteer Force. In N. Gerber & V. Zimmermann (Eds.), *International Symposium on Technikpsychologie (TecPsy)* (pp. 66–81). sciendo. https://doi.org/10.2478/9788366675896-006

Haunschild, J., Kaufhold, M.-A., & Reuter, C. (2020). Sticking with Landlines? Citizens' and Police Social Media Use and Expectation During Emergencies. *Proceedings of the International Conference on Wirtschaftsinformatik (WI) (Best Paper Social Impact Award)*, 1–16. https://doi.org/10.30844/wi_2020_o2-haunschild

Haunschild, J., Kaufhold, M.-A., & Reuter, C. (2022a). Cultural Violence and Fragmentation on Social Media: Interventions and Countermeasures by Humans and Social Bots. In M. D. Cavelty & A. Wenger (Eds.), *Cyber Security Politics: Socio-Technological Transformations and Political Fragmentation* (pp. 48–63). Routledge. https://doi.org/10.4324/9781003110224-5

Haunschild, J., Kaufhold, M.-A., & Reuter, C. (2022b). Perceptions and Use of Warning Apps—Did Recent Crises Lead to Changes in Germany? In M. Mühlhäuser, C. Reuter, B. Pfleging, T. Kosch, A. Matviienko, K. Gerling, S. Mayer, W. Heuten, T. Döring, F. Müller, & M. Schmitz (Eds.), *Proceedings of Mensch und Computer 2022* (pp. 25–40). Association for Computing Machinery. https://doi.org/10.1145/3543758.3543770

Haunschild, J., Pauli, S., & Reuter, C. (2021). Citizens' Perceived Information Responsibilities and Information Challenges During the COVID-19 Pandemic. *Proceedings of the Conference on Information Technology for Social Good*, 151–156. https://doi.org/10.1145/3462203.3475886

Haunschild, J., Pauli, S., & Reuter, C. (2023). Preparedness Nudging for Warning Apps? A Mixed-Method Study Investigating Popularity and Effects of Preparedness Alerts in Warning Apps. *International Journal of Human-Computer Studies, 172*. https://doi.org/10.1016/j.ijhcs.2023.102995

Haunschild, J., & Reuter, C. (2021a). Bridging from Crisis to Everyday Life—An Analysis of User Reviews of the Warning App NINA and the COVID-19 Information Apps CoroBuddy and DarfIchDas. *Companion Publication of the 2021 Conference on Computer Supported Cooperative Work and Social Computing*, 72–78. https://doi.org/10.1145/3462204.3481745

Haunschild, J., & Reuter, C. (2021b). Perceptions of Police Technology Use and Attitudes Towards the Police—A Representative Survey of the German Population. In C. Wienrich, P. Wintersberger, & B. Weyers (Eds.), *Mensch und Computer 2021—Workshopband* (pp. 1–12). Gesellschaft für Informatik e.V. https://doi.org/10.18420/muc2021-mci-ws08-255

Haunschild, Jasmin, Guntrum, Laura Gianna, Cerrillo, Sofía, Bujara, Franziska, & Reuter, Christian. (2024). Towards a Digitally Mediated Transitional Justice Process? An Analysis of Colombian Transitional Justice Organisations' Posting Behaviour on Facebook. *Peace and Conflict Studies, 30*(2). https://nsuworks.nova.edu/pcs/vol30/iss2/4

Hauri, A., Kohler, K., & Scharte, B. (2022). *A Comparative Assessment of Mobile Device-Based Multi-Hazard Warnings: Saving Lives through Public Alerts in Europe*. Center for Security Studies. Zürich. https://doi.org/10.3929/ethz-b-000533908

Hegele, Y., & Schnabel, J. (2021). Federalism and the Management of the COVID-19 Crisis: Centralisation, Decentralisation and (Non-)Coordination. *West European Politics, 0*(0), 1–20. https://doi.org/10.1080/01402382.2021.1873529

Heino, O., & Anttiroiko, A.-V. (2016). Utility-Customer Communication: The Case of Water Utilities. *Public Works Management & Policy, 21*(3), 220–230. https://doi.org/10.1177/1087724X15606738

Helmerichs, J., Klos, S., Heidt, V., Knoch, T., & Groneberg, C. (2018). *Sozialwissenschaftliche Evaluation der Feldübung: Ergebnisse Quantitativer Und Qualitativer Erhebungen der Smarter-Feldübung im September 2017*. https://smarter-projekt.de/wp-content/uploads/2018/01/2018-01-23-Evaluation-der-Feld%C3%BCbung.pdf

Herter, K. (2007). Residential Implementation of Critical-Peak Pricing of Electricity. *Energy Policy, 35*(4), 2121–2130. https://doi.org/10.1016/j.enpol.2006.06.019

Hevner, A., & Chatterjee, S. (2010). Introduction to Design Science Research. In *Design research in information systems: Theory and practice* (pp. 1–8). Springer US. https://doi.org/10.1007/978-1-4419-5653-8_1

Hevner, A. R. (2007). A Three Cycle View of Design Science Research. *Scandinavian Journal of Information Systems, 19*(2).

Hevner, A. R., March, S. T., Park, J., & Ram, S. (2004). Design Sceince in Information Systems Research. *MIS Quarterly, 28*(1), 75–105.

Hewett, T. T., Baecker, R., Card, S., Carey, T., Gasen, J., Mantei, M., Perlman, G., Strong, G., & Verplank, W. (1992). *ACM SIGCHI Curricula for Human-Computer Interaction*. ACM.

Hines, P., Balasubramaniam, K., & Sanchez, E. C. (2009). Cascading Failures in Power Grids. *IEEE Potentials, 28*(5), 24–30. https://doi.org/10.1109/MPOT.2009.933498

Hochheiser, H., & Lazar, J. (2007). HCI and Societal Issues: A Framework for Engagement. *International Journal of Human-Computer Interaction, 23*(3), 339–374. https://doi.org/10.1080/10447310701702717

Hochheiser, H., Wiley, J., Viley, J., & Sussex, W. (2010). *Research Methods in Human-Computer Interactions*. Wiley.

Höchst, J., & Baumgärtner, L. (2020). LoRa-based Device-to-Device Smartphone Communication for Crisis Scenarios. *Proceedings of the 17th International Conference on Information Systems for Crisis Response and Management (ISCRAM)*.

Holden, R. J., & Karsh, B.-T. (2010). The Technology Acceptance Model: Its Past and Its Future in Health Care. *Journal of Biomedical Informatics, 43*(1), 159–172.

Holdener, M., Gut, A., & Angerer, A. (2020). Applicability of the User Engagement Scale to Mobile Health: A Survey-Based Quantitative Study. *JMIR mHealth and uHealth, 8*(1), e13244. https://doi.org/10.2196/13244

Hollick, M., & Katzenbeisser, S. (2019). Resilient Critical Infrastructures. In C. Reuter (Ed.), *Information Technology for Peace and Security* (pp. 305–318). Springer Vieweg. https://doi.org/10.1007/978-3-658-25652-4_14

Hollnagel, E. (2010). *Resilience Engineering in Practice: A Guidebook*. Ashgate Publishing, Ltd.

Holm, S. (1979). A Simple Sequentially Rejective Multiple Test Procedure. *Scandinavian Journal of Statistics, 6*(6), 65–70.

Holzhüter, M., Huhle, G., Reuter-Oppermann, M., Hellriegel, J., & Klafft, M. (2023). Acceptance Study on Application Systems to Improve Situational Incident Management through Bi-Directional Communication between Citizens and Decision-Makers in Emergencies and Crises Situations.

Höppe, P. (2015). *Naturkatastrophen—Immer Häufiger, Heftiger, Tödlicher, Teurer?* https://www.munichre.com/de/risiken/naturkatastrophen.html

Horne, C., Darras, B., Bean, E., Srivastava, A., & Frickel, S. (2015). Privacy, Technology, and Norms: The Case of Smart Meters. *Social Science Research, 51*, 64–76. https://doi.org/10.1016/j.ssresearch.2014.12.003

How, V., Azmi, E. S. B., Mohd Zaki, N. F. B., & Othman, K. B. (2020). Integrating Flood Education Miniature and Interactive E-Learning in a Prototype of Flood Learning Kit for Knowledge Resilience Among School Children. *An Interdisciplinary Approach for Disaster Resilience and Sustainability*, 355–368.

Huang, K.-Y. (2009). Challenges in Human-Computer Interaction Design for Mobile Devices. *Proceedings of the World Congress on Engineering and Computer Science*, 1–6. https://www.iaeng.org/publication/WCECS2009/WCECS2009_pp236-241.pdf

Hughes, A. L., & Palen, L. (2012). The Evolving Role of the Public Information Officer: An Examination of Social Media in Emergency Management. *Journal of Homeland Security and Emergency Management, 9*(1). https://doi.org/10.1515/1547-7355.1976

Hughes, A. L., St. Denis, L. A. A., Palen, L., & Anderson, K. M. (2014). Online Public Communications by Police & Fire Services during the 2012 Hurricane Sandy. *Proceedings of the SIGCHI Conference on Human Factors in Computing Systems*, 1505–1514. https://doi.org/10.1145/2556288.2557227

Hughes, A. L., Palen, L., Sutton, J., Liu, S. B., & Vieweg, S. (2008). "Site-Seeing" in Disaster: An Examination of on-Line Social Convergence. *5th International ISCRAM Conference*, 1–10.

Ingrassia, P. L., Foletti, M., Djalali, A., Scarone, P., Ragazzoni, L., Della Corte, F., Kaptan, K., Lupescu, O., Arculeo, C., & Von Arnim, G. (2014). Education and Training Initiatives for Crisis Management in the European Union: A Web-Based Analysis of Available Programs. *Prehospital and Disaster Medicine, 29*(2), 115–126.

InTradeSys GmbH. (2021). *„Darf Ich Das?"-App*. Köln, Deutschland. https://www.darfichdas.info/

Islam, M. N., & Najmul Islam, A. K. M. (2020). A Systematic Review of the Digital Interventions for Fighting COVID-19: The Bangladesh Perspective. *IEEE Access: Practical Innovations, Open Solutions, 8*, 114078–114087. https://doi.org/10.1109/ACCESS.2020.3002445

Jain, M., Kumar, P., Kota, R., & Patel, S. N. (2018). Evaluating and Informing the Design of Chatbots. *Designing Interactive Systems Conference*, 895–906. https://doi.org/10.1145/3196709.3196735

Janssen, A., Cardona, D. R., Passlick, J., & Breitner, M. H. (2022). How to Make Chatbots Productive—A User-Oriented Implementation Framework. *International Journal of Human-Computer Studies, 168*, 102921. https://doi.org/10.1016/j.ijhcs.2022.102921

Johnson, D., Horton, E., Mulcahy, R., & Foth, M. (2017). Gamification and Serious Games within the Domain of Domestic Energy Consumption: A Systematic Review. *Renewable and Sustainable Energy Reviews, 73*, 249–264. https://doi.org/10.1016/j.rser.2017.01.134

Johnson, E. J., & Goldstein, D. G. (2004). Defaults and Donation Decisions. *Transplantation, 78*(12), 1713–1716. https://doi.org/10.1097/01.TP.0000149788.10382.B2

Johnston, F. H., Wheeler, A. J., Williamson, G. J., Campbell, S. L., Jones, P. J., Koolhof, I. S., Lucani, C., Cooling, N. B., & Bowman, D. M. J. S. (2018). Using Smartphone Technology to Reduce Health Impacts from Atmospheric Environmental Hazards. *Environmental Research Letters, 13*(4). https://doi.org/10.1088/1748-9326/aab1e6

Jones, B., & Jones, R. (2019). Public Service Chatbots: Automating Conversation with BBC News. *Digital Journalism, 7*(8), 1032–1053. https://doi.org/10.1080/21670811.2019.1609371

Jumisko-Pyykkö, S., & Vainio, T. (2010). Framing the Context of Use for Mobile HCI. *International Journal of Mobile Human Computer Interaction, 2*(4), 1–28. https://doi.org/10.4018/jmhci.2010100101

Jung, J. Y., & Mellers, B. A. (2016). American Attitudes toward Nudges. *Judgement and Decision Making, 11*(1), 62–74.

Kahneman, D. (2011). *Thinking, Fast and Slow.* Farrar, Straus and Giroux.

Kahneman, D., Slovic, P., & Tversky, A. (1982, April). *Judgment Under Uncertainty: Heuristics and Biases.* Cambridge University Press.

Kahneman, D., & Tversky, A. (1984). Choices, Values, and Frames. *American Psychologist, 39*(4), 341.

Kangas, E., & Kinnunen, T. (2005). Applying User-Centered Design to Mobile Application Development. *Communications of the ACM, 48*(7), 55–59. https://doi.org/10.1145/1070838.1070866

Karl, I., Rother, K., & Nestler, S. (2015). Crisis-Related Apps: Assistance for Critical and Emergency Situations. *International Journal of Information Systems for Crisis Response and Management (IJISCRAM), 7*(2), 19–35.

Karutz, H., Geier, W., & Mitschke, T. (Eds.). (2017). *Bevölkerungsschutz.* Springer Berlin Heidelberg. https://doi.org/10.1007/978-3-662-44635-5

Kashani, A., & Ozturk, Y. (2017). Residential Energy Consumer Behavior Modification via Gamification. *2017 IEEE 6th International Conference on Renewable Energy Research and Applications (ICRERA)*, 1221–1225. https://doi.org/10.1109/ICRERA.2017.8191247

Kaufhold, M.-A. (2021). *Information Refinement Technologies for Crisis Informatics: User Expectations and Design Principles for Social Media and Mobile Apps.* Springer Vieweg. https://doi.org/10.1007/978-3-658-33341-6

Kaufhold, M.-A., Bäumler, J., & Reuter, C. (2022). The Implementation of Protective Measures and Communication of Cybersecurity Alerts in Germany—A Representative Survey of the Population. https://doi.org/10.18420/muc2022-mci-ws01-228

Kaufhold, M.-A., Bayer, M., & Reuter, C. (2020). Rapid Relevance Classification of Social Media Posts in Disasters and Emergencies: A System and Evaluation Featuring Active, Incremental and Online Learning. *Information Processing & Management, 57*(1), 102132. https://doi.org/10.1016/j.ipm.2019.102132

Kaufhold, M.-A., Fromm, J., Riebe, T., Mirbabaie, M., Kuehn, P., Basyurt, A. S., Bayer, M., Stöttinger, M., Eyilmez, K., Möller, R., Fuchß, C., Stieglitz, S., & Reuter, C. (2021). CYWARN: Strategy and Technology Development for Cross-Platform Cyber Situational Awareness and Actor-Specific Cyber Threat Communication. *Workshop-Proceedings Mensch und Computer.* https://doi.org/10.18420/muc2021-mci-ws08-263

Kaufhold, M.-A., Gizikis, A., Reuter, C., Habdank, M., & Grinko, M. (2019). Avoiding Chaotic Use of Social Media Before, During, and After Emergencies: Design and Evaluation of Citizens' Guidelines. *Journal of Contingencies and Crisis Management (JCCM), 27*(3), 198–213. https://doi.org/10.1111/1468-5973.12249

Kaufhold, M.-A., Haunschild, J., & Reuter, C. (2020). Warning the Public: A Survey on Attitudes, Expectations and Use of Mobile Crisis Apps in Germany. *Proceedings of the 28th European Conference on Information Systems (ECIS)*, 1–16. https://aisel.aisnet.org/ecis2020_rp/84

Kaufhold, M.-A., & Reuter, C. (2016). The Self-Organization of Digital Volunteers across Social Media: The Case of the 2013 European Floods in Germany. *Journal of Homeland Security and Emergency Management (JHSEM), 13*(1), 137–166. https://doi.org/10.1515/jhsem-2015-0063

Kaufhold, M.-A., Reuter, C., Comes, T., Mirbabaie, M., & Stieglitz, S. (2021). 2nd Workshop on Mobile Resilience: Designing Mobile Interactive Systems for Crisis Response, 1–4.

Kaufhold, M.-A., Rupp, N., Reuter, C., Amelunxen, C., & Cristaldi, M. (2018). 112.Social: Design and Evaluation of a Mobile Crisis App for Bidirectional Communication between Emergency Services and Citizens. *ECIS*, 81.

Kaufhold, M.-A., Rupp, N., Reuter, C., & Habdank, M. (2020). Mitigating Information Overload in Social Media during Conflicts and Crises: Design and Evaluation of a Cross-Platform Alerting System. *Behaviour & Information Technology (BIT), 39*(3), 319–342. https://doi.org/10.1080/0144929X.2019.1620334

Keinonen, T. (2008). User-Centered Design and Fundamental Need. *Proceedings of the 5th Nordic Conference on Human-Computer Interaction: Building Bridges*, 211–219. https://doi.org/10.1145/1463160.1463183

Kelley, P. G., Bresee, J., Cranor, L. F., & Reeder, R. W. (2009). A "Nutrition Label" for Privacy. *Proceedings of the 5th Symposium on Usable Privacy and Security*, 1–12. https://doi.org/10.1145/1572532.1572538

Kelley, P. G., Cesca, L., Bresee, J., & Cranor, L. F. (2010). Standardizing Privacy Notices: An Online Study of the Nutrition Label Approach. *Proceedings of the SIGCHI Conference on Human Factors in Computing Systems*, 1573–1582. https://doi.org/10.1145/1753326.1753561

Khalid, M., Shehzaib, U., & Asif, M. (2015). A Case of Mobile App Reviews as a Crowdsource. *International Journal of Information Engineering and Electronic Business, 7*(5), 39–47. https://doi.org/10.5815/ijieeb.2015.05.06

Khine, P. K., Mi, J., & Shahid, R. (2021). A Comparative Analysis of Co-Production in Public Services. *Sustainability, 13*(12), 6730. https://doi.org/10.3390/su13126730

Kim, G., Martel, A., Eisenman, D., Prelip, M., Arevian, A., Johnson, K. L., & Glik, D. (2019). Wireless Emergency Alert Messages: Influences on Protective Action Behaviour. *Journal of Contingencies and Crisis Management, 27*(4), 374–386.

Kim, K., Oglesby-Neal, A., & Mohr, E. (2017). 2016 Law Enforcement Use of Social Media Survey. *Urban Institute*.

Kim, S.-J., Choi, J.-H., & Nam, K.-H. (2021). Serious Game Scenario Design for Earthquake Response Education and Training in the Gyeongsangbuk-do Province. *Journal of the Society of Disaster Information, 17*(4), 769–777.

Kim, S. (2006). Toward an Integrative Framework of Technology Use (IFTU): Alternative Three-Wave Panel Models and Empirical Tests. *DIGIT 2006 Proceedings,*

1–35. http://aisel.aisnet.org/digit2006/8?utm_source=aisel.aisnet.org%2Fdigit2006%2F8&utm_medium=PDF&utm_campaign=PDFCoverPages

Kim, Y. H., Kim, D. J., & Wachter, K. (2013). A Study of Mobile User Engagement (MoEN): Engagement Motivations, Perceived Value, Satisfaction, and Continued Engagement Intention. *Decision Support Systems, 56*, 361–370. https://doi.org/10.1016/j.dss.2013.07.002

Kitchin, R. (2020). Civil Liberties or Public Health, or Civil Liberties and Public Health? Using Surveillance Technologies to Tackle the Spread of COVID-19. *Space and Polity, 24*(3), 362–381. https://doi.org/10.1080/13562576.2020.1770587

Kjeldskov, J., & Graham, C. (2003). A Review of Mobile HCI Research Methods. *International Conference on Mobile Human-Computer Interaction*, 317–335.

Klafft, M. (2013). Diffusion of Emergency Warnings via Multi-Channel Communication Systems an Empirical Analysis. *Eleventh International Symposium on Autonomous Decentralized Systems (ISADS)*, 1–5. https://doi.org/10.1109/ISADS.2013.6513437

Klopfenstein, L. C., Delpriori, S., Malatini, S., & Bogliolo, A. (2017). The Rise of Bots: A Survey of Conversational Interfaces, Patterns, and Paradigms. *Proceedings of the 2017 Conference on Designing Interactive Systems*, 555–565. https://doi.org/10.1145/3064663.3064672

Knowles, B., Blair, L., Walker, S., Coulton, P., Thomas, L., & Mullagh, L. (2014). Patterns of Persuasion for Sustainability. *Proceedings of the Conference on Designing Interactive Systems: Processes, Practices, Methods, and Techniques, DIS*, 1035–1044. https://doi.org/10.1145/2598510.2598536

Knuth, D., Schulz, S., Kietzmann, D., Stumpf, K., & Schmidt, S. (2017). Better Safe than Sorry—Emergency Knowledge and Preparedness in the German Population. *Fire Safety Journal, 93*, 98–101. https://doi.org/10.1016/j.firesaf.2017.08.003

Kohler, K., Hauri, A., Roth, F., & Scharte, B. (2020, September). *Measuring Individual Disaster Preparedness*. ETH Zurich. https://doi.org/10.3929/ETHZ-B-000441285

Korn, O., & Schmidt, A. (2015). Gamification of Business Processes: Re-designing Work in Production and Service Industry. *Procedia Manufacturing, 3*, 3424–3431. https://doi.org/10.1016/j.promfg.2015.07.616

Kotthaus, C., Ludwig, T., & Pipek, V. (2016). Persuasive System Design Analysis of Mobile Warning Apps for Citizens. *11th International Conference on Persuasive Technology*.

Kranenbarg, M. W., Holt, T. J., & van der Ham, J. (2018). Don't Shoot the Messenger! A Criminological and Computer Science Perspective on Coordinated Vulnerability Disclosure. *Crime Science, 7*(16), 1–9. https://doi.org/10.1186/s40163-018-0090-8

Kroese, F. M., Marchiori, D. R., & de Ridder, D. T. (2015). Nudging Healthy Food Choices: A Field Experiment at the Train Station. *Journal of Public Health, 38*(2), e133–e137. https://doi.org/10.1093/pubmed/fdv096

Kuller, M., Schoenholzer, K., & Lienert, J. (2021). Creating Effective Flood Warnings: A Framework from a Critical Review. *Journal of Hydrology, 602*, 126708. https://doi.org/10.1016/j.jhydrol.2021.126708

Kumar, A., Gupta, P. K., & Srivastava, A. (2020). A Review of Modern Technologies for Tackling COVID-19 Pandemic. *Diabetes & Metabolic Syndrome: Clinical Research & Reviews, 14*(4), 569–573.

Kunz, W., & Rittel, H. W. (1972). Information Science: On the Structure of Its Problems. *Information Storage and Retrieval, 8*(2), 95–98.

Kuznetsov, S., & Tomitsch, M. (2018). A Study of Urban Heat: Understanding the Challenges and Opportunities for Addressing Wicked Problems in HCI. *Proceedings of the 2018 CHI Conference on Human Factors in Computing Systems*, 1–13. https://doi.org/10.1145/3173574.3174137

Laato, S., Islam, A. K. M. N., & Islam, M. N. (2020). What Drives Unverified Information Sharing and Cyberchondria during the COVID-19 Pandemic? *European Journal of Information Systems, 29*(3), 288–305. https://doi.org/10.1080/0960085X.2020.1770632

Lai, P. (2017). The Literature Review of Technology Adoption Models and Theories for the Novelty Technology. *Journal of Information Systems and Technology Management, 14*(1), 21–38. https://doi.org/10.4301/S1807-17752017000100002

Landis, J. R., & Koch, G. G. (1977). The Measurement of Observer Agreement for Categorical Data. Biometrics. *Journal of the International Biometric Society, 33*(1), 159–174.

Laspidou, C. (2014). ICT and Stakeholder Participation for Improved Urban Water Management in the Cities of the Future. *Water Utility Journal, 8*, 79–85.

Laughery, K. R., & Wogalter, M. S. (2006). Designing Effective Warnings. *Reviews of Human Factors and Ergonomics, 2*(1), 241–271. https://doi.org/10.1177/1557234x0600200109

Lee, A. J. (2016). *Resilience by Design*. Springer International Publishing. https://doi.org/10.1007/978-3-319-30641-4

Lella, A., & Lipsman, A. (2017). *The 2017 U.S. Mobile App Report*. ComScore. https://www.comscore.com/Insights/Presentations-and-Whitepapers/2017/The-2017-US-Mobile-App-Report

Leonardi, P. M., & Barley, S. R. (2010). What's under Construction Here? Social Action, Materiality, and Power in Constructivist Studies of Technology and Organizing. *The Academy of Management Annals, 4*(1), 1–51.

Leonhart, R. (2008). *Psychologische Methodenlehre/Statistik*. UTB, Ernst Reinhardt Verlag. https://doi.org/10.36198/9783838530642

Lepper, M. R., & Greene, D. (2015). *The Hidden Costs of Reward: New Perspectives on the Psychology of Human Motivation*. Psychology Press.

Leviston, Z., & Uren, H. V. (2020). Overestimating One's "Green" Behavior: Better-than-average Bias May Function to Reduce Perceived Personal Threat from Climate Change. *Journal of Social Issues, 76*(1), 70–85. https://doi.org/10.1111/josi.12365

Lewis, J. R. (2018). The System Usability Scale: Past, Present, and Future. *International Journal of Human-Computer Interaction, 34*(7), 577–590. https://doi.org/10.1080/10447318.2018.1455307

Ley, B., Pipek, V., Reuter, C., & Wiedenhoefer, T. (2012). Supporting Improvisation Work in Inter-Organizational Crisis Management. *Proceedings of the Conference on Human Factors in Computing Systems (CHI)*, 1529. https://doi.org/10.1145/2207676.2208617

Li, C.-H., Yeh, S.-F., Chang, T.-J., Tsai, M.-H., Chen, K., & Chang, Y.-J. (2020). A Conversation Analysis of Non-Progress and Coping Strategies with a Banking Task-Oriented Chatbot. *Proceedings of the 2020 CHI Conference on Human Factors in Computing Systems*, 1–12. https://doi.org/10.1145/3313831.3376209

Li, T., Cobb, C., Yang, J. (., Baviskar, S., Agarwal, Y., Li, B., Bauer, L., & Hong, J. I. (2021). What Makes People Install a COVID-19 Contact-Tracing App? Understanding the Influence of App Design and Individual Difference on Contact-Tracing App Adoption Intention. *Pervasive and Mobile Computing, 75*, 101439. https://doi.org/10.1016/j.pmcj.2021.101439

Limayem, M., Hirt, S. G., & Cheung, C. M. K. (2007). How Habit Limits the Predictive Power of Intention: The Case of Information Systems Continuance. *MIS Quarterly, 31*(4), 705–737. https://doi.org/10.2307/25148817

Lindgren, I., Madsen, C. Ø., Hofmann, S., & Melin, U. (2019). Close Encounters of the Digital Kind: A Research Agenda for the Digitalization of Public Services. *Government Information Quarterly, 36*(3), 427–436. https://doi.org/10.1016/j.giq.2019.03.002

Liu, B. F., Fraustino, J. D., & Jin, Y. (2016). Social Media Use during Disasters: How Information Form and Source Influence Intended Behavioral Responses. *Communication Research, 43*(5), 626–646. https://doi.org/10.1177/0093650214565917

Liu, B. F., Wood, M. M., Egnoto, M., Bean, H., Sutton, J., Mileti, D., & Madden, S. (2017). Is a Picture Worth a Thousand Words? The Effects of Maps and Warning Messages on How Publics Respond to Disaster Information. *Public Relations Review, 43*(3), 493–506. https://doi.org/10.1016/j.pubrev.2017.04.004

Lockton, D., Harrison, D., & Stanton, N. A. (2010). *Design with Intent: 101 Patterns for Influencing Behaviour through Design*. Equifine.

Löhe, F., & Oswald, A. (2021, July 21). *Deutschland Wurde Präzise Gewarnt—Die Bürger Aber Nicht*. https://www.tagesspiegel.de/politik/schwere-vorwuerfe-deutschland-wurde-praezise-gewarnt-die-buerger-aber-nicht/27433034.html

Lou, C., Tandoc, E. C., Hong, L. X., Pong, X. Y., Lye, W. X., & Sng, N. G. (2021). When Motivations Meet Affordances: News Consumption on Telegram. *Journalism Studies, 22*(7), 934–952. https://doi.org/10.1080/1461670X.2021.1906299

Luca, G., Fraternali, P., Chiara, P., Giorgia, B., Santos, A. D. D., Roberto, A., & Valentina, R. (2015). A Gamification Framework for Customer Engagement and Sustainable Water Usage Promotion. *E-Proceedings of the 36th IAHR World Congress*.

Lucivero, F., Marelli, L., Hangel, N., Zimmermann, B. M., Prainsack, B., Galasso, I., Horn, R., Kieslich, K., Lanzing, M., Lievevrouw, E., et al. (2022). Normative Positions towards COVID-19 Contact-Tracing Apps: Findings from a Large-Scale Qualitative Study in Nine European Countries. *Critical Public Health, 32*(1), 5–18. https://doi.org/10.1080/09581596.2021.1925634

Lukoff, K., Lyngs, U., & Alberts, L. (2022). Designing to Support Autonomy and Reduce Psychological Reactance in Digital Self-Control Tools. *Conference on Human Factors in Computing Systems*, 1–6.

Lwin, M. O., Lu, J., Sheldenkar, A., & Schulz, P. J. (2018). Strategic Uses of Facebook in Zika Outbreak Communication: Implications for the Crisis and Emergency Risk Communication Model. *International Journal of Environmental Research and Public Health, 15*(9). https://doi.org/10.3390/ijerph15091974

MacDonald, C. M., & Atwood, M. E. (2014). What Does It Mean for a System to Be Useful? An Exploratory Study of Usefulness. *Proceedings of the 2014 Conference on Designing Interactive Systems*, 885–894. https://doi.org/10.1145/2598510.2598600

MacKay, D., & Robinson, A. (2016). The Ethics of Organ Donor Registration Policies: Nudges and Respect for Autonomy. *The American Journal of Bioethics, 16*(11), 3–12. https://doi.org/10.1080/15265161.2016.1222007

Mackert, M., Mabry-Flynn, A., Champlin, S., Donovan, E. E., & Pounders, K. (2016). Health Literacy and Health Information Technology Adoption: The Potential for a New Digital Divide. *Journal of Medical Internet Research, 18*(10), e264. https://doi.org/10.2196/jmir.6349

Maio, G. R., Olson, J. M., Allen, L., & Bernard, M. M. (2001). Addressing Discrepancies between Values and Behavior: The Motivating Effect of Reasons. *Journal of Experimental Social Psychology, 37*(2), 104–117.

Mair, P., & Wilcox, R. (2020). Robust Statistical Methods in R Using the WRS2 Package. *Behavior Research Methods, 52*(2), 464–488. https://doi.org/10.3758/s13428-019-01246-w

Makasi, T., Nili, A., Desouza, K., & Tate, M. (2022a). Public Service Values and Chatbots in the Public Sector: Reconciling Designer Efforts and User Expectations. *Proceedings of the 55th Hawaii International Conference on System Sciences*, 2334–2343.

Makasi, T., Nili, A., Desouza, K. C., & Tate, M. (2022b). A Typology of Chatbots in Public Service Delivery. *IEEE Software, 39*(3), 58–66. https://doi.org/10.1109/MS.2021.3073674

Marchal, N., & Au, H. (2020). "Coronavirus EXPLAINED": YouTube, COVID-19, and the Socio-Technical Mediation of Expertise. *Social Media and Society, 6*(3), 2–5. https://doi.org/10.1177/2056305120948158

Markwart, H., Vitera, J., Lemanski, S., Kietzmann, D., Brasch, M., & Schmidt, S. (2019). Warning Messages to Modify Safety Behavior during Crisis Situations: A Virtual Reality Study. *International Journal of Disaster Risk Reduction, 38*, 101235. https://doi.org/10.1016/j.ijdrr.2019.101235

Martin, W., Sarro, F., Jia, Y., Zhang, Y., & Harman, M. (2017). A Survey of App Store Analysis for Software Engineering. *IEEE Transactions on Software Engineering, 43*(9), 817–847. https://doi.org/10.1109/TSE.2016.2630689

Martiniello, N., Eisenbarth, W., Lehane, C., Johnson, A., & Wittich, W. (2022). Exploring the Use of Smartphones and Tablets among People with Visual Impairments: Are Mainstream Devices Replacing the Use of Traditional Visual Aids? *Assistive Technology, 34*(1), 34–45. https://doi.org/10.1080/10400435.2019.1682084

Masli, M., & Terveen, L. (2014). Leveraging the Contributory Potential of User Feedback. *Proceedings of the 17th ACM Conference on Computer Supported Cooperative Work & Social Computing*, 956–966.

Matthews, J., Win, K., Oinas-Kukkonen, H., & Freeman, M. (2015). Persuasive Technology in Mobile Applications Promoting Physical Activity: A Systematic Review. *Journal of Medical Systems, 40*(72), 1–13. https://doi.org/10.1007/s10916-015-0425-x

Matthews, J., Win, K. T., Oinas-Kukkonen, H., & Freeman, M. (2016). Persuasive Technology in Mobile Applications Promoting Physical Activity: A Systematic Review. *Journal of Medical Systems, 40*, 1–13.

Mayer, B. (2019). A Review of the Literature on Community Resilience and Disaster Recovery. *Current Environmental Health Reports, 6*, 167–173.

Mayring, P. (2014). *Qualitative Content Analysis: Theoretical Foundation, Basic Procedures and Software Solution*. AUT. https://nbn-resolving.org/urn:nbn:de:0168-ssoar-395173

McKay, D., Zhang, H., & Buchanan, G. (2022). Who Am I, and Who Are You, and Who Are We? A Scientometric Analysis of Gender and Geography in HCI. *Proceedings of the 2022 CHI Conference on Human Factors in Computing Systems*. https://doi.org/10.1145/3491102.3502106

McKay, F. H., Cheng, C., Wright, A., Shill, J., Stephens, H., & Uccellini, M. (2018). Evaluating Mobile Phone Applications for Health Behaviour Change: A Systematic Review. *Journal of Telemedicine and Telecare, 24*(1), 22–30.

Meesters, K., & vande Walle, B. (2013). Disaster in My Backyard: A Serious Game Introduction to Disaster Information Management. *Proceedings of the 10th International ISCRAM Conference.*

Meier, P., Beinke, J. H., Fitte, C., Behne, A., & Teuteberg, F. (2019). FeelFit-Design and Evaluation of a Conversational Agent to Enhance Health Awareness. *ICIS.*

Meijer, A., & Thaens, M. (2013). Social Media Strategies: Understanding the Differences between North American Police Departments. *Government Information Quarterly, 30*(4), 343–350. https://doi.org/10.1016/j.giq.2013.05.023

Meijer, A. J. (2014). New Media and the Coproduction of Safety: An Empirical Analysis of Dutch Practices. *American Review of Public Administration, 44*(1), 17–34. https://doi.org/10.1177/0275074012455843

Meijer, A. J., & Torenvlied, R. (2016). Social Media and the New Organization of Government Communications: An Empirical Analysis of Twitter Usage by the Dutch Police. *The American Review of Public Administration, 46*(2), 143–161. https://doi.org/10.1177/0275074014551381

Menski, U., Wahl, S., Tischer, H., & Braun, J. (2015). Zwischen Anspruch und Wirklichkeit: Die Rolle der Bevölkerung in der Ernährungsnotfallvorsorge. In L. Gerhold, H. Jäckel, J. Schiller, & S. Steiger (Eds.), *Ergebnisse interdisziplinärer Risiko-und Sicherheitsforschung. Eine Zwischenbilanz des Forschungsforum Öffentliche Sicherheit* (pp. 125–142).

Mentler, T., & Herczeg, M. (2016). On the role of User Experience in Mission- or Safety-Critical Systems. *Mensch und Computer.*

Meyer, R. J. (2006). Why We Under-Prepare for Hazards. *On Risk and Disaster: Lessons from Hurricane Katrina, 153–173.*

Michaels, L., & Parag, Y. (2016). Motivations and Barriers to Integrating 'Prosuming' Services into the Future Decentralized Electricity Grid: Findings from Israel. *Energy Research & Social Science, 21,* 70–83. https://doi.org/10.1016/j.erss.2016.06.023

Michalek, G., & Schwarze, R. (2020). *The Strategic Use of Nudging and Behavioural Approaches in Public Health Policy during the Coronavirus Crisis.* Helmholtz-Zentrum für Umweltforschung—UFZ. Leipzig, Germany. https://www.ssoar.info/ssoar/handle/document/68915

Michie, S., Richardson, M., Johnston, M., Abraham, C., Francis, J., Hardeman, W., Eccles, M. P., Cane, J., & Wood, C. E. (2013). The Behavior Change Technique Taxonomy (v1) of 93 Hierarchically Clustered Techniques: Building an International Consensus for the Reporting of Behavior Change Interventions. *Annals of Behavioral Medicine, 46*(1), 81–95. https://doi.org/10.1007/s12160-013-9486-6

Mileti, D. S., & Sorensen, J. H. (1990). Communication of Emergency Public Warnings. *Landslides, 1*(6), 52–70.

Milkman, K. L., Patel, M. S., Gandhi, L., Graci, H. N., Gromet, D. M., Ho, H., Kay, J. S., Lee, T. W., Akinola, M., Beshears, J., Bogard, J. E., Buttenheim, A., Chabris, C. F., Chapman, G. B., Choi, J. J., Dai, H., Fox, C. R., Goren, A., Hilchey, M. D., ... Angela L. Duckworth. (2021). A Megastudy of Text-Based Nudges Encouraging Patients to Get Vaccinated at an Upcoming Doctor's Appointment. In E. U. Weber (Ed.), *Proceedings of the National Academy of Sciences of the United States of America (PNAS)* (pp. 10–12, Vol. 118). https://doi.org/10.1073/pnas.2101165118

Min-Allah, N., Alahmed, B. A., Albreek, E. M., Alghamdi, L. S., Alawad, D. A., Alharbi, A. S., Al-Akkas, N., Musleh, D., & Alrashed, S. (2021). A Survey of COVID-19 Contact-Tracing Apps. *Computers in Biology and Medicine, 137,* 104787. https://doi.org/10.1016/j.compbiomed.2021.104787

Ming, L. C., Untong, N., Aliudin, N. A., Osili, N., Kifli, N., Tan, C. S., Goh, K. W., Ng, P. W., Al-Worafi, Y. M., Lee, K. S., & Goh, H. P. (2020). Mobile Health Apps on COVID-19 Launched in the Early Days of the Pandemic: Content Analysis and Review. *JMIR mHealth and uHealth, 8*(9), 1–17. https://doi.org/10.2196/19796

Mirbabaie, M., Bunker, D., Stieglitz, S., & Deubel, A. (2019). Who Sets the Tone? Determining the Impact of Convergence Behaviour Archetypes in Social Media Crisis Communication. *Information Systems Frontiers.* https://doi.org/10.1007/s10796-019-09917-x

Mirbabaie, M., Bunker, D., Stieglitz, S., Marx, J., & Ehnis, C. (2020). Social Media in Times of Crisis: Learning from Hurricane Harvey for the Coronavirus Disease 2019 Pandemic Response. *Journal of Information Technology,* 1–19. https://doi.org/10.1177/0268396220929258

Mirsch, T., Lehrer, C., & Jung, R. (2017). Digital Nudging: Altering User Behavior in Digital Environments. *Proceedings der 13. Internationalen Tagung Wirtschaftsinformatik (WI 2017),* 634–648.

Mirsch, T., Lehrer, C., & Jung, R. (2018). Making Digital Nudging Applicable: The Digital Nudge Design Method. *Proceedings of the 39th International Conference on Information Systems (ICIS),* 1–16.

Möcker, A. (2021). Notfallinternet per Satellit: Starlink-Einsatz bei der Flutkatastrophe. *Heise.* https://www.heise.de/news/Notfallinternet-per-Satellit-Starlink-Einsatz-bei-der-Flutkatastrophe-6194848.html

Mohsenian-Rad, A.-H., Wong, V. W., Jatskevich, J., Schober, R., & Leon-Garcia, A. (2010). Autonomous Demand-Side Management Based on Game-Theoretic Energy Consumption Scheduling for the Future Smart Grid. *IEEE Transactions on Smart Grid, 1*(3), 320–331. https://doi.org/10.1109/TSG.2010.2089069

Mol, J. M., Botzen, W. J. W., Blasch, J. E., Kranzler, E. C., & Kunreuther, H. C. (2021). All by Myself? Testing Descriptive Social Norm-Nudges to Increase Flood Preparedness among Homeowners. *Behavioural Public Policy,* 1–33. https://doi.org/10.1017/bpp.2021.17

Momani, A. M. (2020). The Unified Theory of Acceptance and Use of Technology: A New Approach in Technology Acceptance. *International Journal of Sociotechnology and Knowledge Development, 12*(3), 79–98. https://doi.org/10.4018/IJSKD.2020070105

Moon, M. J. (2020). Shifting from Old Open Government to New Open Government: Four Critical Dimensions and Case Illustrations. *Public Performance & Management Review, 43*(3), 535–559.

Moore, R. J., & Arar, R. (2019, March). *Conversational UX Design: A Practitioner's Guide to the Natural Conversation Framework.* Association for Computing Machinery.

Moore, R. J., Arar, R., Ren, G.-J., & Szymanski, M. H. (2017). Conversational UX Design. *Proceedings of the 2017 CHI Conference Extended Abstracts on Human Factors in Computing Systems,* 492–497. https://doi.org/10.1145/3027063.3027077

Moore, R. J., Arar, R., Ren, G.-J., & Szymanski, M. H. (Eds.). (2018). *Studies in Conversational UX Design.* Springer Nature.

Moosbrugger, H., & Kelava, A. (2012). *Testtheorie und Fragebogenkonstruktion.* Springer. https://doi.org/10.1007/978-3-642-20072-4

Mulder, F., Ferguson, J., Groenewegen, P., Boersma, K., & Wolbers, J. (2016). Questioning Big Data: Crowdsourcing Crisis Data towards an Inclusive Humanitarian Response. *Big Data and Society, 3*(2), 1–13. https://doi.org/10.1177/2053951716662054

Müller, F., Tomczyk, S., Fathi, R., von Berg, M.-L., Tutt, L., & Fiedrich, F. (2023). Soziale Medien als Psychosoziale Ressource in Krisen und Katastrophen. *Mensch und Computer 2023—Workshopband*. https://doi.org/10.18420/MUC2023-MCI-WS01-348

Mummendey, H. D., & Grau, I. (2014). *Die Fragebogen-Methode. Grundlagen und Anwendung in Persönlichkeits-, Einstellungs- Und Selbstkonzeptforschung* (6th ed.). Hogrefe.

Münscher, R., Vetter, M., & Scheuerle, T. (2016). A Review and Taxonomy of Choice Architecture Techniques. *Journal of Behavioral Decision Making, 29*(5), 511–524. https://doi.org/10.1002/bdm.1897

Munzert, S., Papoutsi, M., & Nowak, H. (2021a). *Ein Jahr digitale Kontaktpersonennachverfolgung mit der Corona-Warn-App: Nutzung—Popularität—Verständnis*. Hertie School. https://opus4.kobv.de/opus4-hsog/frontdoor/deliver/index/docId/3999/file/20210614_covid-apps-report-w2.pdf

Munzert, S., Papoutsi, M., & Nowak, H. (2021b). *Nutzung von Digitalen Tools zur Unterstützung von Covid-19-Kontaktverfolgung: Wie Populär sind Corona-Warn-App Und Luca-App in der Dritten Pandemiewelle?* Hertie School.

Naths, G., Jürgens, C., & Peter, A. (2007). First Responder als Ergänzung des Rettungsdienstes. *Notfall + Rettungsmedizin, 10*(5), 350–356. https://doi.org/10.1007/s10049-007-0929-0

Nestler, S. (2017). Flächendeckende Kommunikation im Stromausfall durch regionale IKT. Krisenszenario: Längerfristiger Stromausfall. In M. Burghardt, R. Wimmer, C. Wolff, & C. Womser-Hacker (Eds.), *Mensch und Computer 2017—Workshopband* (pp. 9–16). Gesellschaft für Informatik e.V.

Neves, B. B., Waycott, J., & Malta, S. (2018). Old and Afraid of New Communication Technologies? Reconceptualising and Contesting the 'Age-Based Digital Divide'. *Journal of Sociology, 54*(2), 236–248. https://doi.org/10.1177/1440783318766119

Newman, L. H. (2017). *Department of Defense's 'Hack the Pentagon' Bug Bounty Program Helps Fix Thousands of Bugs*. https://www.wired.com/story/hack-the-pentagon-bug-bounty-results/

Newman, N., Fletcher, R., Eddy, K., Robertson, C. T., & Nielsen, R. K. (2023). *Digital News Report 2023*. Reuters Institute.

Newman, N., Fletcher, R., Kalogeropoulos, A., & Nielsen, R. K. (2019). *Reuters Institute Digital News Report 2019*. Reuters Institute. https://reutersinstitute.politics.ox.ac.uk/sites/default/files/2019-06/DNR_2019_FINAL_0.pdf

Newman, N., Fletcher, R., Robertson, C. T., Eddy, K., & Nielsen, R. K. (2022). *Digital News Report 2022*. Reuters Institute. https://reutersinstitute.politics.ox.ac.uk/sites/default/files/2022-06/Digital_News-Report_2022.pdf

Nguyen, Q. N., Sidorova, A., & Torres, R. (2022). User Interactions with Chatbot Interfaces vs. Menu-based Interfaces: An Empirical Study. *Computers in Human Behavior, 128*, 107093. https://doi.org/10.1016/j.chb.2021.107093

Niehaves, B., & Plattfaut, R. (2014). Internet Adoption by the Elderly: Employing IS Technology Acceptance Theories for Understanding the Age-Related Digital Divide. *European Journal of Information Systems, 23*(6), 708–726. https://doi.org/10.1057/ejis.2013.19

Nikiforova, A. (2021). Smarter Open Government Data for Society 5.0: Are Your Open Data Smart Enough? *Sensors, 21*(15), 5204. https://doi.org/10.3390/s21155204

Nikou, S., & Mezei, J. (2013). Evaluation of Mobile Services and Substantial Adoption Factors with Analytic Hierarchy Process (AHP). *Telecommunications Policy, 37*(10), 915–929. https://doi.org/10.1016/j.telpol.2012.09.007

Ning, N., Hu, M., Qiao, J., Liu, C., Zhao, X., Xu, W., Xu, W., Zheng, B., Chen, Z., Yu, Y., Hao, Y., & Wu, Q. (2021). Factors Associated With Individual Emergency Preparedness Behaviors: A Cross-Sectional Survey Among the Public in Three Chinese Provinces. *Frontiers in Public Health, 9*, 644421. https://doi.org/10.3389/fpubh.2021.644421

Nkwo, M., Suruliraj, B., & Orji, R. (2021). Persuasive Apps for Sustainable Waste Management: A Comparative Systematic Evaluation of Behavior Change Strategies and State-of-the-Art. *Frontiers in Artificial Intelligence, 4*, 1–18. https://doi.org/10.3389/frai.2021.748454

Nocera, J. A., Dunckley, L., & Sharp, H. (2007). An Approach to the Evaluation of Usefulness as a Social Construct Using Technological Frames. *International Journal of Human-Computer Interaction, 22*(1–2), 153–172. https://doi.org/10.1080/10447310709336959

Nordmann, A., & Ripper, A. (2019). Safety and Security—Their Relation and Transformation. In C. Reuter (Ed.), *Information Technology for Peace and Security: IT Applications and Infrastructures in Conflicts, Crises, War, and Peace* (pp. 343–359). Springer Fachmedien. https://doi.org/10.1007/978-3-658-25652-4_16

Novak, J., Melenhorst, M., Micheel, I., Pasini, C., Fraternali, P., & Rizzoli, A. (2018). Integrating Behavioural Change and Gamified Incentive Modelling for Stimulating Water Saving. *Environmental Modelling & Software, 102*, 120–137. https://doi.org/10.1016/j.envsoft.2017.11.038

Nwafor, O., Singh, R., Collier, C., DeLeon, D., Osborne, J., & DeYoung, J. (2021). Effectiveness of Nudges as a Tool to Promote Adherence to Guidelines in Healthcare and Their Organizational Implications: A Systematic Review. *Social Science & Medicine, 286*, 114321.

Nys, T. R., & Engelen, B. (2017). Judging Nudging: Answering the Manipulation Objection. *Political Studies, 65*(1), 199–214. https://doi.org/10.1177/0032321716629487

Obar, J. A., & Oeldorf-Hirsch, A. (2020). The Biggest Lie on the Internet: Ignoring the Privacy Policies and Terms of Service Policies of Social Networking Services. *Information, Communication & Society, 23*(1), 128–147. https://doi.org/10.1080/1369118X.2018.1486870

O'Brien, H., & Cairns, P. (2015). An Empirical Evaluation of the User Engagement Scale (UES) in Online News Environments. *Information Processing & Management, 51*(4), 413–427. https://doi.org/10.1016/j.ipm.2015.03.003

O'Brien, H. L., Cairns, P., & Hall, M. (2018). A Practical Approach to Measuring User Engagement with the Refined User Engagement Scale (UES) and New UES Short Form. *International Journal of Human-Computer Studies, 112*, 28–39. https://doi.org/10.1016/j.ijhcs.2018.01.004

O'Brien, H. L., & Toms, E. G. (2010). The Development and Evaluation of a Survey to Measure User Engagement. *Journal of the American Society for Information Science and Technology, 61*(1), 50–69. https://doi.org/10.1002/asi.21229

Ohme, J., Vanden Abeele, M. M. P., Van Gaeveren, K., Durnez, W., & De Marez, L. (2020). Staying Informed and Bridging "Social Distance": Smartphone News Use and Mobile Messaging Behaviors of Flemish Adults during the First Weeks of the COVID-19 Pandemic. *Socius: Sociological Research for a Dynamic World, 6*, 237802312095019. https://doi.org/10.1177/2378023120950190

Oinas-Kukkonen, H., & Harjumaa, M. (2009). Persuasive Systems Design: Key Issues, Process Model, and System Features. *Communications of the Association for Information Systems, 24*(1), 485–500. https://doi.org/10.17705/1cais.02428

Olsson, E.-K. (2014). Crisis Communication in Public Organisations: Dimensions of Crisis Communication Revisited. *Journal of Contingencies and Crisis Management, 22*(2), 113–125. https://doi.org/10.1111/1468-5973.12047

Olteanu, A., Vieweg, S., & Castillo, C. (2015). What to Expect When the Unexpected Happens: Social Media Communications across Crises. *Proceedings of the 18th ACM Conference on Computer Supported Cooperative Work & Social Computing*, 994–1009. https://doi.org/10.1145/2675133.2675242

Olubusola, A. O. (2015). User Satisfaction in Mobile Applications. *Research Paper at School of Computer Science, University of Birmingham.*

Onyeji, I., Bazilian, M., & Bronk, C. (2014). Cyber Security and Critical Energy Infrastructure. *The Electricity Journal, 27*(2), 52–60. https://doi.org/10.1016/j.tej.2014.01.011

Orji, R., & Moffatt, K. (2018). Persuasive Technology for Health and Wellness: State-of-the-art and Emerging Trends. *Health Informatics Journal, 24*(1), 66–91. https://doi.org/10.1177/1460458216650979

Ostrom, E. (1996). Crossing the Great Divide: Coproduction, Synergy, and Development. *World Development, 24*(6), 1073–1087. https://doi.org/10.1016/0305-750X(96)00023-X

Oxford English Dictionary. (2020). Motivation.

Pagano, D., & Bruegge, B. (2013). User Involvement in Software Evolution Practice: A Case Study. *Proceedings—2013 35th International Conference on Software Engineering (ICSE)*, 953–962. https://doi.org/10.1109/ICSE.2013.6606645

Pagano, D., & Maalej, W. (2013). User Feedback in the Appstore: An Empirical Study. *2013 21st IEEE International Requirements Engineering Conference, RE 2013—Proceedings*, 125–134. https://doi.org/10.1109/RE.2013.6636712

Page, L. C., Castleman, B. L., & Meyer, K. (2020). Customized Nudging to Improve FAFSA Completion and Income Verification. *Educational Evaluation and Policy Analysis, 42*(1), 3–21. https://doi.org/10.3102/0162373719876916

Page, L. C., Sacerdote, B. I., Goldrick-Rab, S., & Castleman, B. L. (2022). Financial Aid Nudges: A National Experiment with Informational Interventions. *Educational Evaluation and Policy Analysis, 0*(0), 1–25. https://doi.org/10.3102/01623737221111403

Palen, L., & Anderson, K. M. (2016). Crisis Informatics: New Data for Extraordinary Times. *Science (New York, N.Y.), 353*(6296), 224–225.

Palen, L., Anderson, K. M., Mark, G., Martin, J., Sicker, D., Palmer, M., & Grunwald, D. (2010). A Vision for Technology-Mediated Support for Public Participation and Assistance in Mass Emergencies and Disasters. *2010 ACM-BCS Visions of Computer Science Conference*, 87.

Palen, L., Vieweg, S., Liu, S. B., & Hughes, A. L. (2009). Crisis in a Networked World: Features of Computer-Mediated Communication in the April 16, 2007, Virginia Tech Event. *Social Science Computer Review, 27*(4), 467–480. https://doi.org/10.1177/0894439309332302

Paniagua, P., & Rayamajhee, V. (2022). A Polycentric Approach for Pandemic Governance: Nested Externalities and Co-Production Challenges. *Journal of Institutional Economics, 18*(4), 537–552. https://doi.org/10.1017/S1744137421000795

Park, C. S. (2019). Does Too Much News on Social Media Discourage News Seeking? Mediating Role of News Efficacy between Perceived News Overload and News Avoidance on Social Media. *Social Media and Society, 5*(3). https://doi.org/10.1177/2056305119872956

Park, C. H., & Johnston, E. W. (2017). A Framework for Analyzing Digital Volunteer Contributions in Emergent Crisis Response Efforts. *New Media and Society, 19*(8), 1308–1327. https://doi.org/10.1177/1461444817706877

Park, S., & Avery, E. J. (2018). Effects of Media Channel, Crisis Type and Demographics on Audience Intent to Follow Instructing Information During Crisis. *Journal of Contingencies and Crisis Management, 26*(1), 69–78. https://doi.org/10.1111/1468-5973.12137

Park, S., Boatwright, B., & Avery, E. J. (2019). Information Channel Preference in Health Crisis: Exploring the Roles of Perceived Risk, Preparedness, Knowledge, and Intent to Follow Directives. *Public Relations Review, 45*(5), 101794.

Parker, M. J., Fraser, C., Abeler-Dörner, L., & Bonsall, D. (2020). Ethics of Instantaneous Contact Tracing Using Mobile Phone Apps in the Control of the COVID-19 Pandemic. *Journal of Medical Ethics, 46*(7), 427–431. https://doi.org/10.1136/medethics-2020-106314

Paton, D. (2006). Disaster Resilience: An Integrated Approach. Charles C Thomas Pub Ltd.

Paton, D. (2019). Disaster Risk Reduction: Psychological Perspectives on Preparedness. *Australian Journal of Psychology, 71*(4), 327–341. https://doi.org/10.1111/ajpy.12237

Peffers, K., Tuunanen, T., Rothenberger, M. A., & Chatterjee, S. (2007). A Design Science Research Methodology for Information Systems Research. *Journal of Management Information Systems, 24*(3), 45–77.

Petrun Sayers, E. L., Parker, A. M., Seelam, R., & Finucane, M. L. (2021). How Disasters Drive Media Channel Preferences: Tracing News Consumption before, during, and after Hurricane Harvey. *Journal of Contingencies and Crisis Management, 29*(4), 342–356.

Piccolo, L. S. G., Roberts, S., Iosif, A., & Alani, H. (2018, July). Designing Chatbots for Crises: A Case Study Contrasting Potential and Reality. In BCS Learning and Development Ltd. (Ed.), *Proceedings of the 32nd International BCS Human Computer Interaction Conference* (pp. 1–10). https://doi.org/10.14236/ewic/HCI2018.56

Pichert, D., & Katsikopoulos, K. V. (2008). Green Defaults: Information Presentation and pro-Environmental Behaviour. *Journal of Environmental Psychology, 28*(1), 63–73. https://doi.org/10.1016/j.jenvp.2007.09.004

Pine, K. H., Lee, M., Whitman, S. A., Chen, Y., & Henne, K. (2021). Making Sense of Risk Information amidst Uncertainty: Individuals' Perceived Risks Associated with the COVID-19 Pandemic. *Proceedings of the 2021 CHI Conference on Human Factors in Computing Systems (CHI '21)*, 1–15. https://doi.org/10.1145/3411764.3445051

Plotnick, L., & Hiltz, S. R. (2016). Barriers to Use of Social Media by Emergency Managers. *Journal of Homeland Security and Emergency Management, 13*(2), 247–277. https://doi.org/10.1515/jhsem-2015-0068

Polites, G. L., & Karahanna, E. (2012). Shackled to the Status Quo: The Inhibiting Effects of Incumbent System Habit, Switching Costs, and Inertia on New System Acceptance. *MIS Quarterly, 36*(1), 21–42. https://doi.org/10.2307/41410404

Pors, A. S. (2015). Becoming Digital—Passages to Service in the Digitized Bureaucracy. *Journal of Organizational Ethnography, 4*(2), 177–192.

Pradhan, A., Mehta, K., & Findlater, L. (2018). "Accessibility Came by Accident": Use of Voice-Controlled Intelligent Personal Assistants by People with Disabilities. *Proceedings*

of the 2018 CHI Conference on Human Factors in Computing Systems, 1–13. https://doi.
org/10.1145/3173574.3174033

Pratt, B. W., & Erickson, J. D. (2020). Defeat the Peak: Behavioral Insights for Electricity
Demand Response Program Design. *Energy Research & Social Science, 61*, 101352. https://
doi.org/10.1016/j.erss.2019.101352

Rahn, M., Tomczyk, S., & Schmidt, S. (2020). Storms, Fires, and Bombs: Analyzing the
Impact of Warning Message and Receiver Characteristics on Risk Perception in Different
Hazards. *Risk Analysis*. https://doi.org/10.1111/risa.13636

Rahn, M., Tomczyk, S., Schopp, N., Schmidt, S., & Sutton, J. (2021). Warning Messages
in Crisis Communication: Risk Appraisal and Warning Compliance in Severe Weather,
Violent Acts, and the COVID-19 Pandemic. *Frontiers in Psychology, 12*. https://doi.org/
10.3389/fpsyg.2021.557178

Raja, F., Hawkey, K., Hsu, S., Wang, K.-L. C., & Beznosov, K. (2011). A Brick Wall, a Locked
Door, and a Bandit: A Physical Security Metaphor for Firewall Warnings. *Proceedings of
the Seventh Symposium on Usable Privacy and Security*, 1–20. https://doi.org/10.1145/
2078827.2078829

Rana, N. P., Dwivedi, Y. K., & Williams, M. D. (2015). A Meta-Analysis of Existing Research
on Citizen Adoption of e-Government. *Information Systems Frontiers, 17*, 547–563.

Ranchordás, S. (2020). Nudging Citizens through Technology in Smart Cities. *International Review of Law, Computers & Technology, 34*(3), 254–276. https://doi.org/10.1080/
13600869.2019.1590928

Rapp, A., Curti, L., & Boldi, A. (2021). The Human Side of Human-Chatbot Interaction: A
Systematic Literature Review of Ten Years of Research on Text-Based Chatbots. *International Journal of Human-Computer Studies, 151*, 102630. https://doi.org/10.1016/j.ijhcs.
2021.102630

Reisch, L. A., & Sunstein, C. R. (2016). Do Europeans like Nudges? *Lecture Notes in Computer Science (including subseries Lecture Notes in Artificial Intelligence and Lecture Notes
in Bioinformatics), 11 4*, 310–325. https://doi.org/10.2139/ssrn.2739118

Renaud, K., & Zimmermann, V. (2018). Ethical Guidelines for Nudging in Information Security & Privacy. *International Journal of Human Computer Studies, 120*, 22–35. https://doi.
org/10.1016/j.ijhcs.2018.05.011

Renaud, K., & Zimmermann, V. (2019). Nudging Folks towards Stronger Password Choices:
Providing Certainty Is the Key. *Behavioural Public Policy, 3*(2), 228–258. https://doi.org/
10.1017/bpp.2018.3

Reuter, C. (2022). *A European Perspective on Crisis Informatics: Citizens' and Authorities'
Attitudes towards Social Media for Public Safety and Security* (1st ed.). Springer Vieweg.
https://doi.org/10.1007/978-3-658-39720-3

Reuter, C., Haunschild, J., Hollick, M., Mühlhäuser, M., Vogt, J., & Kreutzer, M. (2020).
Towards Secure Urban Infrastructures: Cyber Security Challenges to Information and Communication Technology in Smart Cities. In C. Hansen, A. Nürnberger, & B. Preim (Eds.),
Mensch und Computer 2020—Workshopband (pp. 1–7). Gesellschaft für Informatik e.V.
https://doi.org/10.18420/muc2020-ws117-408

Reuter, C., & Kaufhold, M.-A. (2018). Fifteen Years of Social Media in Emergencies: A Retrospective Review and Future Directions for Crisis Informatics. *Journal of Contingencies
and Crisis Management (JCCM), 26*(1), 41–57. https://doi.org/10.1111/1468-5973.12196

Reuter, C., Kaufhold, M.-A., Leopold, I., & Knipp, H. (2017). KATWARN, NINA, or FEMA? Multi-method Study on Distribution, Use and Public Views on Crisis Apps. *European Conference on Information Systems (ECIS)*, 2187–2201.

Reuter, C., Kaufhold, M.-A., Schmid, S., Spielhofer, T., & Hahne, A. S. (2019). The Impact of Risk Cultures: Citizens' Perception of Social Media Use in Emergencies across Europe. *Technological Forecasting and Social Change, 148*(119724). https://doi.org/10.1016/j.techfore.2019.119724

Reuter, C., Kaufhold, M.-A., Spielhofer, T., & Hahne, A. S. (2017). Social Media in Emergencies: A Representative Study on Citizens' Perception in Germany. *Proceedings of the ACM on Human-Computer Interaction (PACM HCI), 1*, 1–19. https://doi.org/10.1145/3134725

Reuter, C., Ludwig, T., Kaufhold, M.-A., & Spielhofer, T. (2016). Emergency Services Attitudes towards Social Media: A Quantitative and Qualitative Survey across Europe. *International Journal on Human-Computer Studies (IJHCS), 95*, 96–111. https://doi.org/10.1016/j.ijhcs.2016.03.005

Reuter, C., Riebe, T., Haunschild, J., Reinhold, T., & Schmid, S. (2022). Zur Schnittmenge von Informatik mit Friedens- und Sicherheitsforschung: Erfahrungen aus der interdisziplinären Lehre in der Friedensinformatik. *Zeitschrift für Friedens- und Konfliktforschung, 11*(2), 129–140. https://doi.org/10.1007/s42597-022-00078-4

Reuter, C., & Spielhofer, T. (2017). Towards Social Resilience: A Quantitative and Qualitative Survey on Citizens' Perception of Social Media in Emergencies in Europe. *Journal Technological Forecasting and Social Change (TFSC), 121*, 168–180. https://doi.org/10.1016/j.techfore.2016.07.038

Rheinberg, F. (2009). Intrinsische Motivation. In V. Brandstätter & J. H. Otto (Eds.), *Handbuch der Allgemeinen Psychologie—Motivation und Emotion* (pp. 258–265, Vol. 11). Göttingen: Hogrefe.

Rheinberg, F., & Manig, Y. (2003). Was Macht Spaß Am Graffiti-Sprayen? Eine Induktive Anreizanalyse. *Report Psychologie, 28.*

Riebe, T., Haunschild, J., Divo, F., Lang, M., Roitburd, G., Franken, J., & Reuter, C. (2020). Die Veränderung der Vorratsdatenspeicherung in Europa. *Datenschutz und Datensicherheit—DuD, 44*(5), 316–321. https://doi.org/10.1007/s11623-020-1275-3

Rittel, H. W. J. (1973). Dilemmas in a General Theory of Planning. *Policy Sciences*, (4), 155–169. https://doi.org/10.1007/BF01405730

Rizzoli, A. E., Castelletti, A., Fraternali, P., & Novak, J. (2018). SmartH2O, Demonstrating the Impact of Gamification Technologies for Saving Water. *Computer Science—Research and Development, 33*(1), 275–276. https://doi.org/10.1007/s00450-017-0380-5

RKI. (2020, October). *Key Figures for the Corona-Warn-App.* https://www.rki.de/DE/Content/InfAZ/N/Neuartiges_Coronavirus/WarnApp/Archiv_Kennzahlen/Kennzahlen_30102020.pdf?__blob=publicationFile

RKI. (2021). *Coronavirus Disease 2019 (COVID-19)—Weekly Situation Report From the Robert Koch Institute.*

RKI. (2022, March 17). *Todesfälle nach Sterbedatum.* https://www.rki.de/DE/Content/InfAZ/N/Neuartiges_Coronavirus/Projekte_RKI/COVID-19_Todesfaelle.html

Robinson, S. E., Pudlo, J. M., & Wehde, W. (2019). The New Ecology of Tornado Warning Information: A Natural Experiment Assessing Threat Intensity and Citizen-to-Citizen Information Sharing. *Public Administration Review, 79*(6), 905–916. https://doi.org/10.1111/puar.13030

Romero Herrera, N., Rutten, J., & Keyson, D. V. (2017). Designing Ampul: Empowerment to Home Energy Prosumers. In D. V. Keyson, O. Guerra-Santin, & D. Lockton (Eds.), *Living Labs: Design and Assessment of Sustainable Living* (pp. 309–323). Springer International Publishing. https://doi.org/10.1007/978-3-319-33527-8_24

Ross, B., Potthoff, T., Majchrzak, T. A., Chakraborty, N. R., Ben Lazreg, M., & Stieglitz, S. (2018). The Diffusion of Crisis-Related Communication on Social Media: An Empirical Analysis of Facebook Reactions. *Proceedings of the 51st Hawaii International Conference on System Sciences*, 2525–2534. https://doi.org/10.24251/hicss.2018.319

Rovai, A. P., Baker, J. D., & Ponton, M. K. (2013). *Social Science Research Design and Statistics: A Practitioner's Guide to Research Methods and IBM SPSS*. Watertree Press LLC.

Roztocki, N., Strzelczyk, W., & Weistroffer, H. R. (2023). The Role of e Government in Disaster Management: A Review of the Literature. *Journal of Economics and Management, 45*, 1–25.

Ryan, R. M., & Deci, E. L. (2000). When Rewards Compete with Nature: The Undermining of Intrinsic Motivation and Self-Regulation. In C. Sansone & J. M. Harackiewicz (Eds.), *Intrinsic and Extrinsic Motivation* (pp. 13–54). Elsevier. https://doi.org/10.1016/B978-012619070-0/50024-6

Santos, T., Lemmerich, F., Strohmaier, M., & Helic, D. (2019). What's in a Review: Discrepancies between Expert and Amateur Reviews of Video Games on Metacritic. *Proceedings of the ACM on Human-Computer Interaction, 3*. https://doi.org/10.1145/3359242

Sauro, J., & Lewis, J. R. (2016). *Quantifying the User Experience: Practical Statistics for User Research* (2nd ed.). Morgan Kaufmann.

Schlagwein, D., Conboy, K., Feller, J., Leimeister, J. M., & Morgan, L. (2017). "Openness" with and without Information Technology: A Framework and a Brief History. *Journal of Information Technology, 32*(4), 297–305. https://doi.org/10.1057/s41265-017-0049-3

Schminder, E., Ziegler, M., Danay, E., Beyer, L., & Bühner, M. (2010). Is It Really Robust? Reinvestigating the Robustness of ANOVA against Violations of the Normal Distribution. *European Research Journal of Methods for the Behavioral and Social Sciences, 6*, 147–151.

Schmitt, J. B., Debbelt, C. A., & Schneider, F. M. (2018). Too Much Information? Predictors of Information Overload in the Context of Online-News Exposure. *Information, Communication & Society, 21*(8), 1151–1167. https://doi.org/10.1080/1369118X.2017.1305427

Scholta, H., & Lindgren, I. (2019). The Long and Winding Road of Digital Public Services—One Next Step: Proactivity. *ICIS 2019 Proceedings*. https://doi.org/https://aisel.aisnet.org/icis2019/digital_government/digital_government/7

Schoonenboom, J., & Johnson, R. B. (2017). How to Construct a Mixed Methods Research Design. *KZfSS Kölner Zeitschrift für Soziologie und Sozialpsychologie, 69*(S2), 107–131. https://doi.org/10.1007/s11577-017-0454-1

Schrock, A. R. (2015). Communicative Affordances of Mobile Media: Portability, Availability, Locatability, and Multimediality. *International Journal of Communication, 9*, 1–18.

Schroeder, B. L., Whitmer, D. E., & Sims, V. K. (2017). Toward a User-Centered Approach for Emergency Warning Distribution. *Ergonomics in Design, 25*(1), 4–10. https://doi.org/10.1177/1064804616662420

Schubert, C. (2017). Green Nudges: Do They Work? Are They Ethical? *Ecological Economics, 132*, 329–342. https://doi.org/10.1016/j.ecolecon.2016.11.009

Schultz, P. W., Nolan, J. M., Cialdini, R. B., Goldstein, N. J., & Griskevicius, V. (2007). The Constructive, Destructive, and Reconstructive Power of Social Norms. *Psychological Science, 18*(5), 429–434. https://doi.org/10.1111/j.1467-9280.2007.01917.x

Schwartz, S. H., & Howard, J. A. (1984). Internalized Values as Motivators of Altruism. In E. Staub, D. Bar-Tal, J. Karylowski, & J. Reykowski (Eds.), *Development and Maintenance of Prosocial Behavior* (pp. 229–255). Springer US. https://doi.org/10.1007/978-1-4613-2645-8_14

Shahid, A. R., & Elbanna, A. (2015). The Impact of Crowdsourcing on Organisational Practices: The Case of Crowdmapping. *ECIS 2015 Completed Research Papers*. https://doi.org/10.18151/7217474

Shapira, S., Aharonson-Daniel, L., & Bar-Dayan, Y. (2018). Anticipated Behavioral Response Patterns to an Earthquake: The Role of Personal and Household Characteristics, Risk Perception, Previous Experience and Preparedness. *International Journal of Disaster Risk Reduction, 31*, 1–8. https://doi.org/10.1016/j.ijdrr.2018.04.001

Sharma, S. K., Al-Badi, A., Rana, N. P., & Al-Azizi, L. (2018). Mobile Applications in Government Services (mG-App) from User's Perspectives: A Predictive Modelling Approach. *Government Information Quarterly, 35*(4), 557–568. https://doi.org/10.1016/j.giq.2018.07.002

Sheeran, P. (2002). Intention—Behavior Relations: A Conceptual and Empirical Review. *European Review of Social Psychology, 12*(1), 1–36. https://doi.org/10.1080/14792772143000003

Sherman-Morris, K., Pechacek, T., Griffin, D. J., & Senkbeil, J. (2020). Tornado Warning Awareness, Information Needs and the Barriers to Protective Action of Individuals Who Are Blind. *International Journal of Disaster Risk Reduction, 50*, 101709. https://doi.org/10.1016/j.ijdrr.2020.101709

Shi, J., Poorisat, T., & Salmon, C. T. (2018). The Use of Social Networking Sites (SNSs) in Health Communication Campaigns: Review and Recommendations. *Health Communication, 33*(1), 49–56. https://doi.org/10.1080/10410236.2016.1242035

Shklovski, I., Palen, L., & Sutton, J. (2008). Finding Community through Information and Communication Technology during Disaster Events. *Proceedings of the ACM 2008 Conference on Computer Supported Cooperative Work (CSCW '08).*

Shneiderman, B., Plaisant, C., Cohen, M., Jacobs, S., Elmqvist, N., & Diakopoulos, N. (2016). *Designing the User Interface: Strategies for Effective Human-Computer Interaction* (6th ed.). Pearson.

Siebenhaar, K. U., Köther, A. K., & Alpers, G. W. (2020). Dealing with the COVID-19 Infodemic: Distress by Information, Information Avoidance, and Compliance with Preventive Measures. *Frontiers in Psychology, 11*, 1–11. https://doi.org/10.3389/fpsyg.2020.567905

Siegrist, M., & Árvai, J. (2020). Risk Perception: Reflections on 40 Years of Research. *Risk Analysis, 40*(S1), 2191–2206. https://doi.org/10.1111/risa.13599

Simintiras, A. C., Dwivedi, Y. K., & Rana, N. P. (2014). Can Marketing Strategies Enhance the Adoption of Electronic Government Initiatives? *International Journal of Electronic Government Research (IJEGR), 10*(2), 1–7.

Sjöberg, U. (2018). It Is Not about Facts–It Is about Framing. The App Generation's Information-Seeking Tactics: Proactive Online Crisis Communication. *Journal of Contingencies and Crisis Management, 26*(1), 127–137.

Soden, R., & Palen, L. (2018). Informating Crisis: Expanding Critical Perspectives in Crisis Informatics. *Proceedings of the ACM on Human-Computer Interaction, 2*, 1–22. https://doi.org/10.1145/3274431

Soroya, S. H., Farooq, A., Mahmood, K., Isoaho, J., & Zara, S.-E. (2021). From Information Seeking to Information Avoidance: Understanding the Health Information Behavior during a Global Health Crisis. *Information Processing & Management, 58*(2), 1–16. https://doi.org/10.1016/j.ipm.2020.102440

St. Denis, A. L., Hughes, A. L., & Palen, L. (2012). Trial by Fire: The Deployment of Trusted Digital Volunteers in the 2011 Shadow Lake Fire. In L. Rothkrantz, J. Ristvej, & Z. Franco (Eds.), *Proceedings of the information systems for crisis response and management (ISCRAM)* (pp. 1–10). ISCRAM.

Staegemann, D., Volk, M., Daase, C., Pohl, M., & Turowski, K. (2022). A Concept for the Use of Chatbots to Provide the Public with Vital Information in Crisis Situations. *Proceedings of Sixth International Congress on Information and Communication Technology*, Singapore (pp. 281–289).

Starbird, K., Maddock, J., Orand, M., Achterman, P., & Mason, R. M. (2014). Rumors, False Flags, and Digital Vigilantes: Misinformation on Twitter after the 2013 Boston Marathon Bombing. *iConference 2014*, 1–9.

Starbird, K., & Palen, L. (2011). "Voluntweeters": Self-Organizing by Digital Volunteers in Times of Crisis. *Proceedings of the 2011 Annual Conference on Human Factors in Computing Systems—CHI '11*, 1071–1080. https://doi.org/10.1145/1978942.1979102

Statista. (2018). Bevölkerung in Deutschland. https://de.statista.com/statistik/studie/id/11844/dokument/vergleich-der-bundeslaender-statista-dossier/

Statista. (2024a). Most popular messenger apps 2024. https://www.statista.com/statistics/258749/most-popular-global-mobile-messenger-apps/

Statista. (2024b). U.S. Top Messaging Services by Age. Retrieved January 29, 2024, from https://www.statista.com/statistics/1310687/us-top-messaging-services-by-age/

Statista & eMarketer. (2021). Anzahl der Nutzer von Messenger-Apps Weltweit in den Jahren 2018 bis 2020 Sowie eine Prognose bis 2025. https://de.statista.com/statistik/daten/studie/1073385/umfrage/anzahl-der-nutzer-von-messenger-apps-weltweit/

Stieglitz, S., Hofeditz, L., Brünker, F., Ehnis, C., Mirbabaie, M., & Ross, B. (2022). Design Principles for Conversational Agents to Support Emergency Management Agencies. *International Journal of Information Management, 63*, 102469–102469. https://doi.org/10.1016/j.ijinfomgt.2021.102469

Stieglitz, S., Lattemann, C., Robra-Bissantz, S., Zarnekow, R., & Brockmann, T. (Eds.). (2017). *Gamification. Using Game Elements in Serious Contexts*. Springer.

Stieglitz, S., Mirbabaie, M., Fromm, J., & Melzer, S. (2018). The Adoption of Social Media Analytics for Crisis Management—Challenges and Opportunities. *Proceedings of the 26th European Conference on Information Systems (ECIS)*.

Stieglitz, S., Mirbabaie, M., & Milde, M. (2018). Social Positions and Collective Sense-Making in Crisis Communication. *International Journal of Human-Computer Interaction, 34*(4), 328–355. https://doi.org/10.1080/10447318.2018.1427830

Strauss, A. L., & Corbin, J. M. (1998). *Basics of Qualitative Research: Techniques and Procedures for Developing Grounded Theory* (2nd ed.). Sage Publications. https://doi.org/10.1177/1350507600314007

Stuart, A., & Thorsen, E. (2009). *Citizen Journalism: Global Perspectives*. Peter Lang Publ.

Sturm, C., & Nestler, S. (2018). Internationale und Interkulturelle Aspekte. In C. Reuter (Ed.), *Sicherheitskritische Mensch-Computer-Interaktion. Interaktive Technologien und Soziale Medien im Krisen- und Sicherheitsmanagement* (pp. 183–202). Springer.

Su, C.-H., & Cheng, C.-H. (2015). A Mobile Gamification Learning System for Improving the Learning Motivation and Achievements. *Journal of Computer Assisted Learning, 31*(3), 268–286. https://doi.org/10.1111/jcal.12088

Sugerman, D. E., Keir, J. M., Dee, D. L., Lipman, H., Waterman, S. H., Ginsberg, M., & Fishbein, D. B. (2012). Emergency Health Risk Communication During the 2007 San Diego Wildfires: Comprehension, Compliance, and Recall. *Journal of Health Communication, 17*(6), 698–712. https://doi.org/10.1080/10810730.2011.635777

Sugisaki, K., & Bleiker, A. (2020). Usability Guidelines and Evaluation Criteria for Conversational User Interfaces: A Heuristic and Linguistic Approach. *Proceedings of the Conference on Mensch Und Computer*, 309–319. https://doi.org/10.1145/3404983.3405505

Sunstein, C. R. (2013). Behavioral Economics, Consumption, and Environmental Protection. In L. Reisch & J. Thøgersen (Eds.), *Handbook on Research in Sustainable Consumption*. Edward Elgar Publishing. https://doi.org/10.2139/ssrn.2296015

Sunstein, C. R. (2015). Nudges Do Not Undermine Human Agency. *Journal of Consumer Policy, 38*(3), 207–210. https://doi.org/10.1007/s10603-015-9289-1

Sunstein, C. R., & Reisch, L. A. (2016). Climate-Friendly Default Rules. *Discussion Paper No. 87809/2016 Harvard Law School Cambridge, MA 02138.*

Susanto, T. D., Diani, M. M., & Hafidz, I. (2017). User Acceptance of E-Government Citizen Report System (a Case Study of City113 App). *Procedia Computer Science, 124*, 560–568. https://doi.org/10.1016/j.procs.2017.12.190

Susanto, T. D., & Goodwin, R. (2013). User Acceptance of SMS-based e-Government Services: Differences between Adopters and Non-Adopters. *Government Information Quarterly, 30*(4), 486–497. https://doi.org/10.1016/j.giq.2013.05.010

Sutikno, T., Handayani, L., Stiawan, D., Riyadi, M. A., & Subroto, I. M. I. (2016). WhatsApp, Viber and Telegram: Which Is the Best for Instant Messaging? *International Journal of Electrical and Computer Engineering, 6*(3), 909–914. https://doi.org/10.11591/ijece.v6i3.10271

Tackenberg, B., Lukas, T., & Fiedrich, F. (2022). Community Resilience in Krisen und Katastrophen—Nachbarschaftliches Sozialkapital als Bewältigungsressource. In S. Voßschmidt & A. Karsten (Eds.), *Resilienz und Pandemie* (pp. 121–129). Kohlhammer.

Tagesschau. (2020, September 10). *Bilanz Des Innenministeriums: Bundesweiter Warntag "Fehlgeschlagen"*. https://www.tagesschau.de/inland/warntag-115.html

Tagliacozzo, S. (2018). Government Agency Communication during Postdisaster Reconstruction: Insights from the Christchurch Earthquakes Recovery. *Natural Hazards Review, 19*(2), 1–11. https://doi.org/10.1061/(ASCE)NH.1527-6996.0000283

Taj, F., Klein, M. C., & van Halteren, A. (2019). Digital Health Behavior Change Technology: Bibliometric and Scoping Review of Two Decades of Research. *JMIR mHealth and uHealth, 7*(12), e13311.

Takayama, C., Lehdonvirta, V., Shiraishi, M., Washio, Y., Kimura, H., & Nakajima, T. (2009). ECOISLAND: A System for Persuading Users to Reduce CO2 Emissions. *Proceedings of the 2009 Software Technologies for Future Dependable Distributed Systems*, 59–63. https://doi.org/10.1109/STFSSD.2009.8

Tan, M. L., Prasanna, R., Stock, K., Hudson-Doyle, E., Leonard, G., & Johnston, D. (2017). Mobile Applications in Crisis Informatics Literature: A Systematic Review. *International*

Journal of Disaster Risk Reduction, 24, 297–311. https://doi.org/10.1016/j.ijdrr.2017.06.009

Tan, M. L., Prasanna, R., Stock, K., Hudson-Doyle, E., Leonard, G., & Johnston, D. (2019). Enhancing the Usability of a Disaster App: Exploring the Perspectives of the Public as Users. *16th ISCRAM Conference*, 876–886.

Tan, M. L., Prasanna, R., Stock, K., Hudson-Doyle, E., Leonard, G., & Johnston, D. (2020a). Modified Usability Framework for Disaster Apps: A Qualitative Thematic Analysis of User Reviews. *International Journal of Disaster Risk Science, 11*, 615–629. https://doi.org/10.1007/s13753-020-00282-x

Tan, M. L., Prasanna, R., Stock, K., Hudson-Doyle, E., Leonard, G., & Johnston, D. (2020b). Understanding End-Users' Perspectives: Towards Developing Usability Guidelines for Disaster Apps. *Progress in Disaster Science, 7*, 100118. https://doi.org/10.1016/j.pdisas.2020.100118

Tan, M. L., Prasanna, R., Stock, K., Hudson-Doyle, E., Leonard, G., & Johnston, D. (2020c). Usability Factors Influencing the Continuance Intention of Disaster Apps: A Mixed-Methods Study. *International Journal of Disaster Risk Reduction, 50*, 101874. https://doi.org/10.1016/j.ijdrr.2020.101874

Tarute, A., Nikou, S., & Gatautis, R. (2017). Mobile Application Driven Consumer Engagement. *Telematics and Informatics, 34*(4), 145–156. https://doi.org/10.1016/j.tele.2017.01.006

Tatar, D. (2007). The Design Tensions Framework. *Human-Computer Interaction, 22*(4), 413–451.

Telegram. (2022). *Telegram Bot API*. https://core.telegram.org/bots/api

Thaler, R. H., & Sunstein, C. R. (2008). *Nudge: Improving Decisions about Health, Wealth, and Happiness*. Yale University Press.

The British Psychological Society. (2014). Code of Human Research Ethics. https://www.bps.org.uk/news-and-policy/bps-code-human-research-ethics-2nd-edition-2014%20(Accessed%2018%20May%202018)

The European Data Protection Board. (2020). Guidelines 05/2020 on Consent under Regulation 2016/679.

Tikka, M., Ahsanullah, R., Varanasi, U., Härmä, V., Sawhney, N., & Leinonen, T. (2023). Contextual Inquiry of Affordances for Collaboration in Crisis: Lessons from the Finnish Context. *20th International ISCRAM Conference*, 33–42. http://dx.doi.org/10.59297/FCIG6875

Timmermans, S., & Tavory, I. (2012). Theory Construction in Qualitative Research: From Grounded Theory to Abductive Analysis. *Sociological Theory, 30*(3), 167–186. https://doi.org/10.1177/0735275112457914

Toomim, M., Kriplean, T., Pörtner, C., & Landay, J. A. (2011). Utility of Human-Computer Interactions: Toward a Science of Preference Measurement. *Proceedings of the SIGCHI Conference on Human Factors in Computing Systems*, 2275–2284.

Toyoda, Y., Kanegae, H., & Sakai, K. (2014). Gaming Simulation for Community–Based Disaster Reduction. *The Shift from Teaching to Learning: Individual, Collective and Organizational Learning Through Gaming Simulation*, 584–598.

Trang, S., Trenz, M., Weiger, W. H., Tarafdar, M., & Christy M.K. Cheung. (2020). One App to Trace Them All? Examining App Specifications for Mass Acceptance of Contact-Tracing

Apps. *European Journal of Information Systems, 29*(4), 415–428. https://doi.org/10.1080/0960085X.2020.1784046

Treem, J. W., & Leonardi, P. M. (2013). Social Media Use in Organizations: Exploring the Affordances of Visibility, Editability, Persistence, and Association. *Annals of the International Communication Association, 36*(1), 143–189. https://doi.org/10.1080/23808985.2013.11679130

Tsinaraki, C., Mitton, I., Minghini, M., Micheli, M., Kotsev, A., Hernandez Quiros, L., Spinelli, F.-A., Dalla Benetta, A., & Schade, S. (2021). Mobile Apps to Fight the COVID-19 Crisis. *Data, 6*(10), 106. https://doi.org/10.3390/data6100106

Turland, J., Coventry, L., Jeske, D., Briggs, P., & van Moorsel, A. (2015). Nudging towards Security: Developing an Application for Wireless Network Selection for Android Phones. *Proceedings of the 2015 British HCI Conference*, 193–201. https://doi.org/10.1145/2783446.2783588

Tzavella, K., Fekete, A., & Fiedrich, F. (2018). Opportunities Provided by Geographic Information Systems and Volunteered Geographic Information for a Timely Emergency Response during Flood Events in Cologne, Germany. *Natural Hazards, 91*(1), 29–57. https://doi.org/10.1007/s11069-017-3102-1

UN Department of Humanitarian Affairs. (1992). *Internationally Agreed Glossary of Basic Terms Related to Disaster Management* (DHA/93/36). United Nations Department of Humanitarian Affairs. Geneva. https://digitallibrary.un.org/record/793886/files/004DFD3E15B69A67C1256C4C006225C2-dha-glossary-1992.pdf

UN International Strategy for Disaster Reduction. (2009). *Terminology on Disaster Risk Reduction.* UNISDR. Geneva. https://www.unisdr.org/files/7817%7B_%7DDUNISDRTerminologyEnglish.pdf

UN Office for Disaster Risk Reduction. (2015). *Sendai Framework for Disaster Risk Reduction 2015–2030.* United Nations. Geneva, Switzerland. https://www.undrr.org/quick/11409%7D

Utz, C., Becker, S., Schnitzler, T., Farke, F. M., Herbert, F., Schaewitz, L., Degeling, M., & Dürmuth, M. (2021). Apps against the Spread: Privacy Implications and User Acceptance of COVID-19-Related Smartphone Apps on Three Continents. *Proceedings of the 2021 CHI Conference on Human Factors in Computing Systems*, 1–22. https://doi.org/10.1145/3411764.3445517

Van der Meiden, I., Kok, H., & Van der Velde, G. (2019). Nudging Physical Activity in Offices. *Journal of Facilities Management, 17*(4), 317–330. https://doi.org/10.1108/JFM-10-2018-0063

Van Gorp, A. F. (2014). Integration of Volunteer and Technical Communities into the Humanitarian Aid Sector: Barriers to Collaboration. *Proceedings of the 11th International Conference on Information Systems for Crisis Response and Management (ISCRAM 2014)*, 622–631.

VanAhn, V., Auroy, Lola, & Sarradon-Eck, A. (2019). Patients' Perceptions of mHealth Apps: Meta-Ethnographic Review of Qualitative Studies. *JMIR Mhealth and Uhealth, 7*(7), 1–20. https://doi.org/10.2196/13817

van Bavel, R., Rodríguez-Priego, N., Vila, J., & Briggs, P. (2019). Using Protection Motivation Theory in the Design of Nudges to Improve Online Security Behavior. *International Journal of Human Computer Studies, 123*, 29–39. https://doi.org/10.1016/j.ijhcs.2018.11.003

Vargo, D., Zhu, L., Benwell, B., & Yan, Z. (2021). Digital Technology Use during COVID-19 Pandemic: A Rapid Review. *Human Behavior and Emerging Technologies, 3*(1), 13–24. https://doi.org/10.1002/hbe2.242

Venkatesh, Morris, Davis, & Davis. (2003). User Acceptance of Information Technology: Toward a Unified View. *MIS Quarterly, 27*(3), 425. https://doi.org/10.2307/30036540

Venkatesh, V., & Davis, F. D. (2000). A Theoretical Extension of the Technology Acceptance Model: Four Longitudinal Field Studies. *MANAGEMENT SCIENCE, 46*(2), 186–204. https://doi.org/10.1287/mnsc.46.2.186.11926

Venkatesh, V., & Davis, F. D. (1996). A Model of the Antecedents of Perceived Ease of Use: Development and Test. *Decision sciences, 27*(3), 451–481.

Venkatesh, V., Morris, M. G., Sykes, T. A., & Ackerman, P. L. (2004). Individual Reactions to New Technologies in the Workplace: The Role of Gender as a Psychological Construct. *Journal of Applied Social Psychology, 34*(3), 445–467. https://doi.org/10.1111/j.1559-1816.2004.tb02556.x

Verrucci, E., Perez-Fuentes, G., Rossetto, T., Bisby, L., Haklay, M., Rush, D., Rickles, P., Fagg, G., & Joffe, H. (2016). Digital Engagement Methods for Earthquake and Fire Preparedness: A Review. *Natural Hazards, 83*(3), 1583–1604.

Viehmann, C., Ziegele, M., & Quiring, O. (2022). Communication, Cohesion, and Corona: The Impact of People's Use of Different Information Sources on Their Sense of Societal Cohesion in Times of Crises. *Journalism Studies, 23*(5–6), 629–649.

Vieweg, S., Hughes, A. L., Starbird, K., & Palen, L. (2010). Microblogging during Two Natural Hazards Events: What Twitter May Contribute to Situational Awareness. *Proceedings of the SIGCHI Conference on Human Factors in Computing Systems*, 1079–1088. https://doi.org/10.1145/1753326.1753486

Voelsen, D., & Stiftung Wissenschaft und Politik. (2021). Internet from Space: How New Satellite Connections Could Affect Global Internet Governance. https://doi.org/10.18449/2021RP03

vom Brocke, J., Hevner, A., & Maedche, A. (2020). Introduction to Design Science Research. Cases. In J. vom Brocke, A. Hevner, & A. Maedche (Eds.), *Design science research. Cases* (pp. 1–13). Springer International Publishing. https://doi.org/10.1007/978-3-030-46781-4_1

von Wyl, V., Höglinger, M., Sieber, C., Kaufmann, M., Moser, A., Serra-Burriel, M., Ballouz, T., Menges, D., Frei, A., Puhan, M. A., et al. (2021). Drivers of Acceptance of COVID-19 Proximity Tracing Apps in Switzerland: Panel Survey Analysis. *JMIR Public Health and Surveillance, 7*(1), e25701. https://doi.org/10.2196/25701

VOSG. (2019). Virtual Operations Support Group—About. https://vosg.org/about/

Vredenburg, K., Mao, J.-Y., Smith, P. W., & Carey, T. (2002). A Survey of User-Centered Design Practice. *Proceedings of the SIGCHI Conference on Human Factors in Computing Systems*, 471–478.

Vu, P. M., Nguyen, T. T., & Nguyen, T. T. (2019). Why Do App Reviews Get Responded: A Preliminary Study of the Relationship between Reviews and Responses in Mobile Apps. *Proceedings of the 2019 ACM Southeast Conference (ACMSE 2019)*, 237–240. https://doi.org/10.1145/3299815.3314473

Wachinger, G., Renn, O., Begg, C., & Kuhlicke, C. (2013). The Risk Perception Paradox—Implications for Governance and Communication of Natural Hazards. *Risk Analysis, 33*(6), 1049–1065. https://doi.org/10.1111/j.1539-6924.2012.01942.x

Wahl, S., & Gerhold, L. (2021). *Katastrophenkommunikation und Soziale Medien im Bevölkerungsschutz: Kommunikation von Lageinformationen im Bevölkerungsschutz im internationalen Vergleich (KOLIBRI); Abschlussdatum: 30. Juni 2019*. Bundesamt für Bevölkerungsschutz und Katastrophenhilfe. https://www.bbk.bund.de/SharedDocs/Downloads/DE/Mediathek/Publikationen/FiB/FiB-27-kat-kommunikation-soz-medien-bevs.pdf?__blob=publicationFile&v=8

Waidyanatha, N., & Frommberger, L. (2022). Comprehension and Appropriateness of Complex Mobile Pictographs for Crisis Communication. *Natural Hazards, 114*(1), 583–604. https://doi.org/10.1007/s11069-022-05402-y

Ward, N. J., Finley, K., Townsend, A., & Scott, B. G. (2021). The Effects of Message Threat on Psychological Reactance to Traffic Safety Messaging. *Transportation Research Part F: Traffic Psychology and Behaviour, 80*, 250–259. https://doi.org/10.1016/j.trf.2021.04.013

Weichselgartner, J., Guézo, B., Beerlage, I., Després, C., Fekete, A., Hufschmidt, G., Lussignoli, O., Mey-Richters, S., Naumann, J., & Wienand, I. (2018). Urban Resilience and Crisis Management: Perspectives from France and Germany. In *Urban Disaster Resilience and Security* (pp. 473–494). https://doi.org/10.1007/978-3-319-68606-6_27

Weinstein, N. D. (1980). Unrealistic Optimism about Future Life Events. *Journal of Personality and Social Psychology, 39*(5), 806–820. https://doi.org/10.1037//0022-3514.39.5.806

Weinstein, N. D., & Klein, W. M. (1995). Resistance of Personal Risk Perceptions to Debiasing Interventions. *Health Psychology, 14*(2), 132–140. https://doi.org/10.1037/0278-6133.14.2.132

Wendt, A. (1992). Anarchy Is What States Make of It: The Social Construction of Power Politics. *International Organization, 46*(2), 391–425.

Wessel, D., Attig, C., & Franke, T. (2019). ATI-S—An Ultra-Short Scale for Assessing Affinity for Technology Interaction in User Studies. *Proceedings of Mensch Und Computer 2019*, 147–154. https://doi.org/10.1145/3340764.3340766

Wethal, U. (2020). Practices, Provision and Protest: Power Outages in Rural Norwegian Households. *Energy Research & Social Science, 62*, 101388. https://doi.org/10.1016/j.erss.2019.101388

Whyte, K. P., Selinger, E., Caplan, A. L., & Sadowski, J. (2012). Nudge, Nudge or Shove, Shove—the Right Way for Nudges to Increase the Supply of Donated Cadaver Organs. *The American Journal of Bioethics, 12*(2), 32–39. https://doi.org/10.1080/15265161.2011.634484

Williams, B. D., Valero, J. N., & Kim, K. (2018). Social Media, Trust, and Disaster: Does Trust in Public and Nonprofit Organizations Explain Social Media Use during a Disaster? *Quality and Quantity, 52*(2), 537–550. https://doi.org/10.1007/s11135-017-0594-4

Wobbrock, J. O., & Kientz, J. A. (2016). Research Contributions in Human-Computer Interaction. *Interactions, 23*(3), 38–44.

Wogalter, M., & Mayhorn, C. (2005). Providing Cognitive Support with Technology-Based Warning Systems. *Ergonomics, 48*(5), 522–533. https://doi.org/10.1080/00140130400029258

Wood, M. M., Mileti, D. S., Bean, H., Liu, B. F., Sutton, J., & Madden, S. (2018). Milling and Public Warnings. *Environment and Behavior, 50*(5), 535–566.

Wukich, C. (2015). Social Media Use in Emergency Management. *Journal of Emergency Management, 13*(4), 281–294. https://doi.org/10.5055/jem.2015.0242

Xu, X., Zou, T., Xiao, H., Li, Y., Wang, R., Yuan, T., Wang, Y., Shi, Y., Mankoff, J., & Dey, A. K. (2022). TypeOut: Leveraging Just-in-Time Self-Affirmation for Smartphone Overuse Reduction. *Conference on Human Factors in Computing Systems*, 1–17. https://doi.org/10.1145/3491102.3517476

Yeh, S.-F., Wu, M.-H., Chen, T.-Y., Lin, Y.-C., Chang, X., Chiang, Y.-H., & Chang, Y.-J. (2022). How to Guide Task-oriented Chatbot Users, and When: A Mixed-methods Study of Combinations of Chatbot Guidance Types and Timings. *CHI Conference on Human Factors in Computing Systems*, 1–16. https://doi.org/10.1145/3491102.3501941

Yuan, A., Luther, K., Krause, M., Vennix, S., Dow, S. P., & Hartmann, B. (2016). Almost an Expert: The Effects of Rubrics and Expertise on Perceived Value of Crowdsourced Design Critiques. *Proceedings of the 19th ACM Conference on Computer-Supported Cooperative Work & Social Computing, 27*, 1005–1017. https://doi.org/10.1145/2818048.2819953

Zamora, J. (2017). I'm Sorry, Dave, I'm Afraid I Can't Do That: Chatbot Perception and Expectations. *Proceedings of the 5th International Conference on Human Agent Interaction*, 253–260.

Zetterholm, M., Elm, P., & Salavati, S. (2021). Designing for Pandemics—a Design Concept Based on Technology Mediated Nudging for Health Behavior Change. *Proceedings of the 54th Hawaii International Conference on System Sciences*, 3474–3483. https://doi.org/10.24251/hicss.2021.422

Zhao, M., & John, R. (2021). Building Community Resilience Using Gain-Loss Framing to Nudge Homeowner Mitigation and Insurance Decision-Making. *Proceedings of the 54th Hawaii International Conference on System Sciences*, 2206. https://doi.org/10.24251/HICSS.2021.271

Zichermann, G., & Cunningham, C. (2011). *Gamification by Design: Implementing Game Mechanics in Web and Mobile Apps*. O'Reilly Media.

Zimmermann, B. M., Fiske, A., Prainsack, B., Hangel, N., McLennan, S., & Buyx, A. (2021). Early Perceptions of COVID-19 Contact Tracing Apps in German-speaking Countries: Comparative Mixed Methods Study. *Journal of Medical Internet Research, 23*(2), e25525. https://doi.org/10.2196/25525

Zimmermann, S., Hein, A., Schulz, T., Gewald, H., & Krcmar, H. (2021). Digital Nudging Toward Pro-Environmental Behavior: A Literature Review. *Twenty-Fifth Pacific Asia Conference on Information Systems (PACIS)*, 226.

Zimmermann, V., Haunschild, J., Stöver, A., & Gerber, N. (2023). Safe AND Secure Infrastructures?—Studying Human Aspects of Safety and Security Incidents with Experts from both Domains. In P. Fröhlich & V. Cobus (Eds.), *Mensch und Computer 2023—Workshopband*, (pp. 1–8). Gesellschaft für Informatik e.V. https://doi.org/10.18420/MUC2023-MCI-WS01-225

Zimmermann, V., Haunschild, J., Unden, M., Gerber, P., & Gerber, N. (2022). Sicherheitsherausforderungen für Smart City-Infrastrukturen. *Wirtschaftsinformatik & Management, 14*(2), 119–126. https://doi.org/10.1365/s35764-022-00396-5

Zimmermann, V., & Renaud, K. (2021). The Nudge Puzzle: Matching Nudge Interventions to Cybersecurity Decisions. *ACM Transactions on Computer-Human Interaction (TOCHI), 28*(1), 1–45. https://doi.org/10.1145/3429888

Zwilling, M., Klien, G., Lesjak, D., Wiechetek, Ł., Cetin, F., & Basim, H. N. (2022). Cyber Security Awareness, Knowledge and Behavior: A Comparative Study. *Journal of Computer Information Systems, 62*(1), 82–97. https://doi.org/10.1080/08874417.2020.1712269

If you have any concerns about our products,
you can contact us on
ProductSafety@springernature.com

In case Publisher is established outside the EU,
the EU authorized representative is:
Springer Nature Customer Service Center GmbH
Europaplatz 3, 69115 Heidelberg, Germany

Printed by Libri Plureos GmbH
in Hamburg, Germany